OXFORD STATISTICAL SCIENCE SERIES

A. C. Atkinson: *Plots, transformations and regression*

M. Stone: *Coordinate-free multivariate statistic*

W. J. Krzanowski: *Principles of multivariate analysis: a user's perspective*

M. Aitkin, D. Anderson, B. Francis, and J. Hinde: *Statistical modelling in GLIM*

P. J. Diggle: *Time series: a biostatistical introduction*

H. Tong: *Non-linear time series: a dynamical system approach*

V. P. Godambe: *Estimating functions*

A. C. Atkinson and A. N. Donev: *Optimum experimental design*

J. K. Lindsey: *Models for repeated measurements*

N. T. Longford: *Random coefficient models*

P. J. Brown: *Measurement, regression and calibration*

P. J. Diggle, K. Y. Liang, and S. L. Zeger: *The analysis of longitudinal data*

J. I. Ansell and M. J. Phillips: *Practical methods for reliability data analysis*

J. K. Lindsey: *Modelling frequency and count data*

J. L. Jensen: *Saddlepoint approximations*

Steffen L. Lauritzen: *Graphical models*

A. W. Bowman and A. Azzalini: *Applied smoothing methods for data analysis*

J. K. Lindsey: *Models for repeated measurements*, (Second edition)

Michael Evans and Tim Swartz: *Approximating integrals via Monte Carlo and deterministic methods*

D. F. Andrews and J. E. Stafford: *Symbolic computation for statistical inference*

T. A. Severini: *Likelihood methods in statistics*

W. J. Krzanowski: *Principles of multivariate analysis: a user's perspective updated edition*

Time Series Analysis
by State Space Methods

J. DURBIN

Department of Statistics
London School of Economics and Political Science

and

S. J. KOOPMAN

Department of Econometrics
Free University Amsterdam

OXFORD
UNIVERSITY PRESS

OXFORD
UNIVERSITY PRESS

Great Clarendon Street, Oxford OX2 6DP

Oxford University Press is a department of the University of Oxford.
It furthers the University's objective of excellence in research, scholarship,
and education by publishing worldwide in

Oxford New York

Auckland Cape Town Dar es Salaam Hong Kong Karachi
Kuala Lumpur Madrid Melbourne Mexico City Nairobi
New Delhi Shanghai Taipei Toronto

With offices in

Argentina Austria Brazil Chile Czech Republic France Greece
Guatemala Hungary Italy Japan South Korea Poland Portugal
Singapore Switzerland Thailand Turkey Ukraine Vietnam

Oxford is a registered trade mark of Oxford University Press
in the UK and in certain other countries

Published in the United States
by Oxford University Press Inc., New York

A catalogue record for this book is available from the British Library

Library of Congress Cataloging in Publication Data
Durbin, James.
Time series analysis by state space methods / J. Durbin and S. J. Koopman.
(Oxford statistical science series)
Includes bibliographical references and index.
I. Time-series analysis. 2. State-space methods. I. Koopman, S. J. (Siem Jan). II. Title.
III. Series
QA280 .D87 2001 519.5′5 – dc21 00–054845

ISBN 0 19 852354 8

Printed in Great Britain
on acid-free paper by
Biddles Ltd., King's Lynn, Norfolk

To Anne
JD

To my family and friends
SJK

Preface

This book presents a comprehensive treatment of the state space approach to time series analysis. The distinguishing feature of state space time series models is that observations are regarded as made up of distinct components such as trend, seasonal, regression elements and disturbance terms, each of which is modelled separately. The models for the components are put together to form a single model called a state space model which provides the basis for analysis. The techniques that emerge from this approach are very flexible and are capable of handling a much wider range of problems than the main analytical system currently in use for time series analysis, the Box-Jenkins ARIMA system.

The exposition is directed primarily towards students, teachers, methodologists and applied workers in time series analysis and related areas of applied statistics and econometrics. Nevertheless, we hope that the book will also be useful to workers in other fields such as engineering, medicine and biology where state space models are employed. We have made a special effort to make the presentation accessible to readers with no previous knowledge of state space methods. For the development of all important parts of the theory, the only mathematics required that is not elementary is matrix multiplication and inversion, while the only statistical theory required, apart from basic principles, is elementary multivariate normal regression theory.

The techniques that we develop are aimed at practical application to real problems in applied time series analysis. Nevertheless, a surprising degree of beauty is to be found in the elegant way that many of the results drop out; no doubt this is due to the Markovian nature of the models and the recursive structure of the calculations.

State space time series analysis began with the pathbreaking paper of Kalman (1960) and early developments of the subject took place in the field of engineering. The term 'state space' came from engineering, and although it does not strike a natural rapport in statistics and econometrics, we, like others, use it because it is strongly established. We assure beginners that the meaning of the word 'state' will become clear very quickly; however, the attachment of the word 'space' might perhaps remain mysterious to non-engineers.

The book is divided into two parts. Part I discusses techniques of analysis based on the linear Gaussian state space model; the methods we describe represent the core of traditional state space methodology together with some new developments. We have aimed at presenting a state-of-the-art treatment of time series methodology based on this state space model. Although the model has been studied extensively over the past forty years, there is much that is new in our treatment. We use the word 'new' here and below to refer to results derived from original research in our recently published papers or obtained for the first time in this book. In particular, this usage relates to the treatment of disturbance smoothing in Chapter 4, exact initial filtering and smoothing in Chapter 5, the univariate treatment of multivariate observations plus computing algorithms in Chapter 6, aspects of diffuse likelihood, the score vector, the EM algorithm and allowing for the effects of parameter estimation on estimates of variance in Chapter 7, and the use of importance sampling for Bayesian analysis in Chapter 8.

The linear Gaussian model, often after transformation of the observations, provides an adequate basis for the analysis of many of the time series that are encountered in practice. However, situations occur where this model fails to provide an acceptable representation of the data. For example, in the study of road accidents when the numbers observed are small, the Poisson distribution gives an intrinsically better model for the behaviour of the data than the normal distribution. There is therefore a need to extend the scope of state space methodology to cover non-Gaussian observations; it is also important to allow for nonlinearities in the model and for heavy-tailed densities to deal with outliers in the observations and structural shifts in the state.

These extensions are considered in Part II of the book. The treatment given there is new in the sense that it is based partly on the methods developed in our Durbin and Koopman (2000) paper and partly on additional new material. The methodology is based on simulation since exact analytical results are not available for the problems under consideration. Earlier workers had employed Markov chain Monte Carlo simulation methods to study these problems; however, during the research that led to this book, we investigated whether traditional simulation methods based on importance sampling and antithetic variables could be made to work satisfactorily. The results were successful so we have adopted this approach throughout Part II. The simulation methods that we propose are computationally transparent, efficient and convenient for the type of time series applications considered in this book.

An unusual feature of the book is that we provide analyses from both classical and Bayesian perspectives. We start in all cases from the classical standpoint since this is the mode of inference within which each of us normally works. This also makes it easier to relate our treatment to earlier work in the field, most of which has been done from the classical point of view. We discovered as we developed simulation methods for Part II, however, that we could obtain Bayesian solutions by relatively straightforward extensions of the classical treatments. Since our views on statistical inference are eclectic and tolerant, and since we believe that methodology for both approaches should be available to applied workers, we are happy to include solutions from both perspectives in the book.

Both the writing of the book and doing the research on which it was based have been highly interactive and highly enjoyable joint efforts. Subject to this, there have naturally been differences in the contributions that the two authors have made to the work. Most of the new theory in Part I was initiated by SJK while most of the new theory in Part II was initiated by JD. The expository and literary styles of the book were the primary responsibility of JD while the illustrations and computations were the primary responsibility of SJK.

Our collaboration in this work began while we were working in the Statistics Department of the London School of Economics and Political Science (LSE). We were fortunate in having as departmental colleagues three distinguished workers in the field. Our main thanks are to Andrew Harvey who helped us in many ways and whose leadership in the development of the methodology of state space time series analysis at the LSE was an inspiration to us both. We also thank Neil Shephard for many fruitful discussions on various aspects of the statistical treatment of state space models and for his incisive comments on an earlier draft of this book. We are grateful to Piet de Jong for some searching discussions on theoretical points.

We thank Jurgen Doornik of Nuffield College, Oxford for his help over a number of years which has assisted in the development of the computer packages *STAMP* and SsfPack. SJK thanks Nuffield College, Oxford for its hospitality and the Royal Netherlands Academy of Arts and Sciences for its financial support while he was at CentER, Tilburg University.

The book was written in LaTeX using the MiKTeX system (http://www.miktex.org). We thank Jurgen Doornik for assistance in setting up this LaTeX system.

London   J.D.
Amsterdam   S.J.K.
November 2000

Contents

1   Introduction   1
  1.1   Basic ideas of state space analysis   1
  1.2   Linear Gaussian model   1
  1.3   Non-Gaussian and nonlinear models   3
  1.4   Prior knowledge   4
  1.5   Notation   4
  1.6   Other books on state space methods   5
  1.7   Website for the book   5

     I   THE LINEAR GAUSSIAN STATE SPACE MODEL

2   Local level model   9
  2.1   Introduction   9
  2.2   Filtering   11
    2.2.1   The Kalman Filter   11
    2.2.2   Illustration   12
  2.3   Forecast errors   13
    2.3.1   Cholesky decomposition   14
    2.3.2   Error recursions   15
  2.4   State smoothing   16
    2.4.1   Smoothed state   16
    2.4.2   Smoothed state variance   17
    2.4.3   Illustration   18
  2.5   Disturbance smoothing   19
    2.5.1   Smoothed observation disturbances   20
    2.5.2   Smoothed state disturbances   20
    2.5.3   Illustration   21
    2.5.4   Cholesky decomposition and smoothing   22
  2.6   Simulation   22
    2.6.1   Illustration   23
  2.7   Missing observations   23
    2.7.1   Illustration   25
  2.8   Forecasting   25
    2.8.1   Illustration   27

2.9	Initialisation	27
2.10	Parameter estimation	30
	2.10.1 Loglikelihood evaluation	30
	2.10.2 Concentration of loglikelihood	31
	2.10.3 Illustration	32
2.11	Steady state	32
2.12	Diagnostic checking	33
	2.12.1 Diagnostic tests for forecast errors	33
	2.12.2 Detection of outliers and structural breaks	35
	2.12.3 Illustration	35
2.13	Appendix: Lemma in multivariate normal regression	37

3 Linear Gaussian state space models — 38

3.1	Introduction	38
3.2	Structural time series models	39
	3.2.1 Univariate models	39
	3.2.2 Multivariate models	44
	3.2.3 STAMP	45
3.3	ARMA models and ARIMA models	46
3.4	Exponential smoothing	49
3.5	State space versus Box-Jenkins approaches	51
3.6	Regression with time-varying coefficients	54
3.7	Regression with ARMA errors	54
3.8	Benchmarking	54
3.9	Simultaneous modelling of series from different sources	56
3.10	State space models in continuous time	57
	3.10.1 Local level model	57
	3.10.2 Local linear trend model	59
3.11	Spline smoothing	61
	3.11.1 Spline smoothing in discrete time	61
	3.11.2 Spline smoothing in continuous time	62

4 Filtering, smoothing and forecasting — 64

4.1	Introduction	64
4.2	Filtering	65
	4.2.1 Derivation of Kalman filter	65
	4.2.2 Kalman filter recursion	67
	4.2.3 Steady state	68
	4.2.4 State estimation errors and forecast errors	68
4.3	State smoothing	70
	4.3.1 Smoothed state vector	70
	4.3.2 Smoothed state variance matrix	72
	4.3.3 State smoothing recursion	73
4.4	Disturbance smoothing	73

	4.4.1	Smoothed disturbances	73
	4.4.2	Fast state smoothing	75
	4.4.3	Smoothed disturbance variance matrices	75
	4.4.4	Disturbance smoothing recursion	76
4.5	Covariance matrices of smoothed estimators		77
4.6	Weight functions		81
	4.6.1	Introduction	81
	4.6.2	Filtering weights	81
	4.6.3	Smoothing weights	82
4.7	Simulation smoothing		83
	4.7.1	Simulating observation disturbances	84
	4.7.2	Derivation of simulation smoother for observation disturbances	87
	4.7.3	Simulation smoothing recursion	89
	4.7.4	Simulating state disturbances	90
	4.7.5	Simulating state vectors	91
	4.7.6	Simulating multiple samples	92
4.8	Missing observations		92
4.9	Forecasting		93
4.10	Dimensionality of observational vector		94
4.11	General matrix form for filtering and smoothing		95

5	**Initialisation of filter and smoother**		**99**
5.1	Introduction		99
5.2	The exact initial Kalman filter		101
	5.2.1	The basic recursions	101
	5.2.2	Transition to the usual Kalman filter	104
	5.2.3	A convenient representation	105
5.3	Exact initial state smoothing		106
	5.3.1	Smoothed mean of state vector	106
	5.3.2	Smoothed variance of state vector	107
5.4	Exact initial disturbance smoothing		109
5.5	Exact initial simulation smoothing		110
5.6	Examples of initial conditions for some models		110
	5.6.1	Structural time series models	110
	5.6.2	Stationary ARMA models	111
	5.6.3	Nonstationary ARIMA models	112
	5.6.4	Regression model with ARMA errors	114
	5.6.5	Spline smoothing	115
5.7	Augmented Kalman filter and smoother		115
	5.7.1	Introduction	115
	5.7.2	Augmented Kalman filter	115
	5.7.3	Filtering based on the augmented Kalman filter	116

5.7.4   Illustration: the local linear trend model   118
5.7.5   Comparisons of computational efficiency   119
5.7.6   Smoothing based on the augmented Kalman filter   120

6   Further computational aspects   **121**
6.1   Introduction   121
6.2   Regression estimation   121
      6.2.1   Introduction   121
      6.2.2   Inclusion of coefficient vector in state vector   122
      6.2.3   Regression estimation by augmentation   122
      6.2.4   Least squares and recursive residuals   123
6.3   Square root filter and smoother   124
      6.3.1   Introduction   124
      6.3.2   Square root form of variance updating   125
      6.3.3   Givens rotations   126
      6.3.4   Square root smoothing   127
      6.3.5   Square root filtering and initialisation   127
      6.3.6   llustration: local linear trend model   128
6.4   Univariate treatment of multivariate series   128
      6.4.1   Introduction   128
      6.4.2   Details of univariate treatment   129
      6.4.3   Correlation between observation equations   131
      6.4.4   Computational efficiency   132
      6.4.5   Illustration: vector splines   133
6.5   Filtering and smoothing under linear restrictions   134
6.6   The algorithms of SsfPack   134
      6.6.1   Introduction   134
      6.6.2   The SsfPack function   135
      6.6.3   Illustration: spline smoothing   136

7   Maximum likelihood estimation   **138**
7.1   Introduction   138
7.2   Likelihood evaluation   138
      7.2.1   Loglikelihood when initial conditions are known   138
      7.2.2   Diffuse loglikelihood   139
      7.2.3   Diffuse loglikelihood evaluated via augmented Kalman
               filter   140
      7.2.4   Likelihood when elements of initial state vector are
               fixed but unknown   141
7.3   Parameter estimation   142
      7.3.1   Introduction   142
      7.3.2   Numerical maximisation algorithms   142
      7.3.3   The score vector   144
      7.3.4   The EM algorithm   147

	7.3.5	Parameter estimation when dealing with diffuse initial conditions	149
	7.3.6	Large sample distribution of maximum likelihood estimates	150
	7.3.7	Effect of errors in parameter estimation	150
7.4	Goodness of fit		152
7.5	Diagnostic checking		152

8 Bayesian analysis — 155

8.1	Introduction		155
8.2	Posterior analysis of state vector		155
	8.2.1	Posterior analysis conditional on parameter vector	155
	8.2.2	Posterior analysis when parameter vector is unknown	155
	8.2.3	Non-informative priors	158
8.3	Markov chain Monte Carlo methods		159

9 Illustrations of the use of the linear Gaussian model — 161

9.1	Introduction	161
9.2	Structural time series models	161
9.3	Bivariate structural time series analysis	167
9.4	Box-Jenkins analysis	169
9.5	Spline smoothing	172
9.6	Approximate methods for modelling volatility	175

II   NON-GAUSSIAN AND NONLINEAR STATE SPACE MODELS

10 Non-Gaussian and nonlinear state space models — 179

10.1	Introduction		179
10.2	The general non-Gaussian model		179
10.3	Exponential family models		180
	10.3.1	Poisson density	181
	10.3.2	Binary density	181
	10.3.3	Binomial density	181
	10.3.4	Negative binomial density	182
	10.3.5	Multinomial density	182
10.4	Heavy-tailed distributions		183
	10.4.1	t-Distribution	183
	10.4.2	Mixture of normals	184
	10.4.3	General error distribution	184
10.5	Nonlinear models		184
10.6	Financial models		185
	10.6.1	Stochastic volatility models	185

	10.6.2	General autoregressive conditional	
		heteroscedasticity	187
	10.6.3	Durations: exponential distribution	188
	10.6.4	Trade frequencies: Poisson distribution	188

11 Importance sampling   **189**
11.1   Introduction   189
11.2   Basic ideas of importance sampling   190
11.3   Linear Gaussian approximating models   191
11.4   Linearisation based on first two derivatives   193
       11.4.1 Exponential family models   195
       11.4.2 Stochastic volatility model   195
11.5   Linearisation based on the first derivative   195
       11.5.1 t-distribution   197
       11.5.2 Mixture of normals   197
       11.5.3 General error distribution   197
11.6   Linearisation for non-Gaussian state components   198
       11.6.1 t-distribution for state errors   199
11.7   Linearisation for nonlinear models   199
       11.7.1 Multiplicative models   201
11.8   Estimating the conditional mode   202
11.9   Computational aspects of importance sampling   204
       11.9.1 Introduction   204
       11.9.2 Practical implementation of importance sampling   204
       11.9.3 Antithetic variables   205
       11.9.4 Diffuse initialisation   206
       11.9.5 Treatment of t-distribution without importance
               sampling   208
       11.9.6 Treatment of Gaussian mixture distributions without
               importance sampling   210

12 Analysis from a classical standpoint   **212**
12.1   Introduction   212
12.2   Estimating conditional means and variances   212
12.3   Estimating conditional densities and distribution
       functions   213
12.4   Forecasting and estimating with missing observations   214
12.5   Parameter estimation   215
       12.5.1 Introduction   215
       12.5.2 Estimation of likelihood   215
       12.5.3 Maximisation of loglikelihood   216
       12.5.4 Variance matrix of maximum likelihood estimate   217
       12.5.5 Effect of errors in parameter estimation   217

12.5.6   Mean square error matrix due to simulation   217
12.5.7   Estimation when the state disturbances are Gaussian 219
12.5.8   Control variables   219

13 Analysis from a Bayesian standpoint   222
13.1   Introduction   222
13.2   Posterior analysis of functions of the state vector   222
13.3   Computational aspects of Bayesian analysis   225
13.4   Posterior analysis of parameter vector   226
13.5   Markov chain Monte Carlo methods   228

14 Non-Gaussian and nonlinear illustrations   230
14.1   Introduction   230
14.2   Poisson density: van drivers killed in Great Britain   230
14.3   Heavy-tailed density: outlier in gas consumption in UK 233
14.4   Volatility: pound/dollar daily exchange rates   236
14.5   Binary density: Oxford–Cambridge boat race   237
14.6   Non-Gaussian and nonlinear analysis using SsfPack   238

References   241

Author index   249

Subject index   251

1
Introduction

1.1 Basic ideas of state space analysis

State space modelling provides a unified methodology for treating a wide range of problems in time series analysis. In this approach it is assumed that the development over time of the system under study is determined by an unobserved series of vectors $\alpha_1, \ldots, \alpha_n$, with which are associated a series of observations $y_1, \ldots, y_n$; the relation between the α_t's and the y_t's is specified by the state space model. The purpose of state space analysis is to infer the relevant properties of the α_t's from a knowledge of the observations $y_1, \ldots, y_n$. This book presents a systematic treatment of this approach to problems of time series analysis.

Our starting point when deciding the structure of the book was that we wanted to make the basic ideas of state space analysis easy to understand for readers with no previous knowledge of the approach. We felt that if we had begun the book by developing the theory step by step for a general state space model, the underlying ideas would be obscured by the complicated appearance of many of the formulae. We therefore decided instead to devote Chapter 2 of the book to a particularly simple example of a state space model, the local level model, and to develop as many as possible of the basic state space techniques for this model. Our hope is that this will enable readers new to the techniques to gain insights into the ideas behind state space methodology that will help them when working through the greater complexities of the treatment of the general case. With this purpose in mind, we introduce topics such as Kalman filtering, state smoothing, disturbance smoothing, simulation smoothing, missing observations, forecasting, initialisation, maximum likelihood estimation of parameters and diagnostic checking for the local level model. We demonstrate how the basic theory that is needed can be developed very simply from elementary results in regression theory.

1.2 Linear Gaussian model

Before going on to develop the theory for the general model, we present a series of examples that show how the linear Gaussian state space model relates to problems of practical interest. This is done in Chapter 3 where we begin by showing how structural time series models can be put into state space form. By structural time series models we mean models in which the observations are made up of trend, seasonal, cycle and regression components plus error, where the components are generally represented by forms of random walk models. We go on to put Box-Jenkins ARIMA models into state space form, thus demonstrating that these models are special cases of state space models. Next we discuss the history of exponential smoothing and show how it relates to simple forms of state space and ARIMA models. We then compare the relative merits of Box-Jenkins and state space methodologies for applied time series analysis. We follow this by considering various aspects of regression with or without time-varying coefficients or autocorrelated errors. We also show that the benchmarking problem, which is an important problem in official statistics, can be dealt with by state space methods. Further topics discussed are simultaneous modelling series from different sources, continuous time models and spline smoothing in discrete and continuous time.

Chapter 4 begins the development of the theory for the analysis of the general linear Gaussian state space model. This model has provided the basis for most earlier work on state space time series analysis. We give derivations of the Kalman filter and smoothing recursions for the estimation of the state vector and its conditional variance matrix given the data. We also derive recursions for estimating the observation and state disturbances. We derive the simulation smoother which is an important tool in the simulation methods we employ later in the book. We show that forecasting and allowance for missing observations are easily dealt with in the state space framework.

Computational algorithms in state space analyses are mainly based on recursions, that is, formulae in which we calculate the value at time $t + 1$ from earlier values for $t, t - 1, \ldots, 1$. The question of how these recursions are started up at the beginning of the series is called initialisation; it is dealt with in Chapter 5. We give a general treatment in which some elements of the initial state vector have known distributions while others are diffuse, that is, treated as random variables with infinite variance, or are treated as unknown constants to be estimated by maximum likelihood. The treatment is based on recent work and we regard it as more transparent than earlier discussions of the problem.

Chapter 6 discusses computational aspects of filtering and smoothing and begins by considering the estimation of the regression component of the model. It next considers the square root filter and smoother which may be used when the Kalman filter and smoother show signs of numerical instability. It goes on to discuss how multivariate time series can be treated as univariate series by bringing elements of the observational vectors into the system one at a time, with computational savings relative to the multivariate treatment in some cases. The chapter concludes by discussing computer algorithms.

In Chapter 7, maximum likelihood estimation of unknown parameters is considered both for the case where the distribution of the initial state vector is known and for the case where at least some elements of the vector are diffuse or are treated as fixed and unknown. The use of the score vector and the EM algorithm is discussed. The effect of parameter estimation on variance estimation is examined.

Up to this point the exposition has been based on the classical approach to inference in which formulae are worked out on the assumption that parameters are known, while in applications parameter values are replaced by appropriate estimates. In Bayesian analysis the parameters are treated as random variables with a specified or a non-informative prior joint density. In Chapter 8 we consider Bayesian analysis of the linear Gaussian model both for the case where the prior density is proper and for the case where it is non-informative. We give formulae from which the posterior mean can be calculated for functions of the state vector, either by numerical integration or by simulation. We restrict attention to functions which, for given values of the parameters, can be calculated by the Kalman filter and smoother; this however includes a substantial number of quantities of interest.

In Chapter 9 we illustrate the use of the methodology by applying the techniques that have been developed to a number of analyses based on real data. These include a study of the effect of the seat belt law on road accidents in Great Britain, forecasting the number of users logged on to an Internet server, fitting acceleration against time for a simulated motorcycle accident and modelling the volatility of financial time series.

1.3   Non-Gaussian and nonlinear models

Part II of the book extends the treatment to state space models which are not both linear and Gaussian. The analyses are based on simulation using the approach developed by Durbin and Koopman (2000). Chapter 10 illustrates the range of non-Gaussian and nonlinear models that can be analysed using the methods of Part II. This includes exponential family models such as the Poisson distribution for the conditional distribution of the observations given the state. It also includes heavy-tailed distributions for the observational and state disturbances, such as the t-distribution and mixtures of normal densities. Departures from linearity of the models are studied for cases where the basic state space structure is preserved. Financial models such as stochastic volatility models are investigated from the state space point of view.

The simulation techniques employed in this book for handling non-Gaussian models are based on importance sampling and antithetic variables. The methods that we use are described in Chapter 11. We first show how to calculate the conditional mode of the state given the observations for the non-Gaussian model by iterated use of the Kalman filter and smoother. We then find the linear Gaussian model with the same conditional mode given the observations and we use the conditional density of the state given the observations for this model as the importance density. Since this density is Gaussian we can draw random samples from it for the simulation using the simulation smoother described in Chapter 4. To improve efficiency we introduce two antithetic variables intended to balance the simulation sample for location and scale. Finally, we show how cases where the observation or state errors have t-distributions or distributions which are mixtures of normal densities can be treated without using importance sampling.

Chapter 12 discusses the implementation of importance sampling numerically. It describes how to estimate conditional means and variances of functions of elements of the state vector given the observations and how to estimate conditional densities and distributions of these functions. It also shows how variances due to simulation can be estimated by means of simple formulae. It next discusses the estimation of unknown parameters by maximum likelihood. The final topic considered is the effect of errors in the estimation of the parameters on the estimates.

In Chapter 13 we show that by an extension of the importance sampling techniques of Chapter 11 we can obtain a Bayesian treatment for non-Gaussian and nonlinear models. Estimates of posterior means, variances, densities and distribution functions of functions of elements of the state vector are provided. We also estimate the posterior densities of the parameters of the model. Estimates of variances due to simulation are obtained.

We provide examples in Chapter 14 which illustrate the methods that have been developed in Part II for analysing observations using non-Gaussian and nonlinear state space models. We cover both classical and Bayesian analyses. The illustrations include the monthly number of van drivers killed in road accidents in Great Britain, outlying observations in quarterly gas consumption, the volatility of exchange rate returns and analysis of the results of the annual boat race between teams of the universities of Oxford and Cambridge. They demonstrate that the methods we have developed work well in practice.

1.4   Prior knowledge

Only basic knowledge of statistics and matrix algebra is needed in order to understand the theory in this book. In statistics, an elementary knowledge is required of the conditional distribution of a vector y given a vector x in a multivariate normal distribution; the central results needed from this area for much of the theory of the book are stated in the lemma in §2.13. Little previous knowledge of time series analysis is required beyond an understanding of the concepts of a stationary time series and the autocorrelation function. In matrix algebra all that is needed are matrix multiplication and inversion of matrices, together with basic concepts such as rank and trace.

1.5   Notation

Although a large number of mathematical symbols are required for the exposition of the theory in this book, we decided to confine ourselves to the standard English and Greek alphabets. The effect of this is that we occasionally need to use the same symbol more than once; we have aimed however at ensuring that the meaning of the symbol is always clear from the context.

- The same symbol 0 is used to denote zero, a vector of zeros or a matrix of zeros.
- We use the generic notation $p(\cdot)$, $p(\cdot, \cdot)$, $p(\cdot|\cdot)$ to denote a probability density, a joint probability density and a conditional probability density.

- If x is a vector which is normally distributed with mean vector μ and variance matrix V, we write $x \sim N(\mu, V)$.
- If x is a random variable with the chi-squared distribution with ν degrees of freedom, we write $x \sim \chi_\nu^2$.
- We use the same symbol $\mathrm{Var}(x)$ to denote the variance of a scalar random variable x and the variance matrix of a random vector x. Similarly, we use the same symbol $\mathrm{Cov}(x, y)$ to denote the covariance between scalar random variables x and y and between random vectors x and y.

1.6 Other books on state space methods

Without claiming complete coverage, we list here a number of books which contain treatments of state space methods.

First we mention three early books written from an engineering standpoint: Jazwinski (1970), Sage and Melsa (1971) and Anderson and Moore (1979). A later book from a related standpoint is Young (1984).

Books written from the standpoint of statistics and econometrics include Harvey (1989), who gives a comprehensive state space treatment of structural time series models together with related state space material, West and Harrison (1997), who give a Bayesian treatment with emphasis on forecasting, Kitagawa and Gersch (1996) and Kim and Nelson (1999).

More general books on time series analysis and related topics which cover partial treatments of state space topics include Brockwell and Davis (1987) (39 pages on state space out of about 570), Harvey (1993) (48 pages out of about 300), Hamilton (1994) (one chapter of 37 pages on state space out of about 800 pages) and Shumway and Stoffer (2000) (one chapter of 112 pages out of about 545 pages). The monograph of Jones (1993) on longitudinal models has three chapters on state space (66 pages out of about 225). The book by Fahrmeir and Tutz (1994) on multivariate analysis based on generalised linear modelling has a chapter on state space models (48 pages out of about 420).

Books on time series analysis and similar topics with minor treatments of state space analysis include Granger and Newbold (1986) and Mills (1993). We mention finally the book edited by Doucet, deFreitas and Gordon (2000) which contains a collection of articles on Monte Carlo (particle) filtering and the book edited by Akaike and Kitagawa (1999) which contains six chapters (88 pages) on illustrations of state space analysis out of a total of 22 chapters (385 pages).

1.7 Website for the book

We will maintain a website for the book at `http://www.ssfpack.com/dkbook/` for data, code, corrections and other relevant information. We will be grateful to readers if they inform us about their comments and errors in the book so corrections can be placed on the site.

Part I

The linear Gaussian state space model

In Part I we present a state-of-the-art treatment of the construction and analysis of linear Gaussian state space models, and we discuss the software required for implementing the resulting methodology. Methods based on these models, possibly after transformation of the observations, are appropriate for a wide range of problems in practical time series analysis. We also present illustrations of the applications of the methods to real series.

2
Local level model

2.1 Introduction

The purpose of this chapter is to introduce the basic techniques of state space analysis, such as filtering, smoothing, initialisation and forecasting, in terms of a simple example of a state space model, the local level model. This is intended to help beginners grasp the underlying ideas more quickly than they would if we were to begin the book with a systematic treatment of the general case. So far as inference is concerned, we shall limit the discussion to the classical standpoint; a Bayesian treatment of the local level model may be obtained as a special case of the Bayesian analysis of the linear Gaussian model in Chapter 8.

A *time series* is a set of observations $y_1, \ldots, y_n$ ordered in time. The basic model for representing a time series is the additive model

$$y_t = \mu_t + \gamma_t + \varepsilon_t, \qquad t = 1, \ldots, n. \tag{2.1}$$

Here, μ_t is a slowly varying component called the *trend*, γ_t is a periodic component of fixed period called the *seasonal* and ε_t is an irregular component called the *error* or *disturbance*. In general, the observation y_t and the other variables in (2.1) can be vectors but in this chapter we assume they are scalars. In many applications, particularly in economics, the components combine multiplicatively, giving

$$y_t = \mu_t \gamma_t \varepsilon_t. \tag{2.2}$$

By taking logs however and working with logged values model (2.2) reduces to model (2.1), so we can use model (2.1) for this case also.

To develop suitable models for μ_t and γ_t we need the concept of a *random walk*. This is a scalar series α_t determined by the relation $\alpha_{t+1} = \alpha_t + \eta_t$ where the η_t's are independent and identically distributed random variables with zero means and variances σ_η^2.

Consider a simple form of model (2.1) in which $\mu_t = \alpha_t$ where α_t is a random walk, no seasonal is present and all random variables are normally distributed. We assume that ε_t has constant variance σ_ε^2. This gives the model

$$\begin{aligned}
y_t &= \alpha_t + \varepsilon_t, & \varepsilon_t &\sim \mathrm{N}\!\left(0, \sigma_\varepsilon^2\right), \\
\alpha_{t+1} &= \alpha_t + \eta_t, & \eta_t &\sim \mathrm{N}\!\left(0, \sigma_\eta^2\right),
\end{aligned} \tag{2.3}$$

for $t = 1, \ldots, n$ where the ε_t's and η_t's are all mutually independent and are independent of α_1. This model is called the *local level model*. Although it has a simple form, this model is not an artificial special case and indeed it provides the basis for the analysis of important real problems in practical time series analysis; for example, it provides the basis for exponentially weighted moving averages as discussed in §3.4. It exhibits the characteristic structure of state space models in which there is a series of unobserved values $\alpha_1, \ldots, \alpha_n$ which represents the development over time of the system under study, together with a set of observations $y_1, \ldots, y_n$ which are related to the α_t's by the state space model (2.3). The object of the methodology that we shall develop is to infer relevant properties of the α_t's from a knowledge of the observations $y_1, \ldots, y_n$.

We assume initially that $\alpha_1 \sim N(a_1, P_1)$ where a_1 and P_1 are known and that σ_ε^2 and σ_η^2 are known. Since random walks are non-stationary the model is non-stationary. By non-stationary here we mean that distributions of random variables y_t and α_t depend on time t.

For applications of the model to real series using classical inference, we need to compute quantities such as the mean of α_t given $y_1, \ldots, y_{t-1}$ or the mean of α_t given $y_1, \ldots, y_n$, together with their variances; we also need to fit the model to data by calculating maximum likelihood estimates of the parameters σ_ε^2 and σ_η^2. In principle, this could be done by using standard results from multivariate normal theory as described in books such as Anderson (1984). In this approach the observations y_t generated by the local level model are represented as an $n \times 1$ vector y such that

$$y \sim N(1a_1, \Omega), \quad \text{with } y = \begin{pmatrix} y_1 \\ \vdots \\ y_n \end{pmatrix}, \quad 1 = \begin{pmatrix} 1 \\ \vdots \\ 1 \end{pmatrix} \quad \text{and } \Omega = 11'P_1 + \Sigma,$$

(2.4)

where the (i, j)th element of the $n \times n$ matrix Σ is given by

$$\Sigma_{ij} = \begin{cases} (i-1)\sigma_\eta^2, & i < j \\ \sigma_\varepsilon^2 + (i-1)\sigma_\eta^2, & i = j, \\ (j-1)\sigma_\eta^2, & i > j \end{cases} \quad i, j = 1, \ldots, n, \quad (2.5)$$

which follows since the local level model implies that

$$y_t = \alpha_1 + \sum_{j=1}^{t-1} \eta_j + \varepsilon_t, \quad t = 1, \ldots, n. \quad (2.6)$$

Starting from this knowledge of the distribution of y, estimation of conditional means, variances and covariances is in principle a routine matter using standard results in multivariate analysis based on the properties of the multivariate normal distribution. However, because of the serial correlation between the observations y_t, the routine computations rapidly become cumbersome as n increases. This naive approach to estimation can be improved on considerably by using the filtering and smoothing techniques described in the next three sections. In effect, these techniques provide efficient computing algorithms for obtaining the same results as those derived by multivariate analysis theory. The remaining sections of this chapter deal with other important issues such as fitting the local level model and forecasting future observations.

The local level model (2.3) is a simple example of a *linear Gaussian state space model*. The variable α_t is called the *state* and is unobserved. The overall object of the analysis is to study the development of the state over time using the observed values $y_1, \ldots, y_n$. Further examples of these models will be described in Chapter 3. The general form of the linear Gaussian state space model is given in equation (3.1); its properties will be considered in detail from the standpoint of classical inference in Chapters 4-7. A Bayesian treatment of the model will be given in Chapter 8.

2.2 Filtering

2.2.1 THE KALMAN FILTER

The object of filtering is to update our knowledge of the system each time a new observation y_t is brought in. We shall develop the theory of filtering for the local level model (2.3). Since all distributions are normal, conditional joint distributions of one set of obervations given another set are also normal. Let Y_{t-1} be the set of past observations $\{y_1, \ldots, y_{t-1}\}$ and assume that the conditional distribution of α_t given Y_{t-1} is $N(a_t, P_t)$ where a_t and P_t are to be determined. Given that a_t and P_t are known, our object is to calculate a_{t+1} and P_{t+1} when y_t is brought in. We do this by using some results from elementary regression theory.

Since $\quad a_{t+1} = E(\alpha_{t+1}|Y_t) = E(\alpha_t + \eta_t|Y_t) \quad$ and $\quad P_{t+1} = Var(\alpha_{t+1}|Y_t) = Var(\alpha_t + \eta_t|Y_t)$ from (2.3), we have

$$a_{t+1} = E(\alpha_t|Y_t), \qquad P_{t+1} = Var(\alpha_t|Y_t) + \sigma_\eta^2. \qquad (2.7)$$

Define $v_t = y_t - a_t$ and $F_t = Var(v_t)$. Then

$$E(v_t|Y_{t-1}) = E(\alpha_t + \varepsilon_t - a_t|Y_{t-1}) = a_t - a_t = 0.$$

Thus $E(v_t) = E[E(v_t|Y_{t-1})] = 0$ and $Cov(v_t, y_j) = E(v_t y_j) = E[E(v_t|Y_{t-1})y_j] = 0$ so v_t and y_j are independent for $j = 1, \ldots, t - 1$. When Y_t is fixed, Y_{t-1} and y_t are fixed so Y_{t-1} and v_t are fixed and vice versa. Consequently, $E(\alpha_t|Y_t) = E(\alpha_t|Y_{t-1}, v_t)$ and $Var(\alpha_t|Y_t) = Var(\alpha_t|Y_{t-1}, v_t)$. Since all variables are normally distributed, the conditional expectation and variance are given by standard formulae from multivariate normal regression theory. For a general treatment of the results required see the regression lemma in §2.13.

It follows from equation (2.49) of this lemma that

$$E(\alpha_t|Y_t) = E(\alpha_t|Y_{t-1}, v_t) = E(\alpha_t|Y_{t-1}) + Cov(\alpha_t, v_t)Var(v_t)^{-1}v_t, \quad (2.8)$$

where

$$\begin{aligned}
\mathrm{Cov}(\alpha_t, v_t) &= \mathrm{E}[\alpha_t(y_t - a_t)] = \mathrm{E}[\alpha_t(\alpha_t + \varepsilon_t - a_t)] \\
&= \mathrm{E}[\alpha_t(\alpha_t - a_t)] = \mathrm{E}[\mathrm{Var}(\alpha_t|Y_{t-1})] = P_t,
\end{aligned}$$

and

$$\mathrm{Var}(v_t) = F_t = \mathrm{Var}(\alpha_t + \varepsilon_t - a_t) = \mathrm{Var}(\alpha_t|Y_{t-1}) + \mathrm{Var}(\varepsilon_t) = P_t + \sigma_\varepsilon^2.$$

Since $a_t = \mathrm{E}(\alpha_t|Y_{t-1})$ we have from (2.8)

$$\mathrm{E}(\alpha_t|Y_t) = a_t + K_t v_t. \tag{2.9}$$

where $K_t = P_t/F_t$ is the regression coefficient of α_t on v_t. From equation (2.50) of the regression lemma in §2.13 we have

$$\begin{aligned}
\mathrm{Var}(\alpha_t|Y_t) &= \mathrm{Var}(\alpha_t|Y_{t-1}, v_t) \\
&= \mathrm{Var}(\alpha_t|Y_{t-1}) - \mathrm{Cov}(\alpha_t, v_t)^2 \, \mathrm{Var}(v_t)^{-1} \\
&= P_t - P_t^2/F_t \\
&= P_t(1 - K_t). \tag{2.10}
\end{aligned}$$

From (2.7), (2.9) and (2.10), we have the full set of relations for updating from time t to time $t + 1$,

$$\begin{array}{lll}
v_t = y_t - a_t, & F_t = P_t + \sigma_\varepsilon^2, & K_t = P_t/F_t, \\
a_{t+1} = a_t + K_t v_t, & P_{t+1} = P_t(1 - K_t) + \sigma_\eta^2, &
\end{array} \tag{2.11}$$

for $t = 1, \ldots, n$. Note that a_1 and P_1 are assumed known here; however, more general initial specifications will be dealt with in §2.9. Relations (2.11) constitute the celebrated *Kalman filter* for this problem. It should be noted that P_t depends only on σ_ε^2 and σ_η^2 and does not depend on Y_{t-1}. We include the case $t = n$ in (2.11) for convenience even though a_{n+1} and P_{n+1} are not normally needed for anything except forecasting.

The notation employed in (2.11) is not the simplest that could have been used; we have chosen it in order to be compatible with the notation that we consider appropriate for the treatment of the general multivariate model in Chapter 4. A set of relations such as (2.11) which enables us to calculate quantities for $t + 1$ given those for t is called a *recursion*. Formulae (2.9) and (2.10) could be derived in many other ways. For example, a routine Bayesian argument for normal densities could be used in which the prior density is $p(\alpha_t|Y_{t-1}) \sim \mathrm{N}(a_t, P_t)$, the likelihood is $p(y_t|\alpha_t)$ and we obtain the posterior density as $p(\alpha_t|Y_t) \sim \mathrm{N}(a_t^*, P_t^*)$, where a_t^* and P_t^* are given by (2.9) and (2.10), respectively. We have used the regression approach because it is particularly simple and direct for deriving filtering and smoothing recursions for the general state space model in Chapter 4.

2.2.2 ILLUSTRATION

In this chapter we shall illustrate the algorithms using observations from the river Nile. The data set consists of a series of readings of the annual flow volume at

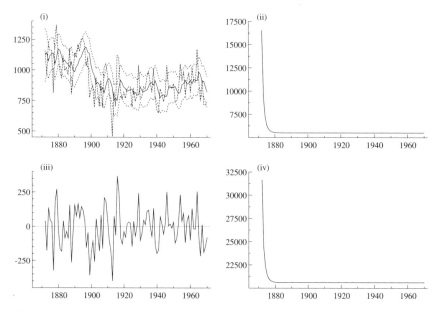

Fig. 2.1. Nile data and output of Kalman filter: (i) filtered state a_t and its 90% confidence intervals; (ii) filtered state variance P_t; (iii) prediction errors v_t; (iv) prediction variance F_t.

Aswan from 1871 to 1970. The series has been analysed by Cobb (1978) and Balke (1993). We analyse the data using the local level model (2.3) with $a_1 = 0$, $P_1 = 10^7$, $\sigma_\varepsilon^2 = 15099$ and $\sigma_\eta^2 = 1469.1$. The values for a_1 and P_1 were chosen arbitrarily for illustrative purposes. The values for σ_ε^2 and σ_η^2 are the maximum likelihood estimates which we obtain in §2.10.3. The output of the Kalman filter (that is v_t, F_t, a_t and P_t for $t = 2, \ldots, n$) is presented graphically together with the raw data in Figure 2.1.

The most obvious feature of the four graphs is that F_t and P_t converge rapidly to constant values which confirms that the local level model has a steady state solution, that is, P_t and F_t converged to fixed values; for discussion of the concept of a steady state see §2.11. However, it was found that this local level model converged numerically to a steady state in around 25 updates of P_t although the graph of P_t seems to suggest that the steady state was obtained after around 10 updates.

2.3 Forecast errors

The Kalman filter residual $v_t = y_t - a_t$ and its variance F_t are the one-step-ahead forecast error and the one-step-ahead forecast error variance of y_t given Y_{t-1} as defined in §2.2. The forecast errors $v_1, \ldots, v_n$ are sometimes called *innovations* because they represent the new part of y_t that cannot be predicted from the past for $t = 1, \ldots, n$. We shall make use of v_t and F_t for a variety of results in the next sections. It is therefore important to study them in detail.

2.3.1 CHOLESKY DECOMPOSITION

First we show that $v_1, \ldots, v_n$ are mutually independent. The joint density of $y_1, \ldots, y_n$ is

$$p(y_1, \ldots, y_n) = p(y_1) \prod_{t=2}^{n} p(y_t | Y_{t-1}). \tag{2.12}$$

Transform from $y_1, \ldots, y_n$ to $v_1, \ldots, v_n$. Since each v_t equals y_t minus a linear function of $y_1, \ldots, y_{t-1}$ for $t = 2, \ldots, n$, the Jacobian is one. From (2.12) on making the substitution we have

$$p(v_1, \ldots, v_n) = \prod_{t=1}^{n} p(v_t), \tag{2.13}$$

since $p(v_1) = p(y_1)$ and $p(v_t) = p(y_t | Y_{t-1})$ for $t = 2, \ldots, n$. Consequently, the v_t's are independently distributed.

We next show that the forecast errors v_t are effectively obtained from a Cholesky decomposition of the observation vector y. The Kalman filter recursions compute the forecast error v_t as a linear function of the initial mean a_1 and the observations $y_1, \ldots, y_t$ since

$$v_1 = y_1 - a_1,$$
$$v_2 = y_2 - a_1 - K_1(y_1 - a_1),$$
$$v_3 = y_3 - a_1 - K_2(y_2 - a_1) - K_1(1 - K_2)(y_1 - a_1), \quad \text{and so on.}$$

It should be noted that K_t does not depend on the initial mean a_1 and the observations $y_1, \ldots, y_n$; it depends only on the initial state variance P_1 and the disturbance variances σ_ε^2 and σ_η^2. Using the definitions in (2.4), we have

$$v = C(y - 1a_1), \quad \text{with} \quad v = \begin{pmatrix} v_1 \\ \vdots \\ v_n \end{pmatrix},$$

where matrix C is the lower triangular matrix

$$C = \begin{bmatrix} 1 & 0 & 0 & & 0 \\ c_{21} & 1 & 0 & & 0 \\ c_{31} & c_{32} & 1 & & 0 \\ & & & \ddots & \vdots \\ c_{n1} & c_{n2} & c_{n3} & \cdots & 1 \end{bmatrix},$$

$$c_{i,i-1} = -K_{i-1},$$
$$c_{ij} = -(1 - K_{i-1})(1 - K_{i-2}) \cdots (1 - K_{j+1})K_j, \tag{2.14}$$

for $i = 2, \ldots, n$ and $j = 1, \ldots, i - 2$. The distribution of v is therefore

$$v \sim N(0, C\Omega C'), \tag{2.15}$$

where $\Omega = \text{Var}(y)$ as given by (2.4). On the other hand we know from (2.11) and (2.13) that $E(v_t) = 0$, $\text{Var}(v_t) = F_t$ and $\text{Cov}(v_t, v_j) = 0$, for $t, j = 1, \dots, n$ and $t \neq j$; therefore,

$$v \sim N(0, F), \quad \text{with} \quad F = \begin{bmatrix} F_1 & 0 & 0 & & 0 \\ 0 & F_2 & 0 & & 0 \\ 0 & 0 & F_3 & & 0 \\ & & & \ddots & \vdots \\ 0 & 0 & 0 & \cdots & F_n \end{bmatrix},$$

where $C\Omega C' = F$. The transformation of a symmetric positive definite matrix (say Ω) into a diagonal matrix (say F) using a lower triangular matrix (say C) by means of the relation $C\Omega C' = F$ is known as the Cholesky decomposition of the symmetric matrix. The Kalman filter can therefore be regarded as essentially a Cholesky decomposition of the variance matrix implied by the local level model (2.3). This result is important for understanding the role of the Kalman filter and it will be used further in §§2.5.4 and 2.10.1. Note also that $F^{-1} = (C')^{-1}\Omega^{-1}C^{-1}$ so we have $\Omega^{-1} = C'F^{-1}C$.

2.3.2 ERROR RECURSIONS

Define the *state estimation error* as

$$x_t = \alpha_t - a_t, \quad \text{with} \quad \text{Var}(x_t) = P_t. \tag{2.16}$$

We now show that the state estimation errors x_t and forecast errors v_t are linear functions of the initial state error x_1 and the disturbances ε_t and η_t analogously to the way that α_t and y_t are linear functions of the initial state and the disturbances, for $t = 1, \dots, n$. It follows directly from the Kalman filter relations (2.11) that

$$\begin{aligned} v_t &= y_t - a_t \\ &= \alpha_t + \varepsilon_t - a_t \\ &= x_t + \varepsilon_t, \end{aligned}$$

and

$$\begin{aligned} x_{t+1} &= \alpha_{t+1} - a_{t+1} \\ &= \alpha_t + \eta_t - a_t - K_t v_t \\ &= x_t + \eta_t - K_t(x_t + \varepsilon_t) \\ &= L_t x_t + \eta_t - K_t \varepsilon_t, \end{aligned}$$

where

$$L_t = 1 - K_t = \sigma_\varepsilon^2 / F_t. \tag{2.17}$$

Thus analogously to the local level model relations

$$y_t = \alpha_t + \varepsilon_t, \qquad \alpha_{t+1} = \alpha_t + \eta_t,$$

we have the error relations

$$v_t = x_t + \varepsilon_t, \qquad x_{t+1} = L_t x_t + \eta_t - K_t \varepsilon_t, \qquad t = 1, \dots, n, \tag{2.18}$$

with $x_1 = \alpha_1 - a_1$. These relations will be used in the next section. We note that P_t, F_t, K_t and L_t do not depend on the initial state mean a_1 or the observations $y_1, \ldots, y_n$ but only on the initial state variance P_1 and the disturbance variances σ_ε^2 and σ_η^2. We note also that the recursion for P_{t+1} in (2.11) can alternatively be derived by

$$
\begin{aligned}
P_{t+1} = \operatorname{Var}(x_{t+1}) &= \operatorname{Cov}(x_{t+1}, \alpha_{t+1}) = \operatorname{Cov}(x_{t+1}, \alpha_t + \eta_t) \\
&= L_t \operatorname{Cov}(x_t, \alpha_t + \eta_t) + \operatorname{Cov}(\eta_t, \alpha_t + \eta_t) - K_t \operatorname{Cov}(\varepsilon_t, \alpha_t + \eta_t) \\
&= L_t P_t + \sigma_\eta^2.
\end{aligned}
$$

2.4   State smoothing

We now consider the estimation of $\alpha_1, \ldots, \alpha_n$ given the entire sample Y_n. We shall find it convenient to use in place of the collective symbol Y_n its representation as the vector $y = (y_1, \ldots, y_n)'$ defined in (2.4). Since all distributions are normal, the conditional density of α_t given y is N$(\hat{\alpha}_t, V_t)$ where $\hat{\alpha}_t = \operatorname{E}(\alpha_t|y)$ and $V_t = \operatorname{Var}(\alpha_t|y)$. We call $\hat{\alpha}_t$ the *smoothed state*, V_t the *smoothed state variance* and the operation of calculating $\hat{\alpha}_1, \ldots, \hat{\alpha}_n$ *state smoothing*.

2.4.1   SMOOTHED STATE

The forecast errors $v_1, \ldots, v_n$ are mutually independent and are a linear transformation of $y_1, \ldots, y_n$, and $v_t, \ldots, v_n$ are independent of $y_1, \ldots, y_{t-1}$ with zero means. Moreover, when $y_1, \ldots, y_n$ are fixed, Y_{t-1} and $v_t, \ldots, v_n$ are fixed and vice versa. By (2.49) of the regression lemma in §2.13 we therefore have

$$
\begin{aligned}
\hat{\alpha}_t = \operatorname{E}(\alpha_t|y) &= \operatorname{E}(\alpha_t|Y_{t-1}, v_t, \ldots, v_n) \\
&= \operatorname{E}(\alpha_t|Y_{t-1}) + \operatorname{Cov}[\alpha_t, (v_t, \ldots, v_n)'] \operatorname{Var}[(v_t, \ldots, v_n)']^{-1}(v_t, \ldots, v_n)' \\
&= a_t + \begin{pmatrix} \operatorname{Cov}(\alpha_t, v_t) \\ \vdots \\ \operatorname{Cov}(\alpha_t, v_n) \end{pmatrix}' \begin{bmatrix} F_t & & 0 \\ & \ddots & \\ 0 & & F_n \end{bmatrix}^{-1} \begin{pmatrix} v_t \\ \vdots \\ v_n \end{pmatrix} \\
&= a_t + \sum_{j=t}^{n} \operatorname{Cov}(\alpha_t, v_j) F_j^{-1} v_j.
\end{aligned} \qquad (2.19)
$$

Now $\operatorname{Cov}(\alpha_t, v_j) = \operatorname{Cov}(x_t, v_j)$ for $j = t, \ldots, n$, and

$$
\operatorname{Cov}(x_t, v_t) = \operatorname{E}[x_t(x_t + \varepsilon_t)] = \operatorname{Var}(x_t) = P_t,
$$
$$
\operatorname{Cov}(x_t, v_{t+1}) = \operatorname{E}[x_t(x_{t+1} + \varepsilon_{t+1})] = \operatorname{E}[x_t(L_t x_t + \eta_t - K_t \varepsilon_t)] = P_t L_t.
$$

Similarly,

$$
\operatorname{Cov}(x_t, v_{t+2}) = P_t L_t L_{t+1},
$$

$$
\vdots \qquad\qquad\qquad (2.20)
$$

$$
\operatorname{Cov}(x_t, v_n) = P_t L_t L_{t+1} \ldots L_{n-1}.
$$

Substituting in (2.19) gives

$$\hat{\alpha}_t = a_t + P_t \frac{v_t}{F_t} + P_t L_t \frac{v_{t+1}}{F_{t+1}} + P_t L_t L_{t+1} \frac{v_{t+2}}{F_{t+2}} + \cdots$$

$$= a_t + P_t r_{t-1},$$

where

$$r_{t-1} = \frac{v_t}{F_t} + L_t \frac{v_{t+1}}{F_{t+1}} + L_t L_{t+1} \frac{v_{t+2}}{F_{t+2}} + L_t L_{t+1} L_{t+2} \frac{v_{t+3}}{F_{t+3}} + \cdots$$

$$+ L_t L_{t+1} \dots L_{n-1} \frac{v_n}{F_n} \tag{2.21}$$

is a weighted sum of innovations after $t - 1$. The value of this at time t is

$$r_t = \frac{v_{t+1}}{F_{t+1}} + L_{t+1} \frac{v_{t+2}}{F_{t+2}} + L_{t+1} L_{t+2} \frac{v_{t+3}}{F_{t+3}} + \cdots$$

$$+ L_{t+1} L_{t+2} \dots L_{n-1} \frac{v_n}{F_n}. \tag{2.22}$$

Obviously, $r_n = 0$ since no observations are available after time n. By substituting from (2.22) into (2.21), it follows that the values of r_{t-1} can be evaluated using the backwards recursion

$$r_{t-1} = \frac{v_t}{F_t} + L_t r_t, \tag{2.23}$$

with $r_n = 0$, for $t = n, n - 1, \dots, 1$. The smoothed state can therefore be calculated by the backwards recursion

$$r_{t-1} = F_t^{-1} v_t + L_t r_t, \qquad \hat{\alpha}_t = a_t + P_t r_{t-1}, \qquad t = n, \dots, 1, \tag{2.24}$$

with $r_n = 0$. The relations in (2.24) are collectively called the *state smoothing recursion*.

2.4.2 SMOOTHED STATE VARIANCE

The error variance of the smoothed state, $V_t = \mathrm{Var}(\alpha_t | Y_n)$, is derived in a similar way. Using the properties of the innovations and the regression lemma in §2.13 with $x = \alpha_t$, $y = (y_1, \dots, y_{t-1})'$ and $z = (v_t, \dots, v_n)'$ in (2.50), we have

$$V_t = \mathrm{Var}(\alpha_t | y) = \mathrm{Var}(\alpha_t | Y_{t-1}, v_t, \dots, v_n)$$

$$= \mathrm{Var}(\alpha_t | Y_{t-1}) - \mathrm{Cov}[\alpha_t, (v_t, \dots, v_n)'] \, \mathrm{Var}[(v_t, \dots, v_n)']^{-1}$$

$$\times \mathrm{Cov}[\alpha_t, (v_t, \dots, v_n)']'$$

$$= P_t - \begin{pmatrix} \mathrm{Cov}(\alpha_t, v_t) \\ \vdots \\ \mathrm{Cov}(\alpha_t, v_n) \end{pmatrix}' \begin{bmatrix} F_t & & 0 \\ & \ddots & \\ 0 & & F_n \end{bmatrix}^{-1} \begin{pmatrix} \mathrm{Cov}(\alpha_t, v_t) \\ \vdots \\ \mathrm{Cov}(\alpha_t, v_n) \end{pmatrix}$$

$$= P_t - \sum_{j=t}^{n} [\mathrm{Cov}(\alpha_t, v_j)]^2 F_j^{-1}, \tag{2.25}$$

where the expressions for $\text{Cov}(\alpha_t, v_j)$ are given by (2.20). Substituting these into (2.25) leads to

$$V_t = P_t - P_t^2 \frac{1}{F_t} - P_t^2 L_t^2 \frac{1}{F_{t+1}} - P_t^2 L_t^2 L_{t+1}^2 \frac{1}{F_{t+2}} - \cdots - P_t^2 L_t^2 L_{t+1}^2 \cdots L_{n-1}^2 \frac{1}{F_n}$$
$$= P_t - P_t^2 N_{t-1}, \tag{2.26}$$

where

$$N_{t-1} = \frac{1}{F_t} + L_t^2 \frac{1}{F_{t+1}} + L_t^2 L_{t+1}^2 \frac{1}{F_{t+2}} + L_t^2 L_{t+1}^2 L_{t+2}^2 \frac{1}{F_{t+3}} + \cdots$$
$$+ L_t^2 L_{t+1}^2 \cdots L_{n-1}^2 \frac{1}{F_n}, \tag{2.27}$$

is a weighted sum of the inverse variances of innovations after time $t - 1$. Its value at time t is

$$N_t = \frac{1}{F_{t+1}} + L_{t+1}^2 \frac{1}{F_{t+2}} + L_{t+1}^2 L_{t+2}^2 \frac{1}{F_{t+3}} + \cdots + L_{t+1}^2 L_{t+2}^2 \cdots L_{n-1}^2 \frac{1}{F_n}, \tag{2.28}$$

and, obviously, $N_n = 0$ since no variances are available after time n. Substituting from (2.28) into (2.27) it follows that the value for N_{t-1} can be calculated using the backwards recursion

$$N_{t-1} = \frac{1}{F_t} + L_t^2 N_t, \tag{2.29}$$

with $N_n = 0$, for $t = n, n-1, \ldots, 1$. We see from (2.22) and (2.28) that $N_t = \text{Var}(r_t)$ since the forecast errors v_t are independent.

Combining these results, the error variance of the smoothed state can be calculated by the backwards recursion

$$N_{t-1} = F_t^{-1} + L_t^2 N_t, \qquad V_t = P_t - P_t^2 N_{t-1}, \qquad t = n, \ldots, 1, \tag{2.30}$$

with $N_n = 0$. The relations in (2.30) are collectively called the *state variance smoothing recursion*. From the standard error $\sqrt{V_t}$ of $\hat{\alpha}_t$ we can construct confidence intervals for α_t for $t = 1, \ldots, n$. It is also possible to derive the smoothed covariances between the states, that is, $\text{Cov}(\alpha_t, \alpha_s | y)$, $t \neq s$, using similar arguments. We shall not give them here but will derive them for the general case in §4.5.

2.4.3   ILLUSTRATION

We now show the results of state smoothing for the Nile data of §2.2.2 using the same local level model. The Kalman filter is applied first and the output v_t, F_t, a_t and P_t is stored for $t = 1, \ldots, n$. Figure 2.2 presents the output of the backwards smoothing recursions (2.24) and (2.30); that is r_t, N_t, $\hat{\alpha}_t$ and V_t. The plot of $\hat{\alpha}_t$ includes the confidence bands for α_t. The graph of $\text{Var}(\alpha_t | y)$ shows that the

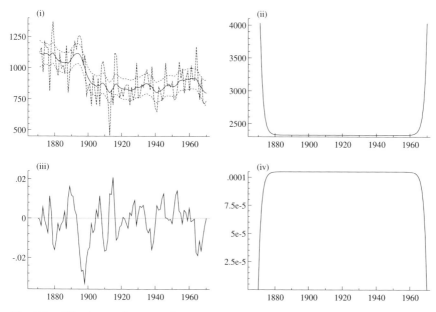

Fig. 2.2. Nile data and output of state smoothing recursion: (i) smoothed state $\hat{\alpha}_t$ and its 90% confidence intervals; (ii) smoothed state variance V_t; (iii) smoothing cumulant r_t; (iv) smoothing variance cumulant N_t.

conditional variance of α_t is larger at the beginning and end of the sample, as it obviously should be on intuitive grounds. Comparing the graphs of a_t and $\hat{\alpha}_t$ in Figures 2.1 and 2.2, we see that the graph of $\hat{\alpha}_t$ is much smoother than that of a_t, except at time points close to the end of the series, as it should be.

2.5 Disturbance smoothing

In this section we consider the calculation of the smoothed observation disturbance $\hat{\varepsilon}_t = \mathrm{E}(\varepsilon_t | y) = y_t - \hat{\alpha}_t$ and the smoothed state disturbance $\hat{\eta}_t = \mathrm{E}(\eta_t | y) = \hat{\alpha}_{t+1} - \hat{\alpha}_t$ together with their error variances. Of course, these could be calculated directly from a knowledge of $\hat{\alpha}_1, \ldots, \hat{\alpha}_n$ and covariances $\mathrm{Cov}(\alpha_t, \alpha_j | y)$ for $j \leq t$. However, it turns out to be computationally advantageous to compute them from r_t and N_t without first calculating $\hat{\alpha}_t$, particularly for the general model discussed in Chapter 4. The merits of smoothed disturbances are discussed in §4.4. For example, the estimates $\hat{\varepsilon}_t$ and $\hat{\eta}_t$ are useful for detecting outliers and structural breaks, respectively; see §2.12.2.

In order to economise on the amount of algebra in this chapter we shall present the required recursions for the local level model without proof, referring the reader to §4.4 for derivations of the analogous recursions for the general model.

2.5.1   SMOOTHED OBSERVATION DISTURBANCES

From (4.36) in §4.4.1, the smoothed observation disturbance $\hat{\varepsilon}_t = \mathrm{E}(\varepsilon_t|y)$ is calculated by

$$\hat{\varepsilon}_t = \sigma_\varepsilon^2 u_t, \qquad t = n, \dots, 1, \tag{2.31}$$

where

$$u_t = F_t^{-1} v_t - K_t r_t, \tag{2.32}$$

and where the recursion for r_t is given by (2.23). The scalar u_t is referred to as the *smoothing error*. Similarly, from (4.44) in §4.4.3, the smoothed variance $\mathrm{Var}(\varepsilon_t|y)$ is obtained by

$$\mathrm{Var}(\varepsilon_t|y) = \sigma_\varepsilon^2 - \sigma_\varepsilon^4 D_t, \qquad t = n, \dots, 1, \tag{2.33}$$

where

$$D_t = F_t^{-1} + K_t^2 N_t, \tag{2.34}$$

and where the recursion for N_t is given by (2.29). Since from (2.22) v_t is independent of r_t, and $\mathrm{Var}(r_t) = N_t$, we have

$$\mathrm{Var}(u_t) = \mathrm{Var}\left(F_t^{-1} v_t - K_t r_t\right) = F_t^{-2} \mathrm{Var}(v_t) + K_t^2 \mathrm{Var}(r_t) = D_t.$$

Consequently, from (2.31) we obtain $\mathrm{Var}(\hat{\varepsilon}_t) = \sigma_\varepsilon^4 D_t$.

Note that the methods for calculating $\hat{\alpha}_t$ and $\hat{\varepsilon}_t$ are consistent since $K_t = P_t F_t^{-1}$, $L_t = 1 - K_t = \sigma_\varepsilon^2 F_t^{-1}$ and

$$\begin{aligned}
\hat{\varepsilon}_t &= y_t - \hat{\alpha}_t \\
&= y_t - a_t - P_t r_{t-1} \\
&= v_t - P_t\left(F_t^{-1} v_t + L_t r_t\right) \\
&= F_t^{-1} v_t (F_t - P_t) - \sigma_\varepsilon^2 P_t F_t^{-1} r_t \\
&= \sigma_\varepsilon^2\left(F_t^{-1} v_t - K_t r_t\right), \qquad t = n, \dots, 1.
\end{aligned}$$

Similar equivalences can be shown for V_t and $\mathrm{Var}(\varepsilon_t|y)$.

2.5.2   SMOOTHED STATE DISTURBANCES

From (4.41) in §4.4.1, the smoothed mean of the disturbance $\hat{\eta}_t = \mathrm{E}(\eta_t|y)$ is calculated by

$$\hat{\eta}_t = \sigma_\eta^2 r_t, \qquad t = n, \dots, 1, \tag{2.35}$$

where the recursion for r_t is given by (2.23). Similarly, from (4.47) in §4.4.3, the smoothed variance $\mathrm{Var}(\eta_t|y)$ is computed by

$$\mathrm{Var}(\eta_t|y) = \sigma_\eta^2 - \sigma_\eta^4 N_t, \qquad t = n, \dots, 1, \tag{2.36}$$

where the recursion for N_t is given by (2.29). Since $\text{Var}(r_t) = N_t$, we have that $\text{Var}(\hat{\eta}_t) = \sigma_\eta^4 N_t$. These results are interesting because they give an interpretation to the values r_t and N_t; they are the scaled smoothed estimator of $\eta_t = \alpha_{t+1} - \alpha_t$ and its unconditional variance, respectively.

The method of calculating $\hat{\eta}_t$ is consistent with the definition $\eta_t = \alpha_{t+1} - \alpha_t$ since

$$
\begin{aligned}
\hat{\eta}_t &= \hat{\alpha}_{t+1} - \hat{\alpha}_t \\
&= a_{t+1} + P_{t+1} r_t - a_t - P_t r_{t-1} \\
&= a_t + K_t v_t - a_t + P_t L_t r_t + \sigma_\eta^2 r_t - P_t \left(F_t^{-1} v_t + L_t r_t \right) \\
&= \sigma_\eta^2 r_t.
\end{aligned}
$$

Similar consistencies can be shown for N_t and $\text{Var}(\eta_t | y)$.

2.5.3 ILLUSTRATION

The smoothed disturbances and their related variances for the Nile data and the local level model of §2.2.2 are calculated by the above recursions and presented in Figure 2.3. We note from the graphs of $\text{Var}(\varepsilon_t | y)$ and $\text{Var}(\eta_t | y)$ the extent that these conditional variances are larger at the beginning and end of the sample. Obviously, the plot of r_t in Figure 2.2 and the plot of $\hat{\eta}_t$ in Figure 2.3 are the same apart from a different scale.

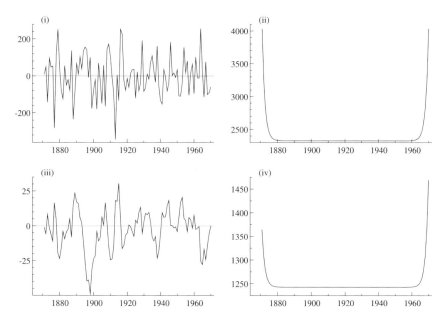

Fig. 2.3. Output of disturbance smoothing recursion: (i) observation error $\hat{\varepsilon}_t$; (ii) observation error variance $\text{Var}(\varepsilon_t | y)$; (iii) state error $\hat{\eta}_t$; (iv) state error variance $\text{Var}(\eta_t | y)$.

2.5.4   CHOLESKY DECOMPOSITION AND SMOOTHING

We now consider the calculation of $\hat{\varepsilon}_t = E(\varepsilon_t|y)$ by direct regression of $\varepsilon = (\varepsilon_1, \ldots, \varepsilon_n)'$ on the observation vector y to obtain $\hat{\varepsilon} = (\hat{\varepsilon}_1, \ldots, \hat{\varepsilon}_n)'$, that is,

$$\hat{\varepsilon} = E(\varepsilon) + \text{Cov}(\varepsilon, y) \, \text{Var}(y)^{-1}[y - E(y)]$$
$$= \text{Cov}(\varepsilon, y)\Omega^{-1}(y - 1a_1).$$

It is obvious from (2.6) that $\text{Cov}(\varepsilon, y) = \sigma_\varepsilon^2 I_n$; also, from the Cholesky decomposition considered in §2.3.1 we have $\Omega^{-1} = C'F^{-1}C$ and $C(y - 1a_1) = v$. We therefore have

$$\hat{\varepsilon} = \sigma_\varepsilon^2 C'F^{-1}v,$$

which, by consulting the definitions of the lower triangular elements of C in (2.14), also leads to the disturbance equations (2.31) and (2.32). Thus

$$\hat{\varepsilon} = \sigma_\varepsilon^2 u, \qquad u = \begin{pmatrix} u_1 \\ \vdots \\ u_n \end{pmatrix},$$

where

$$u = C'F^{-1}v \quad \text{with} \quad v = C(y - 1a_1).$$

It follows that

$$u = C'F^{-1}C(y - 1a_1) = \Omega^{-1}(y - 1a_1), \tag{2.37}$$

where $\Omega = \text{Var}(y)$ and $F = C\Omega C'$, as is consistent with standard regression theory.

2.6   Simulation

It is simple to draw samples generated by the local level model (2.3). We first draw the random normal deviates

$$\varepsilon_t^{(\cdot)} \sim N(0, \sigma_\varepsilon^2), \qquad \eta_t^{(\cdot)} \sim N(0, \sigma_\eta^2), \qquad t = 1, \ldots, n.$$

Then we generate observations using the local level recursion as follows

$$y_t^{(\cdot)} = \alpha_t^{(\cdot)} + \varepsilon_t^{(\cdot)}, \qquad \alpha_{t+1}^{(\cdot)} = \alpha_t^{(\cdot)} + \eta_t^{(\cdot)}, \qquad t = 1, \ldots, n,$$

for some starting value $\alpha_1^{(\cdot)}$.

For certain applications, which will be mainly discussed in Part II of this book, we may require samples generated by the local level model conditional on the observed time series $y_1, \ldots, y_n$. Such samples can be obtained by use of the simulation smoother developed for the general linear Gaussian state space model by de Jong and Shephard (1995) which we derive in §4.7. For the local level model, a simulated sample for the disturbances ε_t, $t = 1, \ldots, n$, given the observations

$y_1, \ldots, y_n$ can be obtained using (4.77) by the backwards recursion

$$\tilde{\varepsilon}_t = d_t + \sigma_\varepsilon^2(v_t/F_t - K_t\tilde{r}_t),$$
$$\tilde{r}_{t-1} = v_t/F_t - \tilde{W}_t d_t/C_t + L_t\tilde{r}_t,$$

where $d_t \sim N(0, C_t)$ with

$$C_t = \sigma_\varepsilon^2 - \sigma_\varepsilon^4(1/F_t + K_t^2\tilde{N}_t),$$
$$\tilde{W}_t = \sigma_\varepsilon^2(1/F_t - K_t\tilde{N}_t L_t),$$
$$\tilde{N}_{t-1} = 1/F_t + \tilde{W}_t^2/C_t + L_t^2\tilde{N}_t,$$

for $t = n, \ldots, 1$ with $\tilde{r}_n = \tilde{N}_n = 0$. The quantities v_t, F_t, K_t and $L_t = 1 - K_t$ are obtained from the Kalman filter and they need to be stored for $t = 1, \ldots, n$. We note that the recursions for $\tilde{r}_t$ and $\tilde{N}_t$ are similar to the recursions for r_t and N_t given by (2.23) and (2.29); in fact, they are equivalent if $\tilde{W}_t = 0$ for $t = 1, \ldots, n$.

Given a sample for ε_t, $t = 1, \ldots, n$, we obtain simulated samples for α_t and η_t via the relations

$$\tilde{\alpha}_t = y_t - \tilde{\varepsilon}_t, \qquad t = 1, \ldots, n,$$
$$\tilde{\eta}_t = \tilde{\alpha}_{t+1} - \tilde{\alpha}_t, \qquad t = 1, \ldots, n - 1.$$

2.6.1 ILLUSTRATION

To illustrate the difference between simulating a sample from the local level model unconditionally and simulating a sample conditional on the observations, we consider the Nile data and the local level model of §2.2.2. In Figure 2.4 (i) we present the smoothed state $\hat{\alpha}_t$ and a sample generated by the local level model unconditionally. The two series have seemingly nothing in common. In the next panel, again the smoothed state is presented but now together with a sample generated conditional on the observations. Here we see that the generated sample is much closer to $\hat{\alpha}_t$. The remaining two panels present the smoothed disturbances together with a sample from the corresponding disturbances conditional on the observations.

2.7 Missing observations

A considerable advantage of the state space approach is the ease with which missing observations can be dealt with. Suppose we have a local level model where observations y_j, with $j = \tau, \ldots, \tau^* - 1$, are missing for $1 < \tau < \tau^* \leq n$. For the filtering stage, the most obvious way to deal with the situation is to define a new series y_t^* where $y_t^* = y_t$ for $t = 1, \ldots, \tau - 1$ and $y_t^* = y_{t+\tau^*-\tau}$ for $t = \tau, \ldots, n^*$ with $n^* = n - (\tau^* - \tau)$. The model for y_t^* with time scale $t = 1, \ldots, n^*$ is then the same as (2.3) with $y_t = y_t^*$ except that $\alpha_\tau = \alpha_{\tau-1} + \eta_{\tau-1}$ where $\eta_{\tau-1} \sim N[0, (\tau^* - \tau)\sigma_\eta^2]$. Filtering for this model can be treated by the methods developed in Chapter 4 for the general state space model. The treatment is readily extended if more than one group of observations is missing.

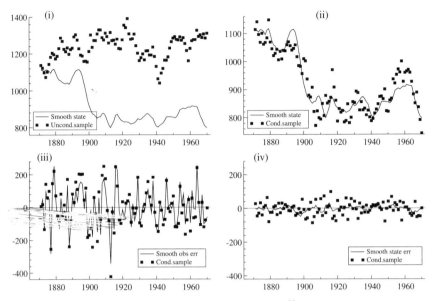

Fig. 2.4. Simulation: (i) smoothed state $\hat{\alpha}_t$ and sample $\alpha_t^{(\cdot)}$; (ii) smoothed state $\hat{\alpha}_t$ and sample $\tilde{\alpha}_t$; (iii) smoothed observation error $\hat{\varepsilon}_t$ and sample $\tilde{\varepsilon}_t$; (iv) smoothed state error $\hat{\eta}_t$ and sample $\tilde{\eta}_t$.

It is, however, easier and more transparent to proceed as follows, using the original time domain. For filtering at times $t = \tau, \ldots, \tau^* - 1$, we have

$$E(\alpha_t | Y_{t-1}) = E(\alpha_t | Y_{\tau-1}) = E\left(\alpha_\tau + \sum_{j=\tau}^{t-1} \eta_j \,\middle|\, Y_{\tau-1}\right) = a_\tau$$

and

$$\mathrm{Var}(\alpha_t | Y_{t-1}) = \mathrm{Var}(\alpha_t | Y_{\tau-1}) = \mathrm{Var}\left(\alpha_\tau + \sum_{j=\tau}^{t-1} \eta_j \,\middle|\, Y_{\tau-1}\right) = P_\tau + (t - \tau)\sigma_\eta^2.$$

giving

$$a_{t+1} = a_t, \qquad P_{t+1} = P_t + \sigma_\eta^2, \qquad t = \tau, \ldots, \tau^* - 1, \qquad (2.38)$$

the remaining values a_t and P_t being given as before by (2.11) for $t = 1, \ldots, \tau$ and $t = \tau^*, \ldots, n$. The consequence is that we can use the original filter (2.11) for all t by taking $v_t = 0$ and $K_t = 0$ at the missing time points. The same procedure is used when more than one group of observations is missing. It follows that allowing for missing observations when using the Kalman filter is extremely simple.

The forecast error recursions from which we derive the smoothing recursions are given by (2.18). These error-updating equations at the missing time points become

$$v_t = x_t + \varepsilon_t, \qquad x_{t+1} = x_t + \eta_t, \qquad t = \tau, \ldots, \tau^* - 1,$$

since $K_t = 0$ and therefore $L_t = 1$. The covariances between the state at the missing time points and the innovations after the missing period are given by

$$\mathrm{Cov}(\alpha_t, v_{\tau^*}) = P_t,$$
$$\mathrm{Cov}(\alpha_t, v_j) = P_t L_{\tau^*} L_{\tau^*+1} \ldots L_{j-1}, \qquad j = \tau^* + 1, \ldots, n, \quad t = \tau, \ldots, \tau^* - 1.$$

By deleting the terms associated with the missing time points, the state smoothing equation (2.19) for the missing time points becomes

$$\hat{\alpha}_t = a_t + \sum_{j=\tau^*}^{n} \mathrm{Cov}(\alpha_t, v_j) F_j^{-1} v_j, \qquad t = \tau, \ldots, \tau^* - 1.$$

Substituting the covariance terms into this and taking into account the definition (2.21) leads directly to

$$r_{t-1} = r_t, \qquad \hat{\alpha}_t = a_t + P_t r_{t-1}, \qquad t = \tau, \ldots, \tau^* - 1. \qquad (2.39)$$

Again, the consequence is that we can use the original state smoother (2.24) for all t by taking $v_t = 0$ and $K_t = 0$, and hence $L_t = 1$, at the missing time points. This device applies to any missing observation within the sample period. In the same way the equations for the variance of the state error and the smoothed disturbances can be obtained by putting $v_t = 0$ and $K_t = 0$ at missing time points.

2.7.1 ILLUSTRATION

Here we consider the Nile data and the same local level model as before; however, we treat the observations at time points $21, \ldots, 40$ and $61, \ldots, 80$ as missing. The Kalman filter is applied first and the output v_t, F_t, a_t and P_t is stored for $t = 1, \ldots, n$. Then, the state smoothing recursions are applied. The first two graphs in Figure 2.5 are the Kalman filter values of a_t and P_t, respectively. The last two graphs are the smoothing output $\hat{\alpha}_t$ and V_t, respectively.

Note that the application of the Kalman filter to missing observations can be regarded as extrapolation of the series to the missing time points, while smoothing at these points is effectively interpolation.

2.8 Forecasting

Let $\bar{y}_{n+j}$ be the minimum mean square error forecast of y_{n+j} given the time series $y_1, \ldots, y_n$ for $j = 1, 2, \ldots, J$ with J as some pre-defined positive integer. By minimum mean square error forecast here we mean the function $\bar{y}_{n+j}$ of $y_1, \ldots, y_n$ which minimises $\mathrm{E}[(y_{n+j} - \bar{y}_{n+j})^2 | Y_n]$. Then $\bar{y}_{n+j} = \mathrm{E}(y_{n+j} | Y_n)$. This follows immediately from the well-known result that if x is a random variable with mean

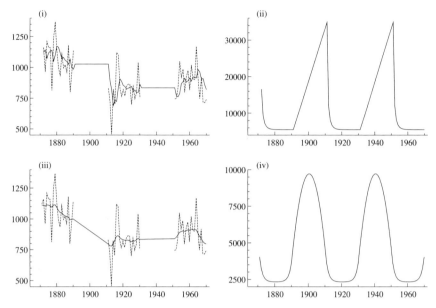

Fig. 2.5. Filtering and smoothing output when observations are missing: (i) filtered state a_t (extrapolation); (ii) filtered state variance P_t; (iii) smoothed state $\hat{\alpha}_t$ (interpolation); (iv) smoothed state variance V_t.

μ the value of λ that minimises $E(x - \lambda)^2$ is $\lambda = \mu$. The variance of the forecast error is denoted by $\bar{F}_{n+j} = \text{Var}(y_{n+j}|Y_n)$. The theory of forecasting for the local level model turns out to be surprisingly simple; we merely regard forecasting as filtering the observations $y_1, \ldots, y_n, y_{n+1}, \ldots, y_{n+J}$ using the recursion (2.11) and treating the last J observations $y_{n+1}, \ldots, y_{n+J}$ as missing.

Letting $\bar{a}_{n+j} = E(\alpha_{n+j}|Y_n)$ and $\bar{P}_{n+j} = \text{Var}(\alpha_{n+j}|Y_n)$, it follows immediately from equation (2.38) with $\tau = n + 1$ and $\tau^* = n + J$ in §2.7 that

$$\bar{a}_{n+j+1} = \bar{a}_{n+j}, \qquad \bar{P}_{n+j+1} = \bar{P}_{n+j} + \sigma_\eta^2, \qquad j = 1, \ldots, J - 1,$$

with $\bar{a}_{n+1} = a_{n+1}$ and $\bar{P}_{n+1} = P_{n+1}$ obtained from the Kalman filter (2.11). Furthermore, we have

$$\bar{y}_{n+j} = E(y_{n+j}|Y_n) = E(\alpha_{n+j}|Y_n) + E(\varepsilon_{n+j}|Y_n) = \bar{a}_{n+j},$$

$$\bar{F}_{n+j} = \text{Var}(y_{n+j}|Y_n) = \text{Var}(\alpha_{n+j}|Y_n) + \text{Var}(\varepsilon_{n+j}|Y_n) = \bar{P}_{n+j} + \sigma_\varepsilon^2,$$

for $j = 1, \ldots, J$. The consequence is that the Kalman filter can be applied for $t = 1, \ldots, n + J$ where we treat the observations at times $n + 1, \ldots, n + J$ as missing. Thus we conclude that forecasts and their error variances are delivered by applying the Kalman filter in a routine way with $v_t = 0$ and $K_t = 0$ for

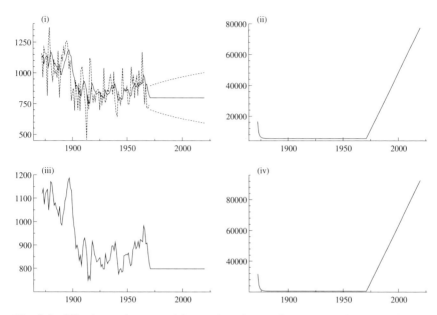

Fig. 2.6. Nile data and output of forecasting: (i) state forecast a_t and 50% confidence intervals; (ii) state variance P_t; (iii) observation forecast $E(y_t|Y_{t-1})$; (iv) observation forecast variance F_t.

$t = n + 1, \ldots, n + J$. The same property holds for the general linear Gaussian state space model as we shall show in §4.9.

2.8.1   ILLUSTRATION

The Nile data set is now extended by 30 missing observations allowing the computation of forecasts for the observations $y_{101}, \ldots, y_{130}$. The Kalman filter only is required. The graphs in Figure 2.6 contain $\hat{y}_{n+j|n} = a_{n+j|n}$, $P_{n+j|n}$, $a_{n+j|n}$ and $F_{n+j|n}$, respectively, for $j = 1, \ldots, J$ with $J = 30$. The confidence interval for $E(y_{n+j}|y)$ is $\hat{y}_{n+j|n} \pm k\sqrt{F_{n+j|n}}$ where k is determined by the required probability of inclusion; in Figure 2.6 this probability is 50%.

2.9   Initialisation

We assumed in previous sections that the distribution of the initial state α_1 is $N(a_1, P_1)$ where a_1 and P_1 are known. We now consider how to start up the filter (2.11) when nothing is known about the distribution of α_1, which is the usual situation in practice. In this situation it is reasonable to represent α_1 as having a *diffuse prior* density, that is, fix a_1 at an arbitrary value and let $P_1 \to \infty$. From (2.11) we have

$$v_1 = y_1 - a_1, \qquad F_1 = P_1 + \sigma_\varepsilon^2,$$

and, by substituting into the equations for a_2 and P_2 in (2.11), it follows that

$$a_2 = a_1 + \frac{P_1}{P_1 + \sigma_\varepsilon^2}(y_1 - a_1), \tag{2.40}$$

$$P_2 = P_1\left(1 - \frac{P_1}{P_1 + \sigma_\varepsilon^2}\right) + \sigma_\eta^2$$

$$= \frac{P_1}{P_1 + \sigma_\varepsilon^2}\sigma_\varepsilon^2 + \sigma_\eta^2. \tag{2.41}$$

Letting $P_1 \to \infty$, we obtain $a_2 = y_1$, $P_2 = \sigma_\varepsilon^2 + \sigma_\eta^2$ and we can then proceed normally with the Kalman filter (2.11) for $t = 2, \ldots, n$. This process is called *diffuse initialisation* of the Kalman filter and the resulting filter is called *the diffuse Kalman filter*. We note the interesting fact that the same values of a_t and P_t for $t = 2, \ldots, n$ can be obtained by treating y_1 as fixed and taking $\alpha_1 \sim N(y_1, \sigma_\varepsilon^2)$. This is intuitively reasonable in the absence of information about the marginal distribution of α_1 since $(y_1 - \alpha_1) \sim N(0, \sigma_\varepsilon^2)$.

We also need to take account of the diffuse distribution of the initial state α_1 in the smoothing recursions. It is shown above that the filtering equations for $t = 2, \ldots, n$ are not affected by letting $P_1 \to \infty$. Therefore, the state and disturbance smoothing equations are also not affected for $t = n, \ldots, 2$ since these only depend on the Kalman filter output. From (2.24), the smoothed mean of the state α_1 is given by

$$\hat{\alpha}_1 = a_1 + P_1\left[\frac{1}{P_1 + \sigma_\varepsilon^2}v_1 + \left(1 - \frac{P_1}{P_1 + \sigma_\varepsilon^2}\right)r_1\right],$$

$$= a_1 + \frac{P_1}{P_1 + \sigma_\varepsilon^2}v_1 + \frac{P_1}{P_1 + \sigma_\varepsilon^2}\sigma_\varepsilon^2 r_1.$$

Letting $P_1 \to \infty$, we obtain $\hat{\alpha}_1 = a_1 + v_1 + \sigma_\varepsilon^2 r_1$ and by substituting for v_1 we have

$$\hat{\alpha}_1 = y_1 + \sigma_\varepsilon^2 r_1.$$

The smoothed conditional variance of the state α_1 given y is, from (2.30)

$$V_1 = P_1 - P_1^2\left[\frac{1}{P_1 + \sigma_\varepsilon^2} + \left(1 - \frac{P_1}{P_1 + \sigma_\varepsilon^2}\right)^2 N_1\right]$$

$$= P_1\left(1 - \frac{P_1}{P_1 + \sigma_\varepsilon^2}\right) - \left(\frac{P_1}{P_1 + \sigma_\varepsilon^2}\right)^2\sigma_\varepsilon^4 N_1$$

$$= \left(\frac{P_1}{P_1 + \sigma_\varepsilon^2}\right)\sigma_\varepsilon^2 - \left(\frac{P_1}{P_1 + \sigma_\varepsilon^2}\right)^2\sigma_\varepsilon^4 N_1.$$

Letting $P_1 \to \infty$, we obtain $V_1 = \sigma_\varepsilon^2 - \sigma_\varepsilon^4 N_1$.

The smoothed means of the disturbances for $t = 1$ are given by

$$\hat{\varepsilon}_1 = \sigma_\varepsilon^2 u_1, \quad \text{with} \quad u_1 = \frac{1}{P_1 + \sigma_\varepsilon^2} v_1 - \frac{P_1}{P_1 + \sigma_\varepsilon^2} r_1,$$

and $\hat{\eta}_1 = \sigma_\eta^2 r_1$. Letting $P_1 \to \infty$, we obtain $\hat{\varepsilon}_1 = -\sigma_\varepsilon^2 r_1$. Note that r_1 depends on the Kalman filter output for $t = 2, \ldots, n$. The smoothed variances of the disturbances for $t = 1$ depend on D_1 and N_1 of which only D_1 is affected by $P_1 \to \infty$; using (2.34),

$$D_1 = \frac{1}{P_1 + \sigma_\varepsilon^2} + \left(\frac{P_1}{P_1 + \sigma_\varepsilon^2} \right)^2 N_1.$$

Letting $P_1 \to \infty$, we obtain $D_1 = N_1$ and therefore $\text{Var}(\hat{\varepsilon}_1) = \sigma_\varepsilon^4 N_1$. The variance of the smoothed estimate of η_1 remains unaltered as $\text{Var}(\hat{\eta}_1) = \sigma_\eta^4 N_1$.

The initial smoothed state $\hat{\alpha}_1$ under diffuse conditions can also be obtained by assuming that y_1 is fixed and $\alpha_1 = y_1 - \varepsilon_1$ where $\varepsilon_1 \sim N(0, \sigma_\varepsilon^2)$. For example, for the smoothed mean of the state at $t = 1$, we have now only $n - 1$ varying y_t's so that

$$\hat{\alpha}_1 = a_1 + \sum_{j=2}^{n} \frac{\text{Cov}(\alpha_1, v_j)}{F_j} v_j$$

with $a_1 = y_1$. It follows from (2.40) that $a_2 = a_1 = y_1$. Further, $v_2 = y_2 - a_2 = \alpha_2 + \varepsilon_2 - y_1 = \alpha_1 + \eta_1 + \varepsilon_2 - y_1 = -\varepsilon_1 + \eta_1 + \varepsilon_2$. Consequently, $\text{Cov}(\alpha_1, v_2) = \text{Cov}(-\varepsilon_1, -\varepsilon_1 + \eta_1 + \varepsilon_2) = \sigma_\varepsilon^2$. We therefore have from (2.19),

$$\hat{\alpha}_1 = a_1 + \frac{\sigma_\varepsilon^2}{F_2} v_2 + \frac{(1 - K_2)\sigma_\varepsilon^2}{F_3} v_3 + \frac{(1 - K_2)(1 - K_3)\sigma_\varepsilon^2}{F_4} v_4 + \cdots$$
$$= y_1 + \sigma_\varepsilon^2 r_1,$$

as before with r_1 as defined in (2.21) for $t = 1$. The equations for the remaining $\hat{\alpha}_t$'s are the same as previously.

Use of a diffuse prior for initialisation is the approach preferred by most time series analysts in the situation where nothing is known about the initial value α_1. However, some workers find the diffuse approach uncongenial because they regard the assumption of an infinite variance as unnatural since all observed time series have finite values. From this point of view an alternative approach is to assume that α_1 is an unknown constant to be estimated from the data by maximum likelihood. The simplest form of this idea is to estimate α_1 by maximum likelihood from the first observation y_1. Denote this maximum likelihood estimate by $\hat{\alpha}_1$ and its variance by $\text{Var}(\hat{\alpha}_1)$. We then initialise the Kalman filter by taking $a_1 = \hat{\alpha}_1$ and $P_1 = \text{Var}(\hat{\alpha}_1)$. Since when α_1 is fixed $y_1 \sim N(\alpha_1, \sigma_\varepsilon^2)$, we have $\hat{\alpha}_1 = y_1$ and $\text{Var}(\hat{\alpha}_1) = \sigma_\varepsilon^2$. We therefore initialise the filter by taking $a_1 = y_1$ and $P_1 = \sigma_\varepsilon^2$. But these are the same values as we obtain by assuming that α_1 is diffuse. It follows that we obtain the same initialisation of the Kalman filter by representing α_1 as a random variable with infinite variance as by assuming that it is fixed and unknown and estimating it from y_1. We shall show that a similar result holds for the general linear Gaussian state space model in §5.7.3.

2.10   Parameter estimation

We now consider the fitting of the local level model to data from the standpoint of classical inference. In effect, this amounts to deriving formulae on the assumption that the parameters are known and then replacing these by their maximum likelihood estimates. Bayesian treatment will be considered for the general linear Gaussian model in Chapter 8. Parameters in state space models are often called *hyperparameters*, possibly to distinguish them from elements of state vectors which can plausibly be thought of as random parameters; however, in this book we shall just call them *parameters*, since with the usual meaning of the word parameter this is what they are. We will discuss methods for calculating the loglikelihood function and the maximisation of it with respect to the parameters, σ_ε^2 and σ_η^2.

2.10.1   LOGLIKELIHOOD EVALUATION

Since

$$p(y_1, \ldots, y_t) = p(Y_{t-1})p(y_t|Y_{t-1}),$$

for $t = 2, \ldots, n$, the joint density of $y_1, \ldots, y_n$ can be expressed as

$$p(y) = \prod_{t=1}^{n} p(y_t|Y_{t-1}),$$

where $p(y_1|Y_0) = p(y_1)$. Now $p(y_t|Y_{t-1}) = \mathrm{N}(a_t, F_t)$ and $v_t = y_t - a_t$ so on taking logs and assuming that a_1 and P_1 are known the loglikelihood is given by

$$\log L = \log p(y) = -\frac{n}{2}\log(2\pi) - \frac{1}{2}\sum_{t=1}^{n}\left(\log F_t + \frac{v_t^2}{F_t}\right). \tag{2.42}$$

The exact loglikelihood can therefore be constructed easily from the Kalman filter (2.11).

Alternatively, let us derive the loglikelihood for the local level model from the representation (2.4). This gives

$$\log L = -\frac{n}{2}\log(2\pi) - \frac{1}{2}\log|\Omega| - \frac{1}{2}(y - a_1 1)'\Omega^{-1}(y - a_1 1),$$

which follows from the multivariate normal distribution $y \sim \mathrm{N}(a_1 1, \Omega)$. Since $\Omega = CFC'$, $|C| = 1$, $\Omega^{-1} = C'F^{-1}C$ and $v = C(y - a_1 1)$, it follows that

$$\log|\Omega| = \log|CFC'| = \log|C||F||C| = \log|F|,$$

and

$$(y - a_1 1)'\Omega^{-1}(y - a_1 1) = v'F^{-1}v.$$

Substitution and using the results $\log|F| = \sum_{t=1}^{n} \log F_t$ and $v'F^{-1}v = \sum_{t=1}^{n} F_t^{-1} v_t^2$ lead directly to (2.42).

The loglikelihood in the diffuse case is derived as follows. All terms in (2.42) remain finite as $P_1 \to \infty$ with y fixed except the term for $t = 1$. It thus seems reasonable to remove the influence of P_1 as $P_1 \to \infty$ by defining the *diffuse loglikelihood* as

$$
\begin{aligned}
\log L_d &= \lim_{P_1 \to \infty} \left(\log L + \frac{1}{2} \log P_1 \right) \\
&= -\frac{1}{2} \lim_{P_1 \to \infty} \left(\log \frac{F_1}{P_1} + \frac{v_1^2}{F_1} \right) - \frac{n}{2} \log(2\pi) - \frac{1}{2} \sum_{t=2}^{n} \left(\log F_t + \frac{v_t^2}{F_t} \right) \\
&= -\frac{n}{2} \log(2\pi) - \frac{1}{2} \sum_{t=2}^{n} \left(\log F_t + \frac{v_t^2}{F_t} \right),
\end{aligned}
\tag{2.43}
$$

since $F_1/P_1 \to 1$ and $v_1^2/F_1 \to 0$ as $P_1 \to \infty$. Note that v_t and F_t remain finite as $P_1 \to \infty$ for $t = 2, \ldots, n$.

Since P_1 does not depend on σ_ε^2 and σ_η^2, the values of σ_ε^2 and σ_η^2 that maximise $\log L$ are identical to the values that maximise $\log L + \frac{1}{2} \log P_1$. As $P_1 \to \infty$, these latter values converge to the values that maximise $\log L_d$ because first and second derivatives with respect to σ_ε^2 and σ_η^2 converge, and second derivatives are finite and strictly negative. It follows that the maximum likelihood estimators of σ_ε^2 and σ_η^2 obtained by maximising (2.42) converge to the values obtained by maximising (2.43) as $P_1 \to \infty$.

We estimate the unknown parameters σ_ε^2 and σ_η^2 by maximising expression (2.42) or (2.43) numerically according to whether a_1 and P_1 are known or unknown. In practice it is more convenient to maximise numerically with respect to the quantities $\psi_\varepsilon = \log \sigma_\varepsilon^2$ and $\psi_\eta = \log \sigma_\eta^2$. An efficient algorithm for numerical maximisation is implemented in the *STAMP* 6.0 package of Koopman, Harvey, Doornik and Shephard (2000). This optimisation procedure is based on the quasi-Newton scheme BFGS for which details are given in §7.3.2.

2.10.2 CONCENTRATION OF LOGLIKELIHOOD

It can be advantageous to re-parameterise the model prior to maximisation in order to reduce the dimensionality of the numerical search. For example, for the local level model we can put $q = \sigma_\eta^2/\sigma_\varepsilon^2$ to obtain the model

$$
\begin{aligned}
y_t &= \alpha_t + \varepsilon_t, & \varepsilon_t &\sim \mathrm{N}\left(0, \sigma_\varepsilon^2\right), \\
\alpha_{t+1} &= \alpha_t + \eta_t, & \eta_t &\sim \mathrm{N}\left(0, q\sigma_\varepsilon^2\right),
\end{aligned}
$$

and estimate the pair σ_ε^2, q in preference to $\sigma_\varepsilon^2, \sigma_\eta^2$. Put $P_t^* = P_t/\sigma_\varepsilon^2$ and $F_t^* = F_t/\sigma_\varepsilon^2$; from (2.11) and §2.9, the diffuse Kalman filter for the re-parameterised

Table 2.1. Estimation of parameters of local level model by maximum likelihood.

Iteration	q	ψ	Score	Log-likelihood
0	1	0	-3.32	-495.68
1	0.0360	-3.32	0.93	-492.53
2	0.0745	-2.60	0.25	-492.10
3	0.0974	-2.32	-0.001	-492.07
4	0.0973	-2.33	0.0	-492.07

local level model is then

$$v_t = y_t - a_t, \qquad F_t^* = P_t^* + 1,$$
$$a_{t+1} = a_t + K_t v_t, \qquad P_{t+1}^* = P_t^*(1 - K_t) + q,$$

where $K_t = P_t^*/F_t^* = P_t/F_t$ for $t = 2, \ldots, n$ and it is initialised with $a_2 = y_1$ and $P_2^* = 1 + q$. Note that F_t^* depends on q but not on σ_ε^2. The loglikelihood (2.43) then becomes

$$\log L_d = -\frac{n}{2} \log(2\pi) - \frac{n-1}{2} \log \sigma_\varepsilon^2 - \frac{1}{2} \sum_{t=2}^{n} \left(\log F_t^* + \frac{v_t^2}{\sigma_\varepsilon^2 F_t^*} \right). \quad (2.44)$$

By maximising (2.44) with respect to σ_ε^2, for given $F_2^*, \ldots, F_n^*$, we obtain

$$\hat{\sigma}_\varepsilon^2 = \frac{1}{n-1} \sum_{t=2}^{n} \frac{v_t^2}{F_t^*}. \quad (2.45)$$

The value of $\log L_d$ obtained by substituting $\hat{\sigma}_\varepsilon^2$ for σ_ε^2 in (2.44) is called the *concentrated diffuse loglikelihood* and is denoted by $\log L_{dc}$, giving

$$\log L_{dc} = -\frac{n}{2} \log(2\pi) - \frac{n-1}{2} - \frac{n-1}{2} \log \hat{\sigma}_\varepsilon^2 - \frac{1}{2} \sum_{t=2}^{n} \log F_t^*. \quad (2.46)$$

This is maximised with respect to q by a one-dimensional numerical search.

2.10.3 ILLUSTRATION

The estimates of the variances σ_ε^2 and $\sigma_\eta^2 = q\sigma_\varepsilon^2$ for the Nile data are obtained by maximising the concentrated diffuse loglikelihood (2.46) with respect to ψ where $q = \exp(\psi)$. In Table 2.1 the iterations of the BFGS procedure are reported starting with $\psi = 0$. The relative percentage change of the loglikelihood goes down very rapidly and convergence is achieved after 4 iterations. The final estimate for ψ is -2.33 and hence the estimate of q is $\hat{q} = 0.097$. The estimate of σ_ε^2 given by (2.45) is 15099 which implies that the estimate of σ_η^2 is $\hat{\sigma}_\eta^2 = \hat{q}\hat{\sigma}_\varepsilon^2 = 0.097 \times 15099 = 1469.1$.

2.11 Steady state

We now consider whether the Kalman filter (2.11) converges to a *steady state* as $n \to \infty$. This will be the case if P_t converges to a positive value, $\bar{P}$ say. Obviously, we would then have $F_t \to \bar{P} + \sigma_\varepsilon^2$ and $K_t \to \bar{P}/(\bar{P} + \sigma_\varepsilon^2)$. To check whether there is a steady state, put $P_{t+1} = P_t = \bar{P}$ in (2.11) and verify whether the resulting equation in $\bar{P}$ has a positive solution. The equation is

$$\bar{P} = \bar{P}\left(1 - \frac{\bar{P}}{\bar{P} + \sigma_\varepsilon^2}\right) + \sigma_\eta^2,$$

which reduces to the quadratic

$$x^2 - xh - h = 0, \qquad\qquad (2.47)$$

where $x = \bar{P}/\sigma_\varepsilon^2$ and $h = \sigma_\eta^2/\sigma_\varepsilon^2$, with the solution

$$x = \left(h + \sqrt{h^2 + 4h}\right)/2.$$

This is positive when $h > 0$ which holds for non-trivial models. The other solution to (2.47) is inapplicable since it is negative for $h > 0$. Thus all non-trivial local level models have a steady state solution.

The practical advantage of knowing that a model has a steady state solution is that, after convergence of P_t to $\bar{P}$ has been verified to be close enough, we can stop computing F_t and K_t and the filter (2.11) reduces to the single relation

$$a_{t+1} = a_t + \bar{K} v_t,$$

with $\bar{K} = \bar{P}/(\bar{P} + \sigma_\varepsilon^2)$ and $v_t = y_t - a_t$. While this has little consequence for the simple local level model we are concerned with, it is a useful property for the more complicated models we shall consider in Chapter 4, where P_t can be a large matrix.

2.12 Diagnostic checking

2.12.1 DIAGNOSTIC TESTS FOR FORECAST ERRORS

The assumptions underlying the local level model are that the disturbances ε_t and η_t are normally distributed and serially independent with constant variances. On these assumptions the standardised one-step forecast errors

$$e_t = \frac{v_t}{\sqrt{F_t}}, \qquad t = 1, \ldots, n, \qquad\qquad (2.48)$$

(or for $t = 2, \ldots, n$ in the diffuse case) are also normally distributed and serially independent with unit variance. We can check that these properties hold by means of the following large-sample diagnostic tests:

- Normality
  The first four moments of the standardised forecast errors are given by

$$m_1 = \frac{1}{n}\sum_{t=1}^{n} e_t,$$

$$m_q = \frac{1}{n}\sum_{t=1}^{n}(e_t - m_1)^q, \qquad q = 2, 3, 4,$$

with obvious modifications in the diffuse case. Skewness and kurtosis are denoted by S and K, respectively, and are defined as

$$S = \frac{m_3}{\sqrt{m_2^3}}, \qquad K = \frac{m_4}{m_2^2},$$

and it can be shown that when the model assumptions are valid they are asymptotically normally distributed as

$$S \sim N\left(0, \frac{6}{n}\right), \qquad K \sim N\left(3, \frac{24}{n}\right);$$

see Bowman and Shenton (1975). Standard statistical tests can be used to check whether the observed values of S and K are consistent with their asymptotic densities. They can also be combined as

$$N = n\left\{\frac{S^2}{6} + \frac{(K-3)^2}{24}\right\},$$

which asymptotically has a χ^2 distribution with 2 degrees of freedom on the null hypothesis that the normality assumption is valid. The *QQ plot* is a graphical display of ordered residuals against their theoretical quantiles. The 45 degree line is taken as a reference line (the closer the residual plot to this line, the better the match).

- Heteroscedasticity
  A simple test for heteroscedasticity is obtained by comparing the sum of squares of two exclusive subsets of the sample. For example, the statistic

$$H(h) = \frac{\sum_{t=n-h+1}^{n} e_t^2}{\sum_{t=1}^{h} e_t^2},$$

is $F_{h,h}$-distributed for some preset positive integer h, under the null hypothesis of homoscedasticity. Here, e_t is defined in (2.48) and the sum of h squared forecast errors in the denominator starts at $t = 2$ in the diffuse case.

- Serial correlation
  When the local level model holds, the standardised forecast errors are serially uncorrelated as we have shown in §2.3.1. Therefore, the correlogram of the forecast errors should reveal serial correlation insignificant. A standard portmanteau test statistic for serial correlation is based on the Box-Ljung statistic suggested by Ljung and Box (1978). This is given by

$$Q(k) = n(n+2)\sum_{j=1}^{k} \frac{c_j^2}{n-j},$$

for some preset positive integer k where c_j is the jth correlogram value

$$c_j = \frac{1}{nm_2}\sum_{t=j+1}^{n} (e_t - m_1)(e_{t-j} - m_1).$$

More details on diagnostic checking will be given in §7.5.

2.12.2 DETECTION OF OUTLIERS AND STRUCTURAL BREAKS

The standardised smoothed residuals are given by

$$u_t^* = \hat{\varepsilon}_t / \sqrt{\mathrm{Var}(\hat{\varepsilon}_t)} = D_t^{-\frac{1}{2}} u_t,$$

$$r_t^* = \hat{\eta}_t / \sqrt{\mathrm{Var}(\hat{\eta}_t)} = N_t^{-\frac{1}{2}} r_t, \qquad t = 1, \ldots, n;$$

see §2.5 for details on computing the quantities u_t, D_t, r_t and N_t. Harvey and Koopman (1992) refer to these standardised residuals as *auxiliary residuals* and they investigate their properties in detail. For example, they show that the auxiliary residuals are autocorrelated and they discuss their autocorrelation function. The auxiliary residuals can be useful in detecting outliers and structural breaks in time series because $\hat{\varepsilon}_t$ and $\hat{\eta}_t$ are estimators of ε_t and η_t. An outlier in a series that we postulate as generated by the local level model is indicated by a large (positive or negative) value for ε_t and a break in the level α_t is indicated by a large (positive or negative) value for η_t. A discussion of use of auxiliary residuals for the general model will be given in §7.5.

2.12.3 ILLUSTRATION

We consider the fitted local level model for the Nile data as obtained in §2.10.3. A plot of e_t is given in Figure 2.7 together with the histogram, the QQ plot and the correlogram. These plots are satisfactory and they suggest that the assumptions

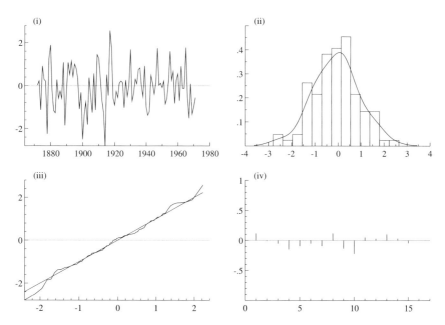

Fig. 2.7. Diagnostic plots for standardised prediction errors: (i) standardised residual; (ii) histogram plus estimated density; (iii) ordered residuals; (iv) correlogram.

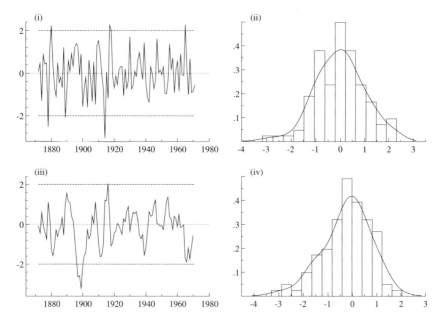

Fig. 2.8. Diagnostic plots for auxiliary residuals: (i) observation residual u_t^*; (ii) histogram and estimated density for u_t^*; (iii) state residual r_t^*; (iv) histogram and estimated density for r_t^*.

underlying the local level model are valid for the Nile data. This is largely confirmed by the following diagnostic test statistics

$$S = -0.03, \quad K = 0.09, \quad N = 0.05, \quad H(33) = 0.61, \quad Q(9) = 8.84.$$

The low value for the heteroscedasticity statistic H indicates a degree of heteroscedasticity in the residuals. This is apparent in the plots of u_t^* and r_t^* together with their histograms in Figure 2.8. These diagnostic plots indicate outliers in 1913 and 1918 and a level break in 1899. The plot of the Nile data confirms these findings.

2.13   Appendix: Lemma in multivariate normal regression

We present here a simple lemma in multivariate normal regression theory which we use extensively in the book to derive results in filtering, smoothing and related problems.

Suppose that x, y and z are random vectors of arbitrary orders that are jointly normally distributed with means μ_p and covariance matrices $\Sigma_{pq} = E[(p - \mu_p)(q - \mu_q)']$ for $p, q = x, y$ and z with $\mu_z = 0$ and $\Sigma_{yz} = 0$. The symbols x, y, z, p and q are employed for convenience and their use here is unrelated to their use in other parts of the book.

Lemma

$$E(x|y, z) = E(x|y) + \Sigma_{xz}\Sigma_{zz}^{-1}z, \tag{2.49}$$

$$\text{Var}(x|y, z) = \text{Var}(x|y) - \Sigma_{xz}\Sigma_{zz}^{-1}\Sigma_{xz}'. \tag{2.50}$$

PROOF. By standard multivariate normal regression theory we have

$$E(x|y) = \mu_x + \Sigma_{xy}\Sigma_{yy}^{-1}(y - \mu_y), \tag{2.51}$$

$$\text{Var}(x|y) = \Sigma_{xx} - \Sigma_{xy}\Sigma_{yy}^{-1}\Sigma_{xy}'; \tag{2.52}$$

see, for example, Anderson (1984, Theorem 2.5.1).

Applying (2.51) to vector $\binom{y}{z}$ in place of y gives

$$E(x|y, z) = \mu_x + [\,\Sigma_{xy} \quad \Sigma_{xz}\,]\begin{bmatrix} \Sigma_{yy}^{-1} & 0 \\ 0 & \Sigma_{zz}^{-1} \end{bmatrix}\begin{pmatrix} y - \mu_y \\ z \end{pmatrix}$$

$$= \mu_x + \Sigma_{xy}\Sigma_{yy}^{-1}(y - \mu_y) + \Sigma_{xz}\Sigma_{zz}^{-1}z,$$

since $\mu_z = 0$ and $\Sigma_{yz} = 0$. This proves (2.49).

Applying (2.52) to vector $\binom{y}{z}$ in place of y gives

$$\text{Var}(x|y, z) = \Sigma_{xx} - [\,\Sigma_{xy} \quad \Sigma_{xz}\,]\begin{bmatrix} \Sigma_{yy}^{-1} & 0 \\ 0 & \Sigma_{zz}^{-1} \end{bmatrix}\begin{bmatrix} \Sigma_{xy}' \\ \Sigma_{xz}' \end{bmatrix}$$

$$= \Sigma_{xx} - \Sigma_{xy}\Sigma_{yy}^{-1}\Sigma_{xy}' - \Sigma_{xz}\Sigma_{zz}^{-1}\Sigma_{xz}',$$

since $\Sigma_{yz} = 0$. This proves (2.50). This simple lemma provides the basis for the treatment of the Kalman filter and smoother in this book.

3
Linear Gaussian state space models

3.1 Introduction

The general linear Gaussian state space model can be written in a variety of ways; we shall use the form

$$y_t = Z_t \alpha_t + \varepsilon_t, \qquad \varepsilon_t \sim N(0, H_t),$$
$$\alpha_{t+1} = T_t \alpha_t + R_t \eta_t, \qquad \eta_t \sim N(0, Q_t), \qquad t = 1, \ldots, n, \tag{3.1}$$

where y_t is a $p \times 1$ vector of observations and α_t is an unobserved $m \times 1$ vector called the *state vector*. The idea underlying the model is that the development of the system over time is determined by α_t according to the second equation of (3.1), but because α_t cannot be observed directly we must base the analysis on observations y_t. The first equation of (3.1) is called the *observation equation* and the second is called the *state equation*. The matrices Z_t, T_t, R_t, H_t and Q_t are initially assumed to be known and the error terms ε_t and η_t are assumed to be serially independent and independent of each other at all time points. Matrices Z_t and T_{t-1} are permitted to depend on $y_1, \ldots, y_{t-1}$. The initial state vector α_1 is assumed to be $N(a_1, P_1)$ independently of $\varepsilon_1, \ldots, \varepsilon_n$ and $\eta_1, \ldots, \eta_n$, where a_1 and P_1 are first assumed known; we will consider in Chapter 5 how to proceed in the absence of knowledge of a_1 and P_1. In practice, some or all of the matrices Z_t, H_t, T_t, R_t and Q_t will depend on elements of an unknown parameter vector ψ, the estimation of which will be considered in Chapter 7.

The first equation of (3.1) has the structure of a linear regression model where the coefficient vector α_t varies over time. The second equation represents a first order vector autoregressive model, the Markovian nature of which accounts for many of the elegant properties of the state space model. The local level model (2.3) considered in the last chapter is a simple special case of (3.1). In many applications R_t is the identity. In others, one could define $\eta_t^* = R_t \eta_t$ and $Q_t^* = R_t Q_t R_t'$ and proceed without explicit inclusion of R_t, thus making the model look simpler. However, if R_t is $m \times r$ with $r < m$ and Q_t is nonsingular, there is an obvious advantage in working with nonsingular η_t rather than singular η_t^*. We assume that R_t is a subset of the columns of I_m; in this case R_t is called a *selection matrix* since it selects the rows of the state equation which have nonzero disturbance terms; however, much of the theory remains valid if R_t is a general $m \times r$ matrix.

Model (3.1) provides a powerful tool for the analysis of a wide range of problems. In this chapter we shall give substance to the general theory to be presented in Chapter 4 by describing a number of important applications of the model to problems in time series analysis and in spline smoothing analysis.

3.2   Structural time series models

3.2.1   UNIVARIATE MODELS

A *structural time series model* is one in which the trend, seasonal and error terms in the basic model (2.1), plus other relevant components, are modelled explicitly. This is in sharp contrast to the philosophy underlying Box-Jenkins ARIMA models where trend and seasonal are removed by differencing prior to detailed analysis. In this section we shall consider structural models for the case where y_t is univariate; we shall extend this to the case where y_t is multivariate in §3.2.2. A detailed discussion of structural time series models, together with further references, has been given by Harvey (1989).

The local level model considered in Chapter 2 is a simple form of a structural time series model. By adding a slope term v_t, which is generated by a random walk, to this we obtain the model

$$
\begin{aligned}
y_t &= \mu_t + \varepsilon_t, & \varepsilon_t &\sim \mathrm{N}(0, \sigma_\varepsilon^2), \\
\mu_{t+1} &= \mu_t + v_t + \xi_t, & \xi_t &\sim \mathrm{N}(0, \sigma_\xi^2), \\
v_{t+1} &= v_t + \zeta_t, & \zeta_t &\sim \mathrm{N}(0, \sigma_\zeta^2).
\end{aligned}
\tag{3.2}
$$

This is called the *local linear trend* model. If $\xi_t = \zeta_t = 0$ then $v_{t+1} = v_t = v$, say, and $\mu_{t+1} = \mu_t + v$ so the trend is exactly linear and (3.2) reduces to the deterministic linear trend plus noise model. The form (3.2) with $\sigma_\xi^2 > 0$ and $\sigma_\zeta^2 > 0$ allows the trend level and slope to vary over time.

Applied workers sometimes complain that the series of values of μ_t obtained by fitting this model does not look smooth enough to represent their idea of what a trend should look like. This objection can be met by setting $\sigma_\xi^2 = 0$ at the outset and fitting the model under this restriction. Essentially the same effect can be obtained by using in place of the second and third equation of (3.2) the model $\Delta^2 \mu_{t+1} = \zeta_t$, i.e. $\mu_{t+1} = 2\mu_t - \mu_{t-1} + \zeta_t$ where Δ is the first difference operator defined by $\Delta x_t = x_t - x_{t-1}$. This and its extension $\Delta^r \mu_t = \zeta_t$ for $r > 2$ have been advocated for modelling trend in state space models in a series of papers by Peter Young and his collaborators under the name *integrated random walk* models; see, for example, Young, Lane, Ng and Palmer (1991). We see that (3.2) can be written in the form

$$
y_t = \begin{pmatrix} 1 & 0 \end{pmatrix} \begin{pmatrix} \mu_t \\ v_t \end{pmatrix} + \varepsilon_t,
$$

$$
\begin{pmatrix} \mu_{t+1} \\ v_{t+1} \end{pmatrix} = \begin{bmatrix} 1 & 1 \\ 0 & 1 \end{bmatrix} \begin{pmatrix} \mu_t \\ v_t \end{pmatrix} + \begin{pmatrix} \xi_t \\ \zeta_t \end{pmatrix},
$$

which is a special case of (3.1).

To model the seasonal term γ_t in (2.1), suppose there are s 'months' per 'year'. Thus for monthly data $s = 12$, for quarterly data $s = 4$ and for daily data, when modelling the weekly pattern, $s = 7$. If the seasonal pattern is constant over time, the seasonal values for months 1 to s can be modelled by the constants $\gamma_1^*, \ldots, \gamma_s^*$ where $\sum_{j=1}^{s} \gamma_j^* = 0$. For the jth 'month' in 'year' i we have $\gamma_t = \gamma_j^*$ where $t = s(i - 1) + j$ for $i = 1, 2, \ldots$ and $j = 1, \ldots, s$. It follows that $\sum_{j=0}^{s-1} \gamma_{t+1-j} = 0$ so $\gamma_{t+1} = -\sum_{j=1}^{s-1} \gamma_{t+1-j}$ with $t = s - 1, s, \ldots$. In practice we often wish to allow the seasonal pattern to change over time. A simple way to achieve this is to add an error term ω_t to this relation giving the model

$$\gamma_{t+1} = -\sum_{j=1}^{s-1} \gamma_{t+1-j} + \omega_t, \qquad \omega_t \sim N(0, \sigma_\omega^2), \qquad (3.3)$$

for $t = 1, \ldots, n$ where initialisation at $t = 1, \ldots, s - 1$ will be taken care of later by our general treatment of the initialisation question in Chapter 5. An alternative suggested by Harrison and Stevens (1976) is to denote the effect of season j at time t by γ_{jt} and then let γ_{jt} be generated by the quasi-random walk

$$\gamma_{j,t+1} = \gamma_{jt} + \omega_{jt}, \qquad t = (i - 1)s + j, \qquad i = 1, 2, \ldots, \qquad j = 1, \ldots, s,$$

with an adjustment to ensure that each successive set of s seasonal components sums to zero; see Harvey (1989, §2.3.4) for details of the adjustment.

It is often preferable to express the seasonal in a trigonometric form, one version of which, for a constant seasonal, is

$$\gamma_t = \sum_{j=1}^{[s/2]} (\tilde{\gamma}_j \cos \lambda_j t + \tilde{\gamma}_j^* \sin \lambda_j t), \qquad \lambda_j = \frac{2\pi j}{s}, \qquad j = 1, \ldots, [s/2], \quad (3.4)$$

where $[a]$ is the largest integer $\leq a$ and where the quantities $\tilde{\gamma}_j$ and $\tilde{\gamma}_j^*$ are given constants. For a time-varying seasonal this can be made stochastic by replacing $\tilde{\gamma}_j$ and $\tilde{\gamma}_j^*$ by the random walks

$$\tilde{\gamma}_{j,t+1} = \tilde{\gamma}_{jt} + \tilde{\omega}_{jt}, \quad \tilde{\gamma}_{j,t+1}^* = \tilde{\gamma}_{jt}^* + \tilde{\omega}_{jt}^*, \quad j = 1, \ldots, [s/2], \quad t = 1, \ldots, n,$$
$$(3.5)$$

where $\tilde{\omega}_{jt}$ and $\tilde{\omega}_{jt}^*$ are independent $N(0, \sigma_\omega^2)$ variables; for details see Young *et al.* (1991). An alternative trigonometric form is the quasi-random walk model

$$\gamma_t = \sum_{j=1}^{[s/2]} \gamma_{jt}, \qquad (3.6)$$

where

$$\gamma_{j,t+1} = \gamma_{jt} \cos \lambda_j + \gamma_{jt}^* \sin \lambda_j + \omega_{jt},$$
$$\gamma_{j,t+1}^* = -\gamma_{jt} \sin \lambda_j + \gamma_{jt}^* \cos \lambda_j + \omega_{jt}^*, \qquad j = 1, \ldots, [s/2], \qquad (3.7)$$

in which the ω_{jt} and ω_{jt}^* terms are independent $N(0, \sigma_\omega^2)$ variables. We can show that when the stochastic terms in (3.7) are zero, the values of γ_t defined by (3.6) are periodic with period s by taking

$$\gamma_{jt} = \tilde{\gamma}_j \cos \lambda_j t + \tilde{\gamma}_j^* \sin \lambda_j t,$$

$$\gamma_{jt}^* = -\tilde{\gamma}_j \sin \lambda_j t + \tilde{\gamma}_j^* \cos \lambda_j t,$$

which are easily shown to satisfy the deterministic part of (3.7). The required result follows since γ_t defined by (3.4) is periodic with period s. In effect, the deterministic part of (3.7) provides a recursion for (3.4).

The advantage of (3.6) over (3.5) is that the contributions of the errors ω_{jt} and ω_{jt}^* are not amplified in (3.6) by the trigonometric functions $\cos \lambda_j t$ and $\sin \lambda_j t$. We regard (3.3) as the main time domain model and (3.6) as the main frequency domain model for the seasonal component in structural time series analysis.

Each of the four seasonal models can be combined with either of the trend models to give a structural time series model and all these can be put in the state space form (3.1). For example, for the local linear trend model (3.2) together with model (3.3) we take

$$\alpha_t = (\mu_t \quad \nu_t \quad \gamma_t \quad \gamma_{t-1} \ldots \gamma_{t-s+2})',$$

with

$$Z_t = (Z_{[\mu]}, Z_{[\gamma]}), \qquad T_t = \mathrm{diag}(T_{[\mu]}, T_{[\gamma]}),$$
$$R_t = \mathrm{diag}(R_{[\mu]}, R_{[\gamma]}), \qquad Q_t = \mathrm{diag}(Q_{[\mu]}, Q_{[\gamma]}), \tag{3.8}$$

where

$$Z_{[\mu]} = (1, 0), \qquad Z_{[\gamma]} = (1, 0, \ldots, 0),$$

$$T_{[\mu]} = \begin{bmatrix} 1 & 1 \\ 0 & 1 \end{bmatrix}, \qquad T_{[\gamma]} = \begin{bmatrix} -1 & -1 & \cdots & -1 & -1 \\ 1 & 0 & & 0 & 0 \\ 0 & 1 & & 0 & 0 \\ & & \ddots & & \\ 0 & 0 & & 1 & 0 \end{bmatrix},$$

$$R_{[\mu]} = I_2, \qquad R_{[\gamma]} = (1, 0, \ldots, 0)',$$

$$Q_{[\mu]} = \begin{bmatrix} \sigma_\xi^2 & 0 \\ 0 & \sigma_\zeta^2 \end{bmatrix}, \qquad Q_{[\gamma]} = \sigma_\omega^2.$$

This model plays a prominent part in the approach of Harvey (1989) to structural time series analysis; he calls it the *basic structural time series model*. The state space form of this basic model with $s = 4$ is therefore

$$\alpha_t = (\mu_t \quad \nu_t \quad \gamma_t \quad \gamma_{t-1} \quad \gamma_{t-2})',$$

$$Z_t = (1 \quad 0 \quad 1 \quad 0 \quad 0), \qquad T_t = \begin{bmatrix} 1 & 1 & 0 & 0 & 0 \\ 0 & 1 & 0 & 0 & 0 \\ 0 & 0 & -1 & -1 & -1 \\ 0 & 0 & 1 & 0 & 0 \\ 0 & 0 & 0 & 1 & 0 \end{bmatrix},$$

$$R_t = \begin{bmatrix} 1 & 0 & 0 \\ 0 & 1 & 0 \\ 0 & 0 & 1 \\ 0 & 0 & 0 \\ 0 & 0 & 0 \end{bmatrix}, \qquad Q_t = \begin{bmatrix} \sigma_\xi^2 & 0 & 0 \\ 0 & \sigma_\zeta^2 & 0 \\ 0 & 0 & \sigma_\omega^2 \end{bmatrix}.$$

Alternative seasonal specifications can also be used within the basic structural model. The Harrison and Stevens (1976) seasonal model referred to below (3.3) has the $(s + 2) \times 1$ state vector

$$\alpha_t = (\mu_t \quad \nu_t \quad \gamma_t \ldots \gamma_{t-s+1})',$$

where the relevant parts of the system matrices for substitution in (3.8) are given by

$$Z_{[\gamma]} = (1, 0, \ldots, 0), \qquad T_{[\gamma]} = \begin{bmatrix} 0 & I_{s-1} \\ 1 & 0 \end{bmatrix},$$

$$R_{[\gamma]} = I_s, \qquad Q_{[\gamma]} = \sigma_\omega^2 (I_s - 11'/s),$$

in which 1 is an $s \times 1$ vector of ones, $\omega_{1t} + \cdots + \omega_{s,t} = 0$ and variance matrix $Q_{[\gamma]}$ has rank $s - 1$. The trigonometric seasonal specification (3.7) has the $(s + 1) \times 1$ state vector

$$\alpha_t = (\mu_t \quad \nu_t \quad \gamma_{1t} \quad \gamma_{1t}^* \quad \gamma_{2t} \ldots)',$$

with the relevant parts of the system matrices given by

$$Z_{[\gamma]} = (1, 0, 1, 0, 1, \ldots, 1, 0, 1) \qquad T_{[\gamma]} = \text{diag}(C_1, \ldots, C_{s^*}, -1),$$

$$R_{[\gamma]} = I_{s-1}, \qquad Q_{[\gamma]} = \sigma_\omega^2 I_{s-1},$$

where we assume that s is even, $s^* = [\frac{s-1}{2}]$ and

$$C_j = \begin{bmatrix} \cos \lambda_j & \sin \lambda_j \\ -\sin \lambda_j & \cos \lambda_j \end{bmatrix}, \qquad \lambda_j = \frac{2\pi j}{s} \qquad j = 1, \ldots, [s/2].$$

When s is odd, we have

$$Z_{[\gamma]} = (1, 0, 1, 0, 1, \ldots, 1, 0) \qquad T_{[\gamma]} = \text{diag}(C_1, \ldots, C_{s^*}),$$

$$R_{[\gamma]} = I_{s-1}, \qquad Q_{[\gamma]} = \sigma_\omega^2 I_{s-1}.$$

Another important component in some time series is the *cycle* c_t which we can introduce by extending the basic time series model (2.2) to

$$y_t = \mu_t + \gamma_t + c_t + \varepsilon_t, \qquad t = 1, \ldots, n. \tag{3.9}$$

In its simplest form c_t is a pure sine wave generated by the relation

$$c_t = \tilde{c} \cos \lambda_c t + \tilde{c}^* \sin \lambda_c t,$$

where λ_c is the frequency of the cycle; the period is $2\pi/\lambda_c$ which is normally substantially greater that the seasonal period s. As with the seasonal, we can allow the cycle to change stochastically over time by means of the relations analogous to (3.7)

$$c_{t+1} = c_t \cos \lambda_c + c_t^* \sin \lambda_c + \tilde{\omega}_t,$$
$$c_{t+1}^* = -c_t \sin \lambda_c + c_t^* \cos \lambda_c + \tilde{\omega}_t^*, \tag{3.10}$$

with

$$c_t^* = -\tilde{c} \sin \lambda_c t + \tilde{c}^* \cos \lambda_c t,$$

where $\tilde{\omega}_t$ and $\tilde{\omega}_t^*$ are independent $N(0, \sigma_{\tilde{w}}^2)$ variables. Cycles of this form fit naturally into the structural time series model framework. The frequency λ_c can be treated as an unknown parameter to be estimated.

Explanatory variables and intervention effects are easily allowed for in the structural model framework. Suppose we have k regressors $x_{1t}, \ldots, x_{kt}$ with regression coefficients $\beta_1, \ldots, \beta_k$ which are constant over time and that we also wish to measure the change in level due to an intervention at time τ. We define an *intervention variable* w_t as follows:

$$\begin{aligned} w_t &= 0, & t < \tau, \\ &= 1, & t \geq \tau. \end{aligned}$$

Adding these to the model (3.9) gives

$$y_t = \mu_t + \gamma_t + c_t + \sum_{j=1}^{k} \beta_j x_{jt} + \lambda w_t + \varepsilon_t, \qquad t = 1, \ldots, n. \tag{3.11}$$

We see that λ measures the change in the level of the series at a known time τ due to an intervention at time τ. The resulting model can readily be put into state space form. For example, if $\gamma_t = c_t = \lambda = 0$, $k = 1$ and if μ_t is determined by a local level model, we can take

$$\alpha_t = (\mu_t \quad \beta_{1t})', \qquad Z_t = (1 \quad x_{1t}),$$

$$T_t = \begin{bmatrix} 1 & 0 \\ 0 & 1 \end{bmatrix}, \qquad R_t = \begin{pmatrix} 1 \\ 0 \end{pmatrix}, \qquad Q_t = \sigma_\xi^2,$$

in (3.1). Here, although we have attached a suffix t to β_1 it is made to satisfy $\beta_{1,t+1} = \beta_{1t}$ so it is constant. Other examples of intervention variables are the *pulse intervention variable* defined by

$$w_t = 0, \qquad t < \tau, \qquad t > \tau,$$
$$\quad = 1, \qquad t = \tau,$$

and the *slope intervention variable* defined by

$$w_t = 0, \qquad\qquad t < \tau,$$
$$\quad = 1 + t - \tau, \qquad t \geq \tau.$$

For other forms of intervention variable designed to represent a more gradual change of level or a transient change see Box and Tiao (1975). Coefficients such as λ which do not change over time can be incorporated into the state vector by setting the corresponding state errors equal to zero. Regression coefficients β_{jt} which change over time can be handled straightforwardly in the state space framework by modelling them by random walks of the form

$$\beta_{j,t+1} = \beta_{jt} + \chi_{jt}, \qquad \chi_{jt} \sim N(0, \sigma_\chi^2), \qquad j = 1, \ldots, k. \qquad (3.12)$$

An example of the use of model (3.11) for intervention analysis is given by Harvey and Durbin (1986) who used it to measure the effect of the British seat belt law on road traffic casualities. Of course, if the cycle term, the regression term or the intervention term are not required, they can be omitted from (3.11). Instead of including regression and intervention coefficients in the state vector, an alternative way of dealing with them is to concentrate them out of the likelihood function and estimate them via regression, as we will show in §6.2.3.

3.2.2   MULTIVARIATE MODELS

The methodology of structural time series models lends itself easily to generalisation to multivariate time series. Consider the local level model for a $p \times 1$ vector of observations y_t, that is

$$y_t = \mu_t + \varepsilon_t,$$
$$\mu_{t+1} = \mu_t + \eta_t, \qquad\qquad (3.13)$$

where μ_t, ε_t and η_t are $p \times 1$ vectors and

$$\varepsilon_t \sim N(0, \Sigma_\varepsilon), \qquad \eta_t \sim N(0, \Sigma_\eta),$$

with $p \times p$ variance matrices Σ_ε and Σ_η. In this so-called *seemingly unrelated time series equations* model, each series in y_t is modelled as in the univariate case, but the disturbances may be correlated instantaneously across series. In the case of a model with other components such as slope, cycle and seasonal, the disturbances associated with the components become vectors which have $p \times p$ variance matrices. The link across the p different time series is through the correlations of the disturbances driving the components.

A seemingly unrelated time series equations model is said to be *homogeneous* when the variance matrices associated with the different disturbances are proportional to each other. For example, the homogeneity restriction for the multivariate local level model is

$$\Sigma_\eta = q \Sigma_\varepsilon,$$

where scalar q is the signal-to-noise ratio. This means that all the series in y_t, and linear combinations thereof, have the same dynamic properties which implies that they have the same autocorrelation function for the stationary form of the model. A homogeneous model is a rather restricted model but it is easy to estimate. For further details we refer to Harvey (1989, Chapter 8).

Consider the multivariate local level model without the homogeneity restriction but with the assumption that the rank of Σ_η is $r < p$. The model then contains only r underlying level components. We may refer to these as *common levels*. Recognition of such common factors yields models which may not only have an interesting interpretation, but may also provide more efficient inferences and forecasts. With an appropriate ordering of the series the model may be written as

$$y_t = a + A\mu_t^* + \varepsilon_t,$$
$$\mu_{t+1}^* = \mu_t^* + \eta_t^*,$$

where μ_t^* and η_t^* are $r \times 1$ vectors, a is a $p \times 1$ vector and A is a $p \times r$ matrix. We further assume that

$$a = \begin{pmatrix} 0 \\ a^* \end{pmatrix}, \qquad A = \begin{bmatrix} I_r \\ A^* \end{bmatrix}, \qquad \eta_t^* \sim N(0, \Sigma_\eta^*),$$

where a^* is a $(p-r) \times 1$ vector and A^* is a $(p-r) \times r$ matrix of nonzero values and where variance matrix Σ_η^* is a $r \times r$ positive definite matrix. The matrix A may be interpreted as a factor loading matrix. When there is more than one common factor $(r > 1)$, the factor loadings are not unique. A factor rotation may give components with a more interesting interpretation.

Further discussion of multivariate extensions of structural time series models are given by Harvey (1989, Chapter 8) and Harvey and Koopman (1997).

3.2.3   STAMP

A wide-ranging discussion of structural time series models can be found in the book of Harvey (1989). A good supplementary source for further applications and later work is Harvey and Shephard (1993). The computer package *STAMP 6.0* of Koopman *et al.* (2000) is designed to analyse, model and forecast time series based on univariate and multivariate structural time series models. The package has implemented the Kalman filter and associated algorithms leaving the user free to concentrate on the important part of formulating a model. More information on *STAMP* can be obtained from the Internet at

http://stamp-software.com/

3.3   ARMA models and ARIMA models

Autoregressive integrated moving average (ARIMA) time series models were introduced by Box and Jenkins in their pathbreaking (1970) book; see Box, Jenkins and Reinsel (1994) for the current version of this book. As with structural time series models considered in the last section, Box and Jenkins typically regarded a univariate time series y_t as made up of trend, seasonal and irregular components. However, instead of modelling the various components separately, their idea was to eliminate the trend and seasonal by differencing at the outset of the analysis. The resulting differenced series are treated as a stationary time series, that is, a series where characteristic properties such as means, covariances, etc. remain invariant under translation through time. Let $\Delta y_t = y_t - y_{t-1}$, $\Delta^2 y_t = \Delta(\Delta y_t)$, $\Delta_s y_t = y_t - y_{t-s}$, $\Delta_s^2 y_t = \Delta_s(\Delta_s y_t)$, and so on, where we are assuming that we have s 'months' per 'year'. Box and Jenkins suggest that differencing is continued until trend and seasonal effects have been eliminated, giving a new variable $y_t^* = \Delta^d \Delta_s^D y_t$ for $d, D = 0, 1, \ldots$, which we model as a stationary autoregressive moving average ARMA(p, q) model given by

$$y_t^* = \phi_1 y_{t-1}^* + \cdots + \phi_p y_{t-p}^* + \zeta_t + \theta_1 \zeta_{t-1} + \cdots + \theta_q \zeta_{t-q}, \qquad \zeta_t \sim N(0, \sigma_\zeta^2),$$
$$(3.14)$$

with non-negative integers p and q and where ζ_t is a serially independent series of N(0, σ_ζ^2) disturbances. This can be written in the form

$$y_t^* = \sum_{j=1}^{r} \phi_j y_{t-j}^* + \zeta_t + \sum_{j=1}^{r-1} \theta_j \zeta_{t-j}, \qquad t = 1, \ldots, n, \qquad (3.15)$$

where $r = \max(p, q + 1)$ and for which some coefficients are zero. Box and Jenkins normally included a constant term in (3.15) but for simplicity we omit this; the modifications needed to include it are straightforward. We use the symbols d, p and q here and elsewhere in their familiar ARIMA context without prejudice to their use in different contexts in other parts of the book.

We now demonstrate how to put these models into state space form, beginning with case where $d = D = 0$, that is, no differencing is needed, so we can model the series by (3.15) with y_t^* replaced by y_t. Take

$$Z_t = (1 \quad 0 \quad 0 \cdots 0),$$

$$\alpha_t = \begin{pmatrix} y_t \\ \phi_2 y_{t-1} + \cdots + \phi_r y_{t-r+1} + \theta_1 \zeta_t + \cdots + \theta_{r-1} \zeta_{t-r+2} \\ \phi_3 y_{t-1} + \cdots + \phi_r y_{t-r+2} + \theta_2 \zeta_t + \cdots + \theta_{r-1} \zeta_{t-r+3} \\ \vdots \\ \phi_r y_{t-1} + \theta_{r-1} \zeta_t \end{pmatrix}, \qquad (3.16)$$

and write the state equation for α_{t+1} as in (3.1) with

$$
T_t = T = \begin{bmatrix} \phi_1 & 1 & & 0 \\ \vdots & & \ddots & \\ \phi_{r-1} & 0 & & 1 \\ \phi_r & 0 & \cdots & 0 \end{bmatrix}, \qquad R_t = R = \begin{pmatrix} 1 \\ \theta_1 \\ \vdots \\ \theta_{r-1} \end{pmatrix}, \qquad \eta_t = \zeta_{t+1}.
$$

$$(3.17)$$

This, together with the observation equation $y_t = Z_t\alpha_t$, is equivalent to (3.15) but is now in the state space form (3.1) with $\varepsilon_t = 0$, implying that $H_t = 0$. For example, with $r = 2$ we have the state equation

$$
\begin{pmatrix} y_{t+1} \\ \phi_2 y_t + \theta_1 \zeta_{t+1} \end{pmatrix} = \begin{bmatrix} \phi_1 & 1 \\ \phi_2 & 0 \end{bmatrix} \begin{pmatrix} y_t \\ \phi_2 y_{t-1} + \theta_1 \zeta_t \end{pmatrix} + \begin{pmatrix} 1 \\ \theta_1 \end{pmatrix} \zeta_{t+1}.
$$

The form given is not the only state space version of an ARMA model but is a convenient one.

We now consider the case of a univariate non-seasonal nonstationary ARIMA model of order p, d and q, with $d > 0$, given by (3.14) with $y_t^* = \Delta^d y_t$. As an example, we first consider the state space form of the ARIMA model with $p = 2$, $d = 1$ and $q = 1$ which is given by

$$
y_t = (1 \quad 1 \quad 0)\alpha_t,
$$

$$
\alpha_{t+1} = \begin{bmatrix} 1 & 1 & 0 \\ 0 & \phi_1 & 1 \\ 0 & \phi_2 & 0 \end{bmatrix} \alpha_t + \begin{pmatrix} 0 \\ 1 \\ \theta_1 \end{pmatrix} \zeta_{t+1},
$$

with the state vector defined as

$$
\alpha_t = \begin{pmatrix} y_{t-1} \\ y_t^* \\ \phi_2 y_{t-1}^* + \theta_1 \zeta_t \end{pmatrix},
$$

and $y_t^* = \Delta y_t = y_t - y_{t-1}$. This example generalises easily to ARIMA models with $d = 1$ with other values for p and q. The ARIMA model with $p = 2$, $d = 2$ and $q = 1$ in state space form is given by

$$
y_t = (1 \quad 1 \quad 1 \quad 0)\alpha_t,
$$

$$
\alpha_{t+1} = \begin{bmatrix} 1 & 1 & 1 & 0 \\ 0 & 1 & 1 & 0 \\ 0 & 0 & \phi_1 & 1 \\ 0 & 0 & \phi_2 & 0 \end{bmatrix} \alpha_t + \begin{pmatrix} 0 \\ 0 \\ 1 \\ \theta_1 \end{pmatrix} \zeta_{t+1},
$$

with

$$\alpha_t = \begin{pmatrix} y_{t-1} \\ \Delta y_{t-1} \\ y_t^* \\ \phi_2 y_{t-1}^* + \theta_1 \zeta_t \end{pmatrix},$$

and $y_t^* = \Delta^2 y_t = \Delta(y_t - y_{t-1})$. The relations between y_t, Δy_t and $\Delta^2 y_t$ follow immediately since

$$\Delta y_t = \Delta^2 y_t + \Delta y_{t-1},$$
$$y_t = \Delta y_t + y_{t-1} = \Delta^2 y_t + \Delta y_{t-1} + y_{t-1}.$$

We deal with the unknown nonstationary values y_0 and Δy_0 in the initial state vector α_1 in §5.6.3 where we describe the initialisation procedure for filtering and smoothing. Instead of estimating y_0 and Δy_0 directly, we treat these elements of α_1 as diffuse random elements while the other elements, including y_t^*, are stationary which have proper unconditional means and variances. The need to facilitate the initialisation procedure explains why we set up the state space model in this form. The state space forms for ARIMA models with other values for p^*, d and q^* can be represented in similar ways. The advantage of the state space formulation is that the array of techniques that have been developed for state space models are made available for ARMA and ARIMA models. In particular, techniques for exact maximum likelihood estimation and for initialisation are available.

As indicated above, for seasonal series both trend and seasonal are eliminated by the differencing operation $y_t^* = \Delta^d \Delta_s^D y_t$ prior to modelling y_t^* by a stationary ARMA model of the form (3.15). The resulting model for y_t^* can be put into state space form by a straightforward extension of the above treatment. A well-known seasonal ARIMA model is the so-called *airline model* which is given by

$$y_t^* = \Delta\Delta_{12} y_t = \zeta_t - \theta_1 \zeta_{t-1} - \theta_{12}\zeta_{t-12} + \theta_1\theta_{12}\zeta_{t-13}, \qquad (3.18)$$

which has a standard ARIMA state space representation.

It is interesting to note that for many state space models an inverse relation holds in the sense that the state space model has an ARIMA representation. For example, if second differences are taken in the local linear trend model (3.2), the terms in μ_t and ν_t disappear and we obtain

$$\Delta^2 y_t = \varepsilon_{t+2} - 2\varepsilon_{t+1} + \varepsilon_t + \xi_{t+1} - \xi_t + \zeta_t.$$

Since the first two autocorrelations of this are nonzero and the rest are zero, we can write it as a moving average series $\zeta_t^* + \theta_1\zeta_{t-1}^* + \theta_2\zeta_{t-2}^*$ where the ζ_t^*'s are independent $N(0, \sigma_{\zeta^*}^2)$ disturbances, and we obtain the representation

$$\Delta^2 y_t = \zeta_t^* + \theta_1\zeta_{t-1}^* + \theta_2\zeta_{t-2}^*. \qquad (3.19)$$

In Box and Jenkins' notation this is an ARIMA(0,2,2) model. It is important to recognise that the model (3.19) is less informative than (3.2) since it has lost the information that exists in the form (3.2) about the level μ_t and the slope v_t. If a seasonal term generated by model (3.3) is added to the local linear trend model, the corresponding ARIMA model has the form

$$\Delta^2 \Delta_s y_t = \zeta_t^* + \sum_{j=1}^{s+2} \theta_j \zeta_{t-j}^*,$$

where $\theta_1, \ldots, \theta_{s+2}$ are determined by the four variances σ_ε^2, σ_ξ^2, σ_ζ^2 and σ_ω^2. In this model, information about the seasonal is lost as well as information about the trend. The fact that structural time series models provide explicit information about trend and seasonal, whereas ARIMA models do not, is an important advantage that the structural modelling approach has over ARIMA modelling. We shall make a detailed comparison of the two approaches to time series analysis in §3.5.

3.4 Exponential smoothing

In this section we consider the development of exponential smoothing in the 1950s and we examine its relation to simple forms of state space and Box-Jenkins models.

Let us start with the introduction in the 1950's of the exponentially weighted moving average (EWMA) for one-step-ahead forecasting of y_{t+1} given a univariate time series $y_t, y_{t-1}, \ldots$. This has the form

$$\hat{y}_{t+1} = (1 - \lambda) \sum_{j=0}^{\infty} \lambda^j y_{t-j}, \qquad 0 < \lambda < 1. \qquad (3.20)$$

From (3.20) we deduce immediately the recursion

$$\hat{y}_{t+1} = (1 - \lambda) y_t + \lambda \hat{y}_t, \qquad (3.21)$$

which is used in place of (3.20) for practical computation. This has a simple structure and requires little storage so it was very convenient for the primitive computers available in the 1950's. As a result, EWMA forecasting became very popular in industry, particularly for sales forecasting of many items simultaneously. We call the operation of calculating forecasts by (3.21) *exponential smoothing*.

Denote the one-step forecast error $y_t - \hat{y}_t$ by u_t and substitute in (3.21) with t replaced by $t - 1$; this gives

$$y_t - u_t = (1 - \lambda) y_{t-1} + \lambda (y_{t-1} - u_{t-1}),$$

that is,

$$\Delta y_t = u_t - \lambda u_{t-1}. \qquad (3.22)$$

Taking u_t to be a series of independent $N(0, \sigma_u^2)$ variables, we see that we have deduced from the EWMA recursion (3.21) the simple ARIMA model (3.22).

An important contribution was made by Muth (1960) who showed that EWMA forecasts produced by the recursion (3.21) are minimum mean square error forecasts in the sense that they minimise $E(\hat{y}_{t+1} - y_{t+1})^2$ for observations

$y_t, y_{t-1}, \dots$ generated by the local level model (2.3), which for convenience we write in the form

$$y_t = \mu_t + \varepsilon_t,$$
$$\mu_{t+1} = \mu_t + \xi_t, \qquad\qquad (3.23)$$

where ε_t and ξ_t are serially independent random variables with zero means and constant variances. Taking first differences of observations y_t generated by (3.23) gives

$$\Delta y_t = y_t - y_{t-1} = \varepsilon_t - \varepsilon_{t-1} + \xi_{t-1}.$$

Since ε_t and ξ_t are serially uncorrelated the autocorrelation coefficient of the first lag for Δy_t is nonzero but all higher autocorrelations are zero. This is the autocorrelation function of a moving average model of order one which, with λ suitably defined, we can write in the form

$$\Delta y_t = u_t - \lambda u_{t-1},$$

which is the same as model (3.22).

We observe the interesting point that these two simple forms of state space and ARIMA models produce the same one-step forecasts and that these can be calculated by the EWMA (3.21) which has proven practical value. We can write this in the form

$$\hat{y}_{t+1} = \hat{y}_t + (1 - \lambda)(y_t - \hat{y}_t),$$

which is the Kalman filter for the simple state space model (3.23).

The EWMA was extended by Holt (1957) and Winters (1960) to series containing trend and seasonal. The extension for trend in the additive case is

$$\hat{y}_{t+1} = m_t + b_t,$$

where m_t and b_t are level and slope terms generated by the EWMA type recursions

$$m_t = (1 - \lambda_1)y_t + \lambda_1(m_{t-1} + b_{t-1}),$$
$$b_t = (1 - \lambda_2)(m_t - m_{t-1}) + \lambda_2 b_{t-1}.$$

In an interesting extension of the results of Muth (1960), Theil and Wage (1964) showed that the forecasts produced by these Holt-Winters recursions are minimum mean square error forecasts for the state space model

$$y_t = \mu_t + \varepsilon_t,$$
$$\mu_{t+1} = \mu_t + v_t + \xi_t, \qquad\qquad (3.24)$$
$$v_{t+1} = v_t + \zeta_t,$$

which is the local linear trend model (3.2). Taking second differences of y_t generated by (3.24), we obtain

$$\Delta^2 y_t = \zeta_{t-2} + \xi_{t-1} - \xi_{t-2} + \varepsilon_t - 2\varepsilon_{t-1} + \varepsilon_{t-2}.$$

This is a stationary series with nonzero autocorrelations at lags 1 and 2 but zero autocorrelations elsewhere. It therefore follows the moving average model

$$\Delta^2 y_t = u_t - \theta_1 u_{t-1} - \theta_2 u_{t-2},$$

which is a simple form of ARIMA model.

Adding the seasonal term $\gamma_{t+1} = -\gamma_t - \cdots - \gamma_{t-s+2} + \omega_t$ from (3.3) to the measurement equation of (3.23) gives the model

$$
\begin{aligned}
y_t &= \mu_t + \gamma_t + \varepsilon_t, \\
\mu_{t+1} &= \mu_t + \xi_t, \\
\gamma_{t+1} &= -\gamma_t - \cdots - \gamma_{t-s+2} + \omega_t,
\end{aligned}
\tag{3.25}
$$

which is a special case of the structural time series models of §3.2. Now take first differences and first seasonal differences of (3.25). We find

$$\Delta\Delta_s y_t = \xi_{t-1} - \xi_{t-s-1} + \omega_{t-1} - 2\omega_{t-2} + \omega_{t-3} + \varepsilon_t - \varepsilon_{t-1} - \varepsilon_{t-s} + \varepsilon_{t-s-1},$$

$$\tag{3.26}$$

which is a stationary time series with nonzero autocorrelations at lags $1, 2, s - 1$, s and $s + 1$. Consider the airline model (3.18) for general s,

$$\Delta\Delta_s y_t = u_t - \theta_1 u_{t-1} - \theta_s u_{t-s} - \theta_1\theta_s u_{t-s-1},$$

which has been found to fit well many economic time series containing trend and seasonal. It has nonzero autocorrelations at lags $1, s - 1, s$ and $s + 1$. Now the autocorrelation at lag 2 from model (3.25) arises only from $\text{Var}(\omega_t)$ which in most cases in practice is small. Thus when we add a seasonal component to the models we find again a close correspondence between state space and ARIMA models. A slope component ν_t can be added to (3.25) as in (3.24) without significantly affecting the conclusions.

A pattern is now emerging. Starting with EWMA forecasting, which in appropriate circumstances has been found to work well in practice, we have found that there are two distinct types of models, the state space models and the Box-Jenkins ARIMA models which appear to be very different conceptually but which both give minimum mean square error forecasts from EWMA recursions. The explanation is that when the time series has an underlying structure which is sufficiently simple, then the appropriate state space and ARIMA models are essentially equivalent. It is when we move towards more complex structures that the differences emerge. In the next section we will make a comparison of the state space and Box-Jenkins approaches to a broader range of problems in times series analysis. The above discussion has been based on Durbin (2000b, §2).

3.5 State space versus Box-Jenkins approaches

In this section we compare the state space and Box-Jenkins approaches to time series analysis. The early development of state space methodology took place in the field of engineering rather than statistics, starting with the pathbreaking paper of Kalman (1960). In this paper Kalman did two crucially important things. He showed that a very wide class of problems could be encapsulated in a simple linear model, essentially the state space model (3.1). Secondly he showed how, due to the Markovian nature of the model, the calculations needed for practical application of the model could be set up in recursive form in a way that was particularly convenient on a computer. A huge amount of work was done in the development of these ideas in the engineering field subsequently. In the 1960s to the early 1980's contributions to state space methodology from statisticians and econometricians were isolated and sporadic. In recent years however there has been a rapid growth of interest in the field in both statistics and econometrics as is indicated by references throughout the book.

The key advantage of the state space approach is that it is based on a structural analysis of the problem. The different components that make up the series, such as trend, seasonal, cycle and calendar variations, together with the effects of explanatory variables and interventions, are modelled separately before being put together in the state space model. It is up to the investigator to identify and model any features in particular situations that require special treatment. In contrast, the Box-Jenkins approach is a kind of 'black box', in which the model adopted depends purely on the data without prior analysis of the structure of the system that generated the data. A second advantage of state space models is that they are flexible. Because of the recursive nature of the models and of the computational techniques used to analyse them, it is straightforward to allow for known changes in the structure of the system over time. On the other hand, Box-Jenkins models are homogeneous through time since they are based on the assumptions that the differenced series is stationary.

State space models are very general. They cover a very wide range including all ARIMA models. Multivariate observations can be handled by straightforward extensions of univariate theory, which is not the case with Box-Jenkins models. It is easy to allow for missing observations with state space models. Explanatory variables can be incorporated into the model without difficulty. Moreover, the associated regression coefficients can be permitted to vary stochastically over time if this seems to be called for in the application. Trading day adjustments and other calendar variations can be readily taken care of. Because of the Markovian nature of state space models, the calculations needed to implement them can be put in recursive form. This enables increasingly large models to be handled effectively without disproportionate increases in the computational burden. No extra theory is required for forecasting. All that is needed is to project the Kalman filter forward into the future. This gives the forecasts required together with their estimated standard errors using the standard formulae used earlier in the series.

It might be asked, if these are all the advantages of state space modelling, what are the disadvantages relative to the Box-Jenkins approach? In our opinion, the only disadvantages are the relative lack in the statistical and econometric communities of information, knowledge and software regarding these models.

ARIMA modelling forms a core part of university courses on time series analysis and there are numerous text books on the subject. Software is widely available in major general packages such as *SAS*, *S-PLUS* and *MINITAB* as well as in many specialist time series packages. In contrast, state space modelling for time series is taught in relatively few universities and on the statistical side, as distinct from the engineering side, very few books are available. There is not much state space software on general statistical packages, and specialist software has only recently become available.

Now let us consider some of the disadvantages of the Box-Jenkins approach. The elimination of trend and seasonal by differencing may not be a drawback if forecasting is the only object of the analysis, but in many contexts, particularly in official statistics and some econometric applications, knowledge about these components has intrinsic importance. It is true that estimates of trend and seasonal can be 'recovered' from the differenced series by maximising the residual mean square as in Burman (1980) but this seems an artificial procedure which is not as appealing as modelling the components directly.

The requirement that the differenced series should be stationary is a weakness of the theory. In the economic and social fields, real series are never stationary however much differencing is done. The investigator has to face the question, how close to stationarity is close enough? This is a hard question to answer.

In the Box-Jenkins system it is relatively difficult to handle matters like missing observations, adding explanatory variables, calendar adjustments and changes in behaviour over time. These are straightforward to deal with in the state space approach. In practice it is found that the airline model and similar ARIMA models fit many data sets quite well, but it can be argued that the reason for this is that they are approximately equivalent to plausible state space models. This point is discussed at length by Harvey (1989, pp. 72, 73). As we move away from airline-type models, the model identification process in the Box-Jenkins system becomes difficult to apply. The main tool is the sample autocorrelation function which is notoriously imprecise due to its high sampling variability. Practitioners in applied time series analysis are familiar with the fact that many examples can be found where the data appear to be explained equally well by models whose specifications look very different.

A final point in favour of structural models is their transparency. One can examine graphs of trend, seasonal and other components to check whether they behave in accordance with expectation. If, for example, a blip was found in the graph of the seasonal component, one could go back to the original data to trace the source and perhaps make an adjustment to the model accordingly.

To sum up, state space models are based on modelling the observed structure of the data. They are more general, more flexible and more transparent than Box-Jenkins models. They can deal with features of the data which are hard to handle in the Box-Jenkins system. Software for state space models is publicly available, as we indicate at appropriate points of the exposition in the chapters to follow. The above discussion has been based on Durbin (2000b, §3).

3.6 Regression with time-varying coefficients

Suppose that in the linear regression model $y_t = Z_t \alpha + \varepsilon_t$ we wish the coefficient vector α to vary over time. A suitable model for this is to put $\alpha = \alpha_t$ and permit each coefficient α_{it} to vary according to a random walk $\alpha_{i,t+1} = \alpha_{it} + \eta_{it}$. This gives a state equation for the vector α_t in the form $\alpha_{t+1} = \alpha_t + \eta_t$. Since the model is a special case of (3.1) it can be handled in a routine fashion by Kalman filter and smoothing techniques.

3.7 Regression with ARMA errors

Consider a regression model of the form

$$y_t = X_t \beta + \xi_t, \qquad t = 1, \dots, n, \tag{3.27}$$

where y_t is a univariate dependent variable, X_t is a $1 \times k$ regressor vector, β is its coefficient vector and ξ_t denotes the error which is assumed to follow an ARMA model of form (3.15); this ARMA model may or may not be stationary and some of the coefficients ϕ_j, θ_j may be zero as long as ϕ_r and θ_{r-1} are not both zero. Let α_t be defined as in (3.16) and let

$$\alpha_t^* = \begin{pmatrix} \beta_t \\ \alpha_t \end{pmatrix},$$

where $\beta_t = \beta$. Writing the state equation implied by (3.17) as $\alpha_{t+1} = T\alpha_t + R\eta_t$, let

$$T^* = \begin{bmatrix} I_k & 0 \\ 0 & T \end{bmatrix}, \qquad R^* = \begin{bmatrix} 0 \\ R \end{bmatrix}, \qquad Z_t^* = (X_t \ 1\ 0 \cdots 0),$$

where T and R are defined in (3.17). Then the model

$$y_t = Z_t^* \alpha_t^*, \qquad \alpha_{t+1}^* = T^* \alpha_t^* + R^* \eta_t,$$

is equivalent to (3.27) and is in state space form (3.1) so Kalman filter and smoothing techniques are applicable; these provide an efficient means of fitting model (3.27). It is evident that the treatment can easily be extended to the case where the regression coefficients are determined by random walks as in §3.6. Moreover, with this approach, unlike some others, it is not necessary for the ARMA model used for the errors to be stationary.

3.8 Benchmarking

A common problem in official statistics is the adjustment of monthly or quarterly observations, obtained from surveys and therefore subject to survey errors, to agree with annual totals obtained from censuses and assumed to be free from error. The annual totals are called *benchmarks* and the process is called *benchmarking*. We shall show how the problem can be handled within a state space framework.

Denote the survey observations, which we take to be monthly ($s = 12$), by y_t and the true values they are intended to estimate by y_t^* for $t = 12(i-1)+j$, $i = 1, \ldots, \ell$ and $j = 1, \ldots, 12$, where ℓ is the number of years. Thus the survey error is $y_t - y_t^*$ which we denote by $\sigma_t^s \xi_t^s$ where σ_t^s is the standard deviation of the survey error at time t. The error ξ_t^s is modelled as an AR(1) model with unit variance. In principle, ARMA models of higher order could be used. We assume that the values of σ_t^s are available from survey experts and that the errors are bias free; we will mention the estimation of bias later. The benchmark values are given by $x_i = \sum_{j=1}^{12} y_{12(i-1)+j}^*$ for $i = 1, \ldots, \ell$. We suppose for simplicity of exposition that we have these annual values for all years in the study though in practice the census values will usually lag a year or two behind the survey observations. We take as the model for the observations

$$ y_t = \mu_t + \gamma_t + \sum_{j=1}^{k} \delta_{jt} w_{jt} + \varepsilon_t + \sigma_t^s \xi_t^s, \qquad t = 1, \ldots, 12\ell, \qquad (3.28) $$

where μ_t is trend, γ_t is seasonal and the term $\sum_{j=1}^{k} \delta_{jt} w_{jt}$ represents systematic effects such as the influence of calendar variations which can have a substantial effect on quantities such as retail sales but which can vary slowly over time.

The series is arranged in the form

$$ y_1, \ldots, y_{12}, x_1, y_{13}, \ldots, y_{24}, x_2, y_{25}, \ldots, y_{12\ell}, x_\ell. $$

Let us regard the time point in the series at which the benchmark occurs as $t = (12i)'$; thus the point $t = (12i)'$ occurs in the series between $t = 12i$ and $t = 12i + 1$. It seems reasonable to update the regression coefficients δ_{jt} only once a year, say in January, so we take for these coefficients the model

$$ \delta_{j,12i+1} = \delta_{j,12i} + \zeta_{j,12i}, \qquad j = 1, \ldots, k, \qquad i = 1, \ldots, \ell, $$

$$ \delta_{j,t+1} = \delta_{j,t}, \qquad \text{otherwise.} $$

Take the integrated random walk model for the trend component and model (3.3) for the seasonal component, that is,

$$ \Delta^2 \mu_t = \xi_t, \qquad \gamma_t = -\sum_{j=1}^{11} \gamma_{t-j} + \omega_t; $$

see §3.2 for alternative trend and seasonal models. It turns out to be convenient to put the observation errors into the state vector, so we take

$$ \alpha_t = \left(\mu_t, \ldots, \mu_{t-11}, \gamma_t, \ldots, \gamma_{t-11}, \delta_{1t}, \ldots, \delta_{kt}, \varepsilon_t, \ldots, \varepsilon_{t-11}, \xi_t^s \right)'. $$

Thus $y_t = Z_t \alpha_t$ where

$$ Z_t = \left(1, 0, \ldots, 0, 1, 0, \ldots, 0, w_{1t}, \ldots, w_{kt}, 1, 0, \ldots, 0, \sigma_t^s \right), \qquad t = 1, \ldots, n, $$

and $x_i = Z_t \alpha_t$ where

$$Z_t = \left(1, \ldots, 1, 0, \ldots, 0, \sum_{s=12i-11}^{12i} w_{1s}, \ldots, \sum_{s=12i-11}^{12i} w_{ks}, 1, \ldots, 1, 0 \right), t = (12i)',$$

for $i = 1, \ldots, \ell$. Using results from §3.2 it is easy to write down the state transition from α_t to α_{t+1} for $t = 12i - 11$ to $t = 12i - 1$, taking account of the
fact that $\delta_{j,t+1} = \delta_{jt}$. From $t = 12i$ to $t = (12i)'$ the transition is the identity.
From $t = (12i)'$ to $t = 12i + 1$, the transition is the same as for $t = 12i - 11$ to
$t = 12i - 1$, except that we take account of the relation $\delta_{j,12i+1} = \delta_{j,12i} + \zeta_{j,12i}$
where $\zeta_{j,12i} \neq 0$.

There are many variants of the benchmarking problem. For example, the annual
totals may be subject to error, the benchmarks maybe values at a particular month,
say December, instead of annual totals, the survey observations may be biased and
the bias needs to be estimated, more complicated models than the AR(1) model can
be used to model y_t^*; finally, the observations may behave multiplicatively whereas
the benchmark constraint is additive, thus leading to a nonlinear model. All these
variants are dealt with in a comprehensive treatment of the benchmarking problem
by Durbin and Quenneville (1997). They also consider a two-step approach to the
problem in which a state space model is first fitted to the survey observations and
the adjustments to satisfy the benchmark constraints takes place in a second stage.

Essentially, this example demonstrates that the state space approach can be
used to deal with situations in which the data come from two different sources.
Another example of such problems will be given in §3.9 where we model different
series which aim at measuring the same phenomenon simultaneously, which are
all subject to sampling error and which are observed at different time intervals.

3.9   Simultaneous modelling of series from different sources

A different problem in which data come from two different sources has been
considered by Harvey and Chung (2000). Here the objective is to estimate the
level of UK unemployment and the month-to-month change of unemployment
given two different series. Of these, the first is a series y_t of observations obtained
from a monthly survey designed to estimate unemployment according to an inter-
nationally accepted standard definition (the so-called ILO definition where ILO
stands for International Labour Office); this estimate is subject to survey error. The
second series consists of monthly counts x_t of the number of individuals claiming
unemployment benefit; although these counts are known accurately, they do not
themselves provide an estimate of unemployment consistent with the ILO definition. Even though the two series are closely related, the relationship is not even
approximately exact and it varies over time. The problem to be considered is how
to use the knowledge of x_t to improve the accuracy of the estimate based on y_t
alone.

The solution suggested by Harvey and Chung (2000) is to model the bivariate series $(y_t, x_t)'$ by the structural time series model

$$
\begin{pmatrix} y_t \\ x_t \end{pmatrix} = \mu_t + \varepsilon_t, \qquad \varepsilon_t \sim N(0, \Sigma_\varepsilon),
$$
$$
\mu_{t+1} = \mu_t + \nu_t + \xi_t, \qquad \xi_t \sim N(0, \Sigma_\xi),
$$
$$
\nu_{t+1} = \nu_t + \zeta_t, \qquad \zeta_t \sim N(0, \Sigma_\zeta),
$$

(3.29)

for $t = 1, \ldots, n$. Here, $\mu_t, \varepsilon_t, \nu_t, \xi_t$ and ζ_t are 2×1 vectors and $\Sigma_\varepsilon, \Sigma_\xi$ and Σ_ζ are 2×2 variance matrices. Seasonals can also be incorporated. Many complications are involved in implementing the analysis based on this model, particularly those arising from design features of the survey such as overlapping samples. A point of particular interest is that the claimant count x_t is available one month ahead of the survey value y_t. This extra value of x_t can be easily and efficiently made use of by the missing observations technique discussed in §4.8. For a discussion of the details we refer the reader to Harvey and Chung (2000).

This is not the only way in which the information available in x_t can be utilised. For example, in the published discussion of the paper, Durbin (2000a) suggested two further possibilities, the first of which is to model the series $y_t - x_t$ by a structural time series model of one of the forms considered in §3.2.1; unemployment level could then be estimated by $\hat{\mu}_t + x_t$ where $\hat{\mu}_t$ is the forecast of the trend μ_t in the model using information up to time $t - 1$, while the month-to-month change could be estimated by $\hat{\nu}_t + x_t - x_{t-1}$ where $\hat{\nu}_t$ is the forecast of the slope ν_t. Alternatively, x_t could be incorporated as an explanatory variable into an appropriate form of model (3.11) with coefficient β_j replaced by β_{jt} which varies over time according to (3.12). In an obvious notation, trend and change would then be estimated by $\hat{\mu}_t + \hat{\beta}_t x_t$ and $\hat{\nu}_t + \hat{\beta}_t x_t - \hat{\beta}_{t-1} x_{t-1}$.

3.10   State space models in continuous time

In contrast to all the models that we have considered so far, suppose that the observation $y(t)$ is a continuous function of time for t in an interval which we take to be $0 \le t \le T$. We shall aim at constructing state space models for $y(t)$ which are the analogues in continuous time for models that we have already studied in discrete time. Such models are useful not only for studying phenomena which genuinely operate in continuous time, but also for providing a convenient theoretical base for situations where the observations take place at time points $t_1 \le \cdots \le t_n$ which are not equally spaced.

3.10.1   LOCAL LEVEL MODEL

We begin by considering a continuous version of the local level model (2.3). To construct this, we need a continuous analogue of the Gaussian random walk. This is the *Brownian motion process*, defined as the continuous stochastic process $w(t)$ such that $w(0) = 0$, $w(t) \sim N(0, t)$ for $0 < t < \infty$, where increments $w(t_2) - w(t_1)$, $w(t_4) - w(t_3)$ for $0 \le t_1 \le t_2 \le t_3 \le t_4$ are independent. We sometimes need to consider increments $dw(t)$, where $dw(t) \sim N(0, dt)$ for dt infinitesimally small. Analogously to the random walk $\alpha_{t+1} = \alpha_t + \eta_t$, $\eta_t \sim N(0, \sigma_\eta^2)$ for the discrete model, we define $\alpha(t)$ by the continuous time relation $d\alpha(t) = \sigma_\eta dw(t)$ where σ_η is an appropriate positive scale parameter. This suggests that as the continuous analogue of the local level model we adopt the continuous time state space model

$$
\begin{aligned}
y(t) &= \alpha(t) + \varepsilon(t), \\
\alpha(t) &= \alpha(0) + \sigma_\eta w(t), \qquad 0 \le t \le T,
\end{aligned}
\tag{3.30}
$$

where $T > 0$.

The nature of $\varepsilon(t)$ in (3.30) requires careful thought. It must first be recognised that for any analysis that is performed digitally, which is all that we consider in this book, $y(t)$ cannot be admitted into the calculations as a continuous record; we can only deal with it as a series of values observed at a discrete set of time points $0 \le t_1 < t_2 < \cdots < t_n \le T$. Secondly, $\text{Var}[\varepsilon(t)]$ must be bounded significantly away from zero; there is no point in carrying out an analysis when $y(t)$ is indistinguishably close to $\alpha(t)$. Thirdly, in order to obtain a continuous analogue of the local level model we need to assume that $\text{Cov}[\varepsilon(t_i), \varepsilon(t_j)] = 0$ for observational points t_i, t_j $(i \ne j)$. It is obvious that if the observational points are close together it may be advisable to set up an autocorrelated model for $\varepsilon(t)$, for example a low-order autoregressive model; however, the coefficients of this would have to be put into the state vector and the resulting model would not be a continuous local level model. In order to allow $\text{Var}[\varepsilon(t)]$ to vary over time we assume that $\text{Var}[\varepsilon(t)] = \sigma^2(t)$ where $\sigma^2(t)$ is a non-stochastic function of t that may depend on unknown parameters. We conclude that in place of (3.30) a more appropriate form of the model is

$$
\begin{aligned}
y(t) &= \alpha(t) + \varepsilon(t), & t &= t_1, \ldots, t_n, & \varepsilon(t_i) &\sim N[0, \sigma^2(t_i)], \\
\alpha(t) &= \alpha(0) + \sigma_\eta w(t), & 0 &\le t \le T.
\end{aligned}
\tag{3.31}
$$

We next consider the estimation of unknown parameters by maximum likelihood. Since by definition the likelihood is equal to

$$
p[y(t_1)]p[y(t_2)|y(t_1)] \cdots p[y(t_n)|y(t_1), \ldots, y(t_{n-1})],
$$

it depends on $\alpha(t)$ only at values $t_1, \ldots, t_n$. Thus for estimation of parameters we can employ the reduced model

$$
\begin{aligned}
y_i &= \alpha_i + \varepsilon_i, \\
\alpha_{i+1} &= \alpha_i + \eta_i, \qquad i = 1, \ldots, n,
\end{aligned}
\tag{3.32}
$$

where $y_i = y(t_i)$, $\alpha_i = \alpha(t_i)$, $\varepsilon_i = \varepsilon(t_i)$, $\eta_i = \sigma_\eta[w(t_{i+1}) - w(t_i)]$ and where the ε_i's are assumed to be independent. This is a discrete local level model which differs from (2.3) only because the variances of the ε_i's can be unequal; consequently we can calculate the loglikelihood by a slight modification of the method of §2.10.1 which allows for the variance inequality.

Having estimated the model parameters, suppose that we wish to estimate $\alpha(t)$ at values $t = t_{j_*}$ between t_j and t_{j+1} for $1 \le j < n$. We adjust and extend equations

(3.32) to give

$$\begin{aligned}
\alpha_{j_*} &= \alpha_j + \eta_j^*, \\
y_{j_*} &= \alpha_{j_*} + \varepsilon_{j_*}, \\
\alpha_{j+1} &= \alpha_{j_*} + \eta_{j_*}^*,
\end{aligned} \tag{3.33}$$

where $y_{j_*} = y(t_{j_*})$ is treated as missing, $\eta_j^* = \sigma_\eta[w(t_{j_*}) - w(t_j)]$ and $\eta_{j_*}^* = \sigma_\eta[w(t_{j+1}) - w(t_{j_*})]$. We can now calculate $\mathrm{E}[\alpha_{j_*} | y(t_1), \ldots, y(t_n)]$ and $\mathrm{Var}[\alpha_{j_*} | y(t_1), \ldots, y(t_n)]$ by routine applications of the Kalman filter and smoother for series with missing observations, as described in §2.7, with a slight modification to allow for unequal observational error variances.

3.10.2   LOCAL LINEAR TREND MODEL

Now let us consider the continuous analogue of the local linear trend model (3.2) for the case where $\sigma_\xi^2 = 0$, so that, in effect, the trend term μ_t is modelled by the relation $\Delta^2 \mu_{t+1} = \zeta_t$. For the continuous case, denote the trend by $\mu(t)$ and the slope by $v(t)$ by analogy with (3.2). The natural model for the slope is then $dv(t) = \sigma_\zeta dw(t)$, where $w(t)$ is standard Brownian motion and $\sigma_\zeta > 0$, which gives

$$v(t) = v(0) + \sigma_\zeta w(t), \qquad 0 \le t \le T. \tag{3.34}$$

By analogy with (3.2) with $\sigma_\xi^2 = 0$, the model for the trend level is $d\mu(t) = v(t)dt$, giving

$$\begin{aligned}
\mu(t) &= \mu(0) + \int_0^t v(s)\, ds \\
&= \mu(0) + v(0)t + \sigma_\zeta \int_0^t w(s)\, ds.
\end{aligned} \tag{3.35}$$

As before, suppose that $y(t)$ is observed at times $t_1 \le \cdots \le t_n$. Analogously to (3.31), the observation equation for the continuous model is

$$y(t) = \alpha(t) + \varepsilon(t), \qquad t = t_1, \ldots, t_n, \qquad \varepsilon(t_i) \sim \mathrm{N}[0, \sigma^2(t_i)], \tag{3.36}$$

and the state equation can be written in the form

$$d \begin{bmatrix} \mu(t) \\ v(t) \end{bmatrix} = \begin{bmatrix} 0 & 1 \\ 0 & 0 \end{bmatrix} \begin{bmatrix} \mu(t) \\ v(t) \end{bmatrix} dt + \sigma_\zeta \begin{bmatrix} 0 \\ dw(t) \end{bmatrix}. \tag{3.37}$$

For maximum likelihood estimation we employ the discrete state space model,

$$y_i = \mu_i + \varepsilon_i,$$

$$\begin{pmatrix} \mu_{i+1} \\ v_{i+1} \end{pmatrix} = \begin{bmatrix} 1 & \delta_i \\ 0 & 1 \end{bmatrix} \begin{pmatrix} \mu_i \\ v_i \end{pmatrix} + \begin{pmatrix} \xi_i \\ \zeta_i \end{pmatrix}, \qquad i = 1, \ldots, n, \tag{3.38}$$

where $\mu_i = \mu(t_i)$, $v_i = v(t_i)$, $\varepsilon_i = \varepsilon(t_i)$ and $\delta_i = t_{i+1} - t_i$; also

$$\xi_i = \sigma_\zeta \int_{t_i}^{t_{i+1}} [w(s) - w(t_i)]\, ds,$$

and

$$\zeta_i = \sigma_\zeta [w(t_{i+1}) - w(t_i)],$$

as can be verified from (3.34) and (3.35). From (3.36), $\text{Var}(\varepsilon_i) = \sigma^2(t_i)$. Since $E[w(s) - w(t_i)] = 0$ for $t_i \le s \le t_{i+1}$, $E(\xi_i) = E(\zeta_i) = 0$. To calculate $\text{Var}(\xi_i)$, approximate ξ_i by the sum

$$\frac{\delta_i}{M} \sum_{j=0}^{M-1} (M - j) w_j$$

where $w_j \sim N(0, \sigma_\zeta^2 \delta_i / M)$ and $E(w_j w_k) = 0$ $(j \ne k)$. This has variance

$$\sigma_\zeta^2 \frac{\delta_i^3}{M} \sum_{j=0}^{M-1} \left(1 - \frac{j}{M}\right)^2,$$

which converges to

$$\sigma_\zeta^2 \delta_i^3 \int_0^1 x^2\, dx = \frac{1}{3} \sigma_\zeta^2 \delta_i^3$$

as $M \to \infty$. Also,

$$E(\xi_i \zeta_i) = \sigma_\zeta^2 \int_{t_i}^{t_{i+1}} E[\{w(s) - w(t_i)\}\{w(t_{i+1}) - w(t_i)\}]\, ds$$
$$= \sigma_\zeta^2 \int_0^{\delta_i} x\, dx$$
$$= \frac{1}{2} \sigma_\zeta^2 \delta_i^2,$$

and $E(\zeta_i^2) = \sigma_\zeta^2 \delta_i$. Thus the variance matrix of the disturbance term in the state equation (3.38) is

$$Q_i = \text{Var}\begin{pmatrix} \xi_i \\ \zeta_i \end{pmatrix} = \sigma_\zeta^2 \delta_i \begin{bmatrix} \frac{1}{3}\delta_i^2 & \frac{1}{2}\delta_i \\ \frac{1}{2}\delta_i & 1 \end{bmatrix}. \tag{3.39}$$

The loglikelihood is then calculated by means of the Kalman filter as in §7.3.

As with model (3.31), adjustments (3.33) can also be introduced to model (3.36) and (3.37) in order to estimate the conditional mean and variance matrix of the state vector $[\mu(t), v(t)]'$ at values of t other than $t_1, \ldots, t_n$.

Chapter 9 of Harvey (1989) may be consulted for extensions to more general models.

3.11 Spline smoothing

3.11.1 SPLINE SMOOTHING IN DISCRETE TIME

Suppose we have a univariate series $y_1, \ldots, y_n$ of values which are equispaced in time and we wish to approximate the series by a relatively smooth function $\mu(t)$. A standard approach is to choose $\mu(t)$ by minimising

$$\sum_{t=1}^{n} [y_t - \mu(t)]^2 + \lambda \sum_{t=1}^{n} [\Delta^2 \mu(t)]^2 \tag{3.40}$$

with respect to $\mu(t)$ for given $\lambda > 0$. It is important to note that we are considering $\mu(t)$ here to be a discrete function of t at time points $t = 1, \ldots, n$, in contrast to the situation considered in the next section where $\mu(t)$ is a continuous function of time. If λ is small, the values of $\mu(t)$ will be close to the y_t's but $\mu(t)$ may not be smooth enough. If λ is large the $\mu(t)$ series will be smooth but the values of $\mu(t)$ may not be close enough to the y_t's. The function $\mu(t)$ is called a *spline*. Reviews of methods related to this idea are given in Silverman (1985), Wahba (1990) and Green and Silverman (1994). Note that in this book we usually take t as the time index but it can also refer to other sequentially ordered measures such as temperature, earnings and speed.

Let us now consider this problem from a state space standpoint. Let $\alpha_t = \mu(t)$ for $t = 1, \ldots, n$ and assume that y_t and α_t obey the state space model

$$y_t = \alpha_t + \varepsilon_t, \qquad \Delta^2 \alpha_t = \zeta_t, \qquad t = 1, \ldots, n, \tag{3.41}$$

where $\text{Var}(\varepsilon_t) = \sigma^2$ and $\text{Var}(\zeta_t) = \sigma^2 / \lambda$ with $\lambda > 0$. We observe that the second equation of (3.41) is one of the smooth models for trend considered in §3.2. For simplicity suppose that α_{-1} and α_0 are fixed and known. The log of the joint density of $\alpha_1, \ldots, \alpha_n, y_1, \ldots, y_n$ is then, apart from irrelevant constants,

$$-\frac{\lambda}{2\sigma^2} \sum_{t=1}^{n} \left(\Delta_t^2 \alpha_t \right)^2 - \frac{1}{2\sigma^2} \sum_{t=1}^{n} (y_t - \alpha_t)^2. \tag{3.42}$$

Now suppose that our objective is to smooth the y_t series by estimating α_t by $\hat{\alpha}_t = E(\alpha_t | Y_n)$. We shall employ a technique that we shall use extensively later so we state it in general terms. Suppose $\alpha = (\alpha_1', \ldots, \alpha_n')'$ and $y = (y_1', \ldots, y_n')'$ are jointly normally distributed stacked vectors with density $p(\alpha, y)$ and we wish to calculate $\hat{\alpha} = E(\alpha | y)$. Then $\hat{\alpha}$ is the solution of the equations

$$\frac{\partial \log p(\alpha, y)}{\partial \alpha} = 0.$$

This follows since $\log p(\alpha | y) = \log p(\alpha, y) - \log p(y)$ so $\partial \log p(\alpha | y) / \partial \alpha = \partial \log p(\alpha, y) / \partial \alpha$. Now the solution of the equations $\partial \log p(\alpha | y) / \partial \alpha = 0$ is the mode of the density $p(\alpha | y)$ and since the density is normal the mode is equal to the mean vector $\hat{\alpha}$. The conclusion follows. Since $p(\alpha | y)$ is the conditional distribution of α given y, we call this technique *conditional mode estimation* of $\alpha_1, \ldots, \alpha_n$.

Applying this technique to (3.42), we see that $\hat{\alpha}_1, \ldots, \hat{\alpha}_n$ can be obtained by minimising

$$\sum_{t=1}^{n}(y_t - \alpha_t)^2 + \lambda \sum_{t=1}^{n}(\Delta^2\alpha_t)^2.$$

Comparing this with (3.40), and ignoring for the moment the initialisation question, we see that the spline smoothing problem can be solved by finding $E(\alpha_t | Y_n)$ for model (3.41). This is achieved by a standard extension of the smoothing technique of §2.4 that will be given in §4.3. It follows that state space techniques can be used for spline smoothing. Treatments along these lines have been given by Kohn, Ansley and Wong (1992). This approach has the advantages that the models can be extended to include extra features such as explanatory variables, calendar variations and intervention effects in the ways indicated earlier in this chapter; moreover, unknown quantities, for example λ in (3.40), can be estimated by maximum likelihood using methods that we shall describe in Chapter 7.

3.11.2   SPLINE SMOOTHING IN CONTINUOUS TIME

Let us now consider the smoothing problem where the observation $y(t)$ is a continuous function of time t for t in an interval which for simplicity we take to be $0 \leq t \leq T$. Suppose that we wish to smooth $y(t)$ by a function $\mu(t)$ given a sample of values $y(t_i)$ for $i = 1, \ldots, n$ where $0 < t_1 < \cdots < t_n < T$. A traditional approach to the problem is to choose $\mu(t)$ to be the twice-differentiable function on $(0, T)$ which minimises

$$\sum_{i=1}^{n}[y(t_i) - \mu(t_i)]^2 + \lambda \int_0^T \left[\frac{\partial^2\mu(t)}{\partial t^2}\right]^2 dt, \tag{3.43}$$

for given $\lambda > 0$. We observe that (3.43) is the analogue in continuous time of (3.40) in discrete time. This is a well-known problem, a standard treatment to which is presented in Chapter 2 of Green and Silverman (1994). Their approach is to show that the resulting $\mu(t)$ must be a *cubic spline*, which is defined as a cubic polynomial function in t between each pair of time points t_i, t_{i+1} for $i = 0, 1, \ldots, n$ with $t_0 = 0$ and $t_{n+1} = T$, such that $\mu(t)$ and its first two derivatives are continuous at each t_i for $i = 1, \ldots, n$. The properties of the cubic spline are then used to solve the minimisation problem. In contrast, we shall present a solution based on a continuous time state space model of the kind considered in §3.10.

We begin by adopting a model for $\mu(t)$ in the form (3.35), which for convenience we reproduce here as

$$\mu(t) = \mu(0) + \nu(0)t + \sigma_\zeta \int_0^t w(s)\,ds, \qquad 0 \leq t \leq T. \tag{3.44}$$

This is a natural model to consider since it is the simplest model in continuous time for a trend with smoothly varying slope. As the observation equation we take

$$y(t_i) = \mu(t_i) + \varepsilon_i, \qquad \varepsilon_i \sim N(0, \sigma_\varepsilon^2), \qquad i = 1, \ldots, n,$$

where ε_i's are independent of each other and of $w(t)$ for $0 < t \leq T$. We have taken Var(ε_i) to be constant since this is a reasonable assumption for many smoothing problems, and also since it leads to the same solution to the problem of minimising (3.43) as the Green-Silverman approach.

Since $\mu(0)$ and $\nu(0)$ are normally unknown, we represent them by diffuse priors. On these assumptions, Wahba (1978) has shown that on taking

$$\lambda = \frac{\sigma_\varepsilon^2}{\sigma_\zeta^2},$$

the conditional mean $\hat{\mu}(t)$ of $\mu(t)$ defined by (3.44), given the observations $y(t_1), \ldots, y(t_n)$, is the solution to the problem of minimising (3.43) with respect to $\mu(t)$. We shall not give details of the proof here but will instead refer to discussions of the result by Wecker and Ansley (1983, §2) and Green and Silverman (1994, §3.8.3). The result is important since it enables problems in spline smoothing to be solved by state space methods. We note that Wahba and Wecker and Ansley in the papers cited consider the more general problem in which the second term of (3.43) is replaced by the more general form

$$\lambda \int_0^T \left[\frac{d^m \mu(t)}{dt^m} \right] dt,$$

for $m = 2, 3, \ldots$.

We have reduced the problem of minimising (3.43) to the treatment of a special case of the state space model (3.36) and (3.37) in which $\sigma^2(t_i) = \sigma_\varepsilon^2$ for all i. We can therefore compute $\hat{\mu}(t)$ and Var$[\mu(t)|y(t_1), \ldots, y(t_n)]$ by routine Kalman filtering and smoothing. We can also compute the loglikelihood and, consequently, estimate λ by maximum likelihood; this can be done efficiently by concentrating out σ_ε^2 by a straightforward extension of the method described in §2.10.2 and then maximising the concentrated loglikelihood with respect to λ in a one-dimensional search. The implication of these results is that the flexibility and computational power of state space methods can be employed to solve problems in spline smoothing.

4
Filtering, smoothing and forecasting

4.1 Introduction

In this chapter and the following three chapters we provide a general treatment from the standpoint of classical inference of the linear Gaussian state space model (3.1). The observations y_t will be treated as multivariate. For much of the theory, the development is a straightforward extension to the general case of the treatment of the simple local level model in Chapter 2. In this chapter we will consider filtering, smoothing, simulation, missing observations and forecasting. Filtering is aimed at updating our knowledge of the system as each observation y_t comes in. Smoothing enables us to base our estimates of quantities of interest on the entire sample $y_1, \ldots, y_n$. As with the local level model, we shall show that when state space models are used, it is easy to allow for missing observations. Forecasting is of special importance in many applications of time series analysis; we will demonstrate that by using our approach, we can obtain the required forecasting results merely by treating the future values $y_{n+1}, y_{n+2}, \ldots$ as missing values. Chapter 5 will discuss the initialisation of the Kalman filter in cases where some or all of the elements of the initial state vector are unknown or have unknown distributions. Chapter 6 will discuss various computational aspects of filtering and smoothing. In the classical approach to inference, the parameter vector ψ is regarded as fixed but unknown. In Chapter 7 we will consider the estimation of ψ by maximum likelihood together with the related questions of goodness of fit of the model and diagnostic checking. In the Bayesian approach to inference, ψ is treated as a random vector which has a specified prior density or a non-informative prior. We will show how the linear Gaussian state space model can be dealt with from the Bayesian standpoint by simulation methods in Chapter 8. The use in practice of the techniques presented in these four chapters is illustrated in detail in Chapter 9 by describing their application to five real time series.

Denote the set of observations $y_1, \ldots, y_t$ by Y_t. In §4.2 we will derive the Kalman filter, which is a recursion for calculating $a_{t+1} = \mathrm{E}(\alpha_{t+1}|Y_t)$ and $P_{t+1} = \mathrm{Var}(\alpha_{t+1}|Y_t)$ given a_t and P_t. The derivation requires only some elementary properties of multivariate normal regression theory. In the following section we investigate some properties of one-step forecast errors which we shall need for subsequent work. In §4.3 we use the output of the Kalman filter and the properties of forecast errors to obtain recursions for smoothing the series, that is, calculating the conditional mean and variance matrix of α_t given all the observations $y_1, \ldots, y_n$ for $t = 1, \ldots, n$. Estimates of the disturbance vectors ε_t and η_t and their error variance matrices given all the data are investigated in §4.4. The weights associated with filtered and smoothed estimates of functions of the state and disturbance vectors are discussed in §4.6. Section 4.7 describes how to generate random samples for purposes of simulation from the smoothed density of the state and disturbance vectors given the observations. The problem of missing observations is considered in §4.8 where we show that with the state space approach the matter is easily dealt with by means of simple modifications of the Kalman filter and the smoothing recursions. Section 4.9 discusses forecasting by using the results of §4.8. A comment on varying dimensions of the observation vector is given in §4.10. Finally, in §4.11 we consider a general matrix formulation of the state space model.

4.2   Filtering

4.2.1   DERIVATION OF KALMAN FILTER

For convenience we restate the linear Gaussian state space model (3.1) here as

$$
\begin{aligned}
y_t &= Z_t \alpha_t + \varepsilon_t, & \varepsilon_t &\sim N(0, H_t), & & \\
\alpha_{t+1} &= T_t \alpha_t + R_t \eta_t, & \eta_t &\sim N(0, Q_t), & t &= 1, \ldots, n, \\
& & \alpha_1 &\sim N(a_1, P_1),
\end{aligned}
\tag{4.1}
$$

where details are given below (3.1). Let Y_{t-1} denote the set of past observations $y_1, \ldots, y_{t-1}$. Starting at $t = 1$ and building up the distributions of α_t and y_t recursively, it is easy to show that $p(y_t | \alpha_1, \ldots, \alpha_t, Y_{t-1}) = p(y_t | \alpha_t)$ and $p(\alpha_{t+1} | \alpha_1, \ldots, \alpha_t, Y_t) = p(\alpha_{t+1} | \alpha_t)$. In Table 4.1 we give the dimensions of the vectors and matrices of the state space model.

In this section we derive the Kalman filter for model (4.1) for the case where the initial state α_1 is $N(a_1, P_1)$ where a_1 and P_1 are known. Our object is to obtain the conditional distribution of α_{t+1} given Y_t for $t = 1, \ldots, n$ where $Y_t = \{y_1, \ldots, y_t\}$. Since all distributions are normal, conditional distributions of subsets of variables given other subsets of variables are also normal; the required

Table 4.1.   Dimensions of state space model (4.1).

Vector		Matrix	
y_t	$p \times 1$	Z_t	$p \times m$
α_t	$m \times 1$	T_t	$m \times m$
ε_t	$p \times 1$	H_t	$p \times p$
η_t	$r \times 1$	R_t	$m \times r$
		Q_t	$r \times r$
a_1	$m \times 1$	P_1	$m \times m$

distribution is therefore determined by a knowledge of $a_{t+1} = \mathrm{E}(\alpha_{t+1}|Y_t)$ and $P_{t+1} = \mathrm{Var}(\alpha_{t+1}|Y_t)$. Assume that α_t given Y_{t-1} is $\mathrm{N}(a_t, P_t)$. We now show how to calculate a_{t+1} and P_{t+1} from a_t and P_t recursively.

Since $\alpha_{t+1} = T_t\alpha_t + R_t\eta_t$, we have

$$
\begin{aligned}
a_{t+1} &= \mathrm{E}(T_t\alpha_t + R_t\eta_t|Y_t) \\
&= T_t\,\mathrm{E}(\alpha_t|Y_t),
\end{aligned}
\tag{4.2}
$$

$$
\begin{aligned}
P_{t+1} &= \mathrm{Var}(T_t\alpha_t + R_t\eta_t|Y_t) \\
&= T_t\,\mathrm{Var}(\alpha_t|Y_t)T_t' + R_t Q_t R_t',
\end{aligned}
\tag{4.3}
$$

for $t = 1, \ldots, n$. Let

$$
v_t = y_t - \mathrm{E}(y_t|Y_{t-1}) = y_t - \mathrm{E}(Z_t\alpha_t + \varepsilon_t|Y_{t-1}) = y_t - Z_t a_t.
\tag{4.4}
$$

Then v_t is the one-step forecast error of y_t given Y_{t-1}. When Y_{t-1} and v_t are fixed then Y_t is fixed and vice versa. Thus $\mathrm{E}(\alpha_t|Y_t) = \mathrm{E}(\alpha_t|Y_{t-1}, v_t)$. But $\mathrm{E}(v_t|Y_{t-1}) = \mathrm{E}(y_t - Z_t a_t|Y_{t-1}) = \mathrm{E}(Z_t\alpha_t + \varepsilon_t - Z_t a_t|Y_{t-1}) = 0$. Consequently, $\mathrm{E}(v_t) = 0$ and $\mathrm{Cov}(y_j, v_t) = \mathrm{E}[y_j\mathrm{E}(v_t|Y_{t-1})'] = 0$ with $j = 1, \ldots, t - 1$. By (2.49) in the regression lemma in §2.13 we therefore have

$$
\begin{aligned}
\mathrm{E}(\alpha_t|Y_t) &= \mathrm{E}(\alpha_t|Y_{t-1}, v_t) \\
&= \mathrm{E}(\alpha_t|Y_{t-1}) + \mathrm{Cov}(\alpha_t, v_t)[\mathrm{Var}(v_t)]^{-1}v_t \\
&= a_t + M_t F_t^{-1}v_t,
\end{aligned}
\tag{4.5}
$$

where $M_t = \mathrm{Cov}(\alpha_t, v_t)$, $F_t = \mathrm{Var}(v_t)$ and $\mathrm{E}(\alpha_t|Y_{t-1}) = a_t$ by definition of a_t. Here,

$$
\begin{aligned}
M_t = \mathrm{Cov}(\alpha_t, v_t) &= \mathrm{E}[\mathrm{E}\{\alpha_t(Z_t\alpha_t + \varepsilon_t - Z_t a_t)'|Y_{t-1}\}] \\
&= \mathrm{E}[\mathrm{E}\{\alpha_t(\alpha_t - a_t)'Z_t'|Y_{t-1}\}] = P_t Z_t',
\end{aligned}
\tag{4.6}
$$

and

$$
F_t = \mathrm{Var}(Z_t\alpha_t + \varepsilon_t - Z_t a_t) = Z_t P_t Z_t' + H_t.
\tag{4.7}
$$

We assume that F_t is nonsingular; this assumption is normally valid in well-formulated models, but in any case it is relaxed in §6.4. Substituting in (4.2) and (4.5) gives

$$
\begin{aligned}
a_{t+1} &= T_t a_t + T_t M_t F_t^{-1}v_t \\
&= T_t a_t + K_t v_t, \qquad t = 1, \ldots, n,
\end{aligned}
\tag{4.8}
$$

with

$$
K_t = T_t M_t F_t^{-1} = T_t P_t Z_t' F_t^{-1}.
\tag{4.9}
$$

We observe that a_{t+1} has been obtained as a linear function of the previous value a_t and v_t, the forecast error of y_t given Y_{t-1}.

By (2.50) of the regression lemma in §2.13 we have

$$
\begin{aligned}
\mathrm{Var}(\alpha_t | Y_t) &= \mathrm{Var}(\alpha_t | Y_{t-1}, v_t) \\
&= \mathrm{Var}(\alpha_t | Y_{t-1}) - \mathrm{Cov}(\alpha_t, v_t)[\mathrm{Var}(v_t)]^{-1} \mathrm{Cov}(\alpha_t, v_t)' \\
&= P_t - M_t F_t^{-1} M_t' \\
&= P_t - P_t Z_t' F_t^{-1} Z_t P_t.
\end{aligned}
\tag{4.10}
$$

Substituting in (4.3) gives

$$
P_{t+1} = T_t P_t L_t' + R_t Q_t R_t', \qquad t = 1, \ldots, n,
\tag{4.11}
$$

with

$$
L_t = T_t - K_t Z_t.
\tag{4.12}
$$

The recursions (4.8) and (4.11) constitute the celebrated Kalman filter for model (4.1). They enable us to update our knowledge of the system each time a new observation comes in. It is noteworthy that we have derived these recursions by simple applications of standard results of multivariate normal regression theory. The key advantage of the recursions is that we do not have to invert a $(pt \times pt)$ matrix to fit the model each time the tth observation comes in for $t = 1, \ldots, n$; we only have to invert the $(p \times p)$ matrix F_t and p is generally much smaller than n; indeed, in the most important case in practice, $p = 1$. Although relations (4.8) and (4.11) constitute the forms in which the multivariate Kalman filter recursions are usually presented, we shall show in §6.4 that variants of them in which elements of the observational vector y_t are brought in one at a time, rather than the entire vector y_t, are in general computationally superior.

It can be shown that when the observations are not normally distributed and we restrict attention to estimates which are linear in the y_t's, and when also matrices Z_t and T_t do not depend on previous y_t's, then under appropriate assumptions the value of a_{t+1} given by the filter minimises the mean square error of the estimate of each component of α_{t+1}. See, for example, Duncan and Horn (1972) or Anderson and Moore (1979) for details of this approach. A similar result holds for arbitrary linear functions of the elements of the a_t's. This emphasises the point that although our results are obtained under the assumption of normality, they have a wider validity in the sense of minimum mean square errors when the variables involved are not normally distributed.

4.2.2   KALMAN FILTER RECURSION

For convenience we collect together the filtering equations

$$
\begin{aligned}
v_t &= y_t - Z_t a_t, & F_t &= Z_t P_t Z_t' + H_t, & & \\
K_t &= T_t P_t Z_t' F_t^{-1}, & L_t &= T_t - K_t Z_t, & t &= 1, \ldots, n, \\
a_{t+1} &= T_t a_t + K_t v_t, & P_{t+1} &= T_t P_t L_t' + R_t Q_t R_t', & &
\end{aligned}
\tag{4.13}
$$

Table 4.2. Dimensions of Kalman filter.

Vector		Matrix	
v_t	$p \times 1$	F_t	$p \times p$
		K_t	$m \times p$
		L_t	$m \times m$
		M_t	$m \times p$
a_t	$m \times 1$	P_t	$m \times m$
$a_{t|t}$	$m \times 1$	$P_{t|t}$	$m \times m$

with a_1 and P_1 as the mean vector and variance matrix of the initial state vector α_1, respectively. The recursion (4.13) is called the *Kalman filter*.

The so-called *contemporaneous filtering equations* incorporate the computation of the state vector estimator $\mathrm{E}(\alpha_t|Y_t)$ and its associated error variance matrix, which we denote by $a_{t|t}$ and $P_{t|t}$, respectively. These equations are just a re-formulation of the Kalman filter and are given by

$$
\begin{aligned}
v_t &= y_t - Z_t a_t, & F_t &= Z_t P_t Z_t' + H_t, \\
& & M_t &= P_t Z_t', \\
a_{t|t} &= a_t + M_t F_t^{-1} v_t, & P_{t|t} &= P_t - M_t F_t^{-1} M_t', & t &= 1, \ldots, n, & (4.14) \\
a_{t+1} &= T_t a_{t|t}, & P_{t+1} &= T_t P_{t|t} T_t' + R_t Q_t R_t',
\end{aligned}
$$

for given a_1 and P_1. In Table 4.2 we give the dimensions of the vectors and matrices of the Kalman filter equations.

4.2.3   STEADY STATE

When dealing with a time-invariant state space model in which the system matrices are constant over time, the Kalman recursion for P_{t+1} converges to a constant matrix $\bar{P}$ which is the solution to the matrix equation

$$
\bar{P} = T \bar{P} T' - T \bar{P} Z' \bar{F}^{-1} Z \bar{P} T' + R Q R',
$$

where $\bar{F} = Z \bar{P} Z' + H$. The solution that is reached after convergence to $\bar{P}$ is referred to as the *steady state solution* of the Kalman filter. Use of the steady state after convergence leads to considerable computational savings because the computations for F_t, $K_t = M_t F_t^{-1}$ and P_{t+1} are no longer required.

4.2.4   STATE ESTIMATION ERRORS AND FORECAST ERRORS

Define the *state estimation error* as

$$
x_t = \alpha_t - a_t, \quad \text{with} \quad \mathrm{Var}(x_t) = P_t, \tag{4.15}
$$

as for the local level model in §2.3.2. We now investigate how these errors are related to each other and to the one-step forecast errors $v_t = y_t - \mathrm{E}(y_t|Y_{t-1}) = y_t - Z_t a_t$. Since v_t is the part of y_t that cannot be predicted from the past we shall sometimes refer to the v_t's as *innovations*. It follows immediately from the Kalman filter relations and the definition of x_t that

$$
\begin{aligned}
v_t &= y_t - Z_t a_t \\
&= Z_t \alpha_t + \varepsilon_t - Z_t a_t \\
&= Z_t x_t + \varepsilon_t,
\end{aligned}
\tag{4.16}
$$

and

$$
\begin{aligned}
x_{t+1} &= \alpha_{t+1} - a_{t+1} \\
&= T_t \alpha_t + R_t \eta_t - T_t a_t - K_t v_t \\
&= T_t x_t + R_t \eta_t - K_t Z_t x_t - K_t \varepsilon_t \\
&= L_t x_t + R_t \eta_t - K_t \varepsilon_t,
\end{aligned}
\tag{4.17}
$$

where we note that these recursions are similar to (2.18) for the local level model in Chapter 2. Analogously to the state space relations

$$
y_t = Z_t \alpha_t + \varepsilon_t, \qquad \alpha_{t+1} = T_t \alpha_t + R_t \eta_t,
$$

we obtain the *innovation analogue* of the state space model, that is,

$$
v_t = Z_t x_t + \varepsilon_t, \qquad x_{t+1} = L_t x_t + R_t \eta_t - K_t \varepsilon_t,
\tag{4.18}
$$

with $x_1 = \alpha_1 - a_1$, for $t = 1, \dots, n$. The recursion for P_{t+1} can be derived more easily than in §4.2.1 by the steps

$$
\begin{aligned}
P_{t+1} &= \mathrm{Var}(x_{t+1}) = \mathrm{E}[(\alpha_{t+1} - a_{t+1})x_{t+1}'] \\
&= \mathrm{E}(\alpha_{t+1}x_{t+1}') \\
&= \mathrm{E}[(T_t \alpha_t + R_t \eta_t)(L_t x_t + R_t \eta_t - K_t \varepsilon_t)'] \\
&= T_t P_t L_t' + R_t Q_t R_t',
\end{aligned}
$$

since $\mathrm{Cov}(x_t, \eta_t) = 0$. Relations (4.18) will be used for deriving the smoothing recursions in the next section.

We finally show that the forecast errors are independent of each other using the same arguments as in §2.3.1. The joint density of the observational vectors $y_1, \dots, y_n$ is

$$
p(y_1, \dots, y_n) = p(y_1) \prod_{t=2}^{n} p(y_t | Y_{t-1}).
$$

Transforming from y_t to $v_t = y_t - Z_t a_t$ we have

$$
p(v_1, \dots, v_n) = \prod_{t=1}^{n} p(v_t),
$$

since $p(y_1) = p(v_1)$ and the Jacobian of the transformation is unity because each v_t is y_t minus a linear function of $y_1, \dots, y_{t-1}$ for $t = 2, \dots, n$. Consequently $v_1, \dots, v_n$ are independent of each other, from which it also follows that $v_t, \dots, v_n$ are independent of Y_{t-1}.

4.3 State smoothing

We now consider the estimation of α_t given the entire series $y_1, \ldots, y_n$. Let us denote the stacked vector $(y_1', \ldots, y_n')'$ by y; thus y is Y_n represented as a vector. We shall estimate α_t by its conditional mean $\hat{\alpha}_t = \mathrm{E}(\alpha_t|y)$ and we shall also calculate the error variance matrix $V_t = \mathrm{Var}(\alpha_t - \hat{\alpha}_t) = \mathrm{Var}(\alpha_t|y)$, for $t = 1, \ldots, n$. Our approach is to construct recursions for $\hat{\alpha}_t$ and V_t on the assumption that $\alpha_1 \sim N(a_1, P_1)$ where a_1 and P_1 are known, deferring consideration of the case a_1 and P_1 unknown until Chapter 5. We emphasise the point that the derivations are elementary since the only theoretical background they require is the regression lemma in §2.13.

4.3.1 SMOOTHED STATE VECTOR

The vector y is fixed when Y_{t-1} and $v_t, \ldots, v_n$ are fixed. By (2.49) in the regression lemma in §2.13 and the fact that $v_t, \ldots, v_n$ are independent of Y_{t-1} and of each other with zero means, we therefore have

$$\hat{\alpha}_t = \mathrm{E}(\alpha_t|y) = \mathrm{E}(\alpha_t|Y_{t-1}, v_t, \ldots, v_n)$$

$$= a_t + \sum_{j=t}^{n} \mathrm{Cov}(\alpha_t, v_j)F_j^{-1}v_j, \qquad (4.19)$$

for $t = 1, \ldots, n$, with $\mathrm{Cov}(\alpha_t, v_j) = \mathrm{E}(\alpha_t v_j')$. It follows from (4.18) that

$$\mathrm{E}(\alpha_t v_j') = \mathrm{E}[\alpha_t(Z_j x_j + \varepsilon_j)'] = \mathrm{E}(\alpha_t x_j')Z_j', \qquad j = t, \ldots, n. \quad (4.20)$$

Moreover,

$$\mathrm{E}(\alpha_t x_t') = \mathrm{E}[\mathrm{E}(\alpha_t x_t'|y)] = \mathrm{E}[\mathrm{E}\{\alpha_t(\alpha_t - a_t)'|y\}] = P_t,$$
$$\mathrm{E}(\alpha_t x_{t+1}') = \mathrm{E}[\mathrm{E}\{\alpha_t(L_t x_t + R_t \eta_t - K_t \varepsilon_t)'|y\}] = P_t L_t',$$
$$\mathrm{E}(\alpha_t x_{t+2}') = P_t L_t' L_{t+1}', \qquad (4.21)$$

$$\vdots$$

$$\mathrm{E}(\alpha_t x_n') = P_t L_t' \cdots L_{n-1}'.$$

Note that here and elsewhere we interpret $L_t' \cdots L_{n-1}'$ as I_m when $t = n$ and as L_{n-1}' when $t = n - 1$. Substituting into (4.19) gives

$$\hat{\alpha}_n = a_n + P_n Z_n' F_n^{-1} v_n,$$
$$\hat{\alpha}_{n-1} = a_{n-1} + P_{n-1} Z_{n-1}' F_{n-1}^{-1} v_{n-1} + P_{n-1} L_n' Z_n' F_n^{-1} v_n,$$
$$\hat{\alpha}_t = a_t + P_t Z_t' F_t^{-1} v_t + P_t L_t' Z_{t+1}' F_{t+1}^{-1} v_{t+1}$$
$$+ \cdots + P_t L_t' \cdots L_{n-1}' Z_n' F_n^{-1} v_n,$$

for $t = n - 2, n - 3, \ldots, 1$. We can express the smoothed state vector as

$$\hat{\alpha}_t = a_t + P_t r_{t-1}, \qquad (4.22)$$

where $r_{n-1} = Z'_n F_n^{-1} v_n$, $r_{n-2} = Z'_{n-1} F_{n-1}^{-1} v_{n-1} + L'_{n-1} Z'_n F_n^{-1} v_n$ and

$$r_{t-1} = Z'_t F_t^{-1} v_t + L'_t Z'_{t+1} F_{t+1}^{-1} v_{t+1} + \cdots + L'_t L'_{t+1} \cdots L'_{n-1} Z'_n F_n^{-1} v_n, \quad (4.23)$$

for $t = n - 2, n - 3, \ldots, 1$. The vector r_{t-1} is a weighted sum of innovations v_j occurring after time $t - 1$, that is, $j = t, \ldots, n$. The value at time t is

$$r_t = Z'_{t+1} F_{t+1}^{-1} v_{t+1} + L'_{t+1} Z'_{t+2} F_{t+2}^{-1} v_{t+2} + \cdots + L'_{t+1} \cdots L'_{n-1} Z'_n F_n^{-1} v_n, \quad (4.24)$$

and $r_n = 0$ since no innovations are available after time n. Substituting (4.24) into (4.23) we obtain the backwards recursion

$$r_{t-1} = Z'_t F_t^{-1} v_t + L'_t r_t, \qquad t = n, \ldots, 1, \qquad (4.25)$$

with $r_n = 0$.

Collecting these results together gives the recursion for state smoothing,

$$r_{t-1} = Z'_t F_t^{-1} v_t + L'_t r_t, \qquad \hat{\alpha}_t = a_t + P_t r_{t-1}, \qquad t = n, \ldots, 1, \quad (4.26)$$

with $r_n = 0$; this provides an efficient algorithm for calculating $\hat{\alpha}_1, \ldots, \hat{\alpha}_n$. The smoother, together with the recursion for computing the variance matrix of the smoothed state vector which we present in §4.3.2, is sometimes referred to as the *fixed interval smoother* and was proposed in the forms (4.26) and (4.31) below by de Jong (1988a), de Jong (1989) and Kohn and Ansley (1989) although the earlier treatments in the engineering literature by Bryson and Ho (1969) and Young (1984) are similar.

Alternative algorithms for state smoothing have also been proposed. For example, Anderson and Moore (1979) present the so-called *classical fixed interval smoother* which for our state space model is given by

$$\hat{\alpha}_t = a_{t|t} + P_{t|t} T'_t P_{t+1}^{-1} (\hat{\alpha}_{t+1} - a_{t+1}), \qquad t = n, \ldots, 1, \qquad (4.27)$$

where

$$a_{t|t} = E(\alpha_t | Y_t) = a_t + P_t Z'_t F_t^{-1} v_t, \qquad P_{t|t} = \text{Var}(\alpha_t | Y_t) = P_t - P_t Z'_t F_t^{-1} Z_t P_t;$$

see equations (4.5) and (4.10). Notice that $T_t P_{t|t} = L_t P_t$.

Following Koopman (1998), we now show that (4.26) can be derived from (4.27). Substituting for $a_{t|t}$ and $T_t P_{t|t}$ into (4.27) we have

$$\hat{\alpha}_t = a_t + P_t Z'_t F_t^{-1} v_t + P_t L'_t P_{t+1}^{-1} (\hat{\alpha}_{t+1} - a_{t+1}).$$

By defining $r_t = P_{t+1}^{-1} (\hat{\alpha}_{t+1} - a_{t+1})$ and re-ordering the terms, we obtain

$$P_t^{-1} (\hat{\alpha}_t - a_t) = Z'_t F_t^{-1} v_t + L'_t P_{t+1}^{-1} (\hat{\alpha}_{t+1} - a_{t+1}),$$

and hence

$$r_{t-1} = Z'_t F_t^{-1} v_t + L'_t r_t,$$

which is (4.25). Note that the alternative definition of r_t also implies that $r_n = 0$. Finally, it follows immediately from the definitional relation $r_{t-1} = P_t^{-1} (\hat{\alpha}_t - a_t)$ that $\hat{\alpha}_t = a_t + P_t r_{t-1}$.

A comparison of the two different algorithms shows that the Anderson and Moore smoother requires inversion of $n - 1$ possibly large matrices P_t whereas the smoother (4.26) requires no inversion other than of F_t which has been inverted during the Kalman filter. This is a considerable advantage for large models. For both smoothers the Kalman filter vector a_t and matrix P_t need to be stored together with v_t, F_t^{-1} and K_t, for $t = 1, \ldots, n$. The state smoothing equation of Koopman (1993), which we consider in §4.4.2, does not involve a_t and P_t and it therefore leads to further computational savings.

4.3.2 SMOOTHED STATE VARIANCE MATRIX

A recursion for calculating $V_t = \text{Var}(\alpha_t | y)$ will now be derived. Using (2.50) in the regression lemma in §2.13.2 with $x = \alpha_t$, $y = (y_1', \ldots, y_{t-1}')'$ and $z = (v_t', \ldots, v_n')'$, we obtain

$$V_t = \text{Var}(\alpha_t | Y_{t-1}, v_t, \ldots, v_n) = P_t - \sum_{j=t}^{n} \text{Cov}(\alpha_t, v_j) F_j^{-1} \text{Cov}(\alpha_t, v_j)',$$

since $v_t, \ldots, v_n$ are independent of each other and of Y_{t-1} with zero means. Using (4.20) and (4.21) we obtain immediately

$$V_t = P_t - P_t Z_t' F_t^{-1} Z_t P_t - P_t L_t' Z_{t+1}' F_{t+1}^{-1} Z_{t+1} L_t P_t - \cdots$$
$$- P_t L_t' \cdots L_{n-1}' Z_n' F_n^{-1} Z_n L_{n-1} \cdots L_t P_t$$
$$= P_t - P_t N_{t-1} P_t,$$

where

$$N_{t-1} = Z_t' F_t^{-1} Z_t + L_t' Z_{t+1}' F_{t+1}^{-1} Z_{t+1} L_t + \cdots$$
$$+ L_t' \cdots L_{n-1}' Z_n' F_n^{-1} Z_n L_{n-1} \cdots L_t. \tag{4.28}$$

We note that here, as in the previous subsection, we interpret $L_t' \cdots L_{n-1}'$ as I_m when $t = n$ and as L_{n-1}' when $t = n - 1$. The value at time t is

$$N_t = Z_{t+1}' F_{t+1}^{-1} Z_{t+1} + L_{t+1}' Z_{t+2}' F_{t+2}^{-1} Z_{t+2} L_{t+1} + \cdots$$
$$+ L_{t+1}' \cdots L_{n-1}' Z_n' F_n^{-1} Z_n L_{n-1} \cdots L_{t+1}. \tag{4.29}$$

Substituting (4.29) into (4.28) we obtain the backwards recursion

$$N_{t-1} = Z_t' F_t^{-1} Z_t + L_t' N_t L_t, \qquad t = n, \ldots, 1. \tag{4.30}$$

Noting from (4.29) that $N_{n-1} = Z_n' F_n^{-1} Z_n$ we deduce that recursion (4.30) is initialised with $N_n = 0$. Collecting these results, we find that V_t can be efficiently calculated by the recursion

$$N_{t-1} = Z_t' F_t^{-1} Z_t + L_t' N_t L_t, \qquad V_t = P_t - P_t N_{t-1} P_t, \qquad t = n, \ldots, 1. \tag{4.31}$$

Table 4.3. Dimensions of smoothing recursions of §§4.3.3 and 4.4.4.

Vector		Matrix	
r_t	$m \times 1$	N_t	$m \times m$
$\hat{\alpha}_t$	$m \times 1$	V_t	$m \times m$
u_t	$p \times 1$	D_t	$p \times p$
$\hat{\varepsilon}_t$	$p \times 1$		
$\hat{\eta}_t$	$r \times 1$		

with $N_n = 0$. Since $v_{t+1}, \ldots, v_n$ are independent it follows from (4.24) and (4.29) that $N_t = \text{Var}(r_t)$.

4.3.3 STATE SMOOTHING RECURSION

For convenience we collect together the smoothing equations for the state vector,

$$
\begin{aligned}
r_{t-1} &= Z_t' F_t^{-1} v_t + L_t' r_t, & N_{t-1} &= Z_t' F_t^{-1} Z_t + L_t' N_t L_t, \\
\hat{\alpha}_t &= a_t + P_t r_{t-1}, & V_t &= P_t - P_t N_{t-1} P_t,
\end{aligned}
\tag{4.32}
$$

for $t = n, \ldots, 1$ initialised with $r_n = 0$ and $N_n = 0$. We refer to these collectively as the *state smoothing recursion*. Taken together, the recursions (4.13) and (4.32) will be referred to as the *Kalman filter and smoother*. We see that the way the filtering and smoothing is performed is that we proceed forwards through the series using (4.13) and backwards through the series using (4.32) to obtain $\hat{\alpha}_t$ and V_t for $t = 1, \ldots, n$. During the forwards pass we need to store the quantities v_t, F_t, K_t, a_t and P_t for $t = 1, \ldots, n$. Alternatively we can store a_t and P_t only and re-calculate v_t, F_t and K_t using a_t and P_t but this is usually not done since the dimensions of v_t, F_t and K_t are usually small relatively to a_t and P_t, so the extra storage required is small. In Table 4.3 we present the dimensions of the vectors and matrices of the smoothing equations of this section and §4.4.4.

4.4 Disturbance smoothing

In this section we will derive recursions for computing the smoothed estimates $\hat{\varepsilon}_t = \text{E}(\varepsilon_t | y)$ and $\hat{\eta}_t = \text{E}(\eta_t | y)$ of the disturbance vectors ε_t and η_t given all the observations $y_1, \ldots, y_n$. These estimates have a variety of uses, particularly for parameter estimation and diagnostic checking, as will be indicated in §§7.3 and 7.5.

4.4.1 SMOOTHED DISTURBANCES

Let $\hat{\varepsilon}_t = \text{E}(\varepsilon_t | y)$. By (2.49) of the regression lemma at the end of Chapter 2 we have

$$
\hat{\varepsilon}_t = \text{E}(\varepsilon_t | Y_{t-1}, v_t, \ldots, v_n) = \sum_{j=t}^{n} \text{E}(\varepsilon_t v_j') F_j^{-1} v_j, \qquad t = 1, \ldots, n, \tag{4.33}
$$

since $E(\varepsilon_t|Y_{t-1}) = 0$. It follows from (4.18) that $E(\varepsilon_t v'_j) = E(\varepsilon_t x'_j)Z'_j + E(\varepsilon_t \varepsilon'_j)$ with $E(\varepsilon_t x'_t) = 0$ for $t = 1, \ldots, n$ and $j = t, \ldots, n$. Therefore

$$E(\varepsilon_t v'_j) = \begin{cases} H_t, & j = t, \\ E(\varepsilon_t x'_j)Z'_j, & j = t+1, \ldots, n, \end{cases} \tag{4.34}$$

with

$$E(\varepsilon_t x'_{t+1}) = -H_t K'_t,$$
$$E(\varepsilon_t x'_{t+2}) = -H_t K'_t L'_{t+1},$$
$$\vdots \tag{4.35}$$
$$E(\varepsilon_t x'_n) = -H_t K'_t L'_{t+1} \cdots L'_{n-1},$$

which follow from (4.16) and (4.18), for $t = 1, \ldots, n - 1$. Note that here as elsewhere we interpret $L'_{t+1} \cdots L'_{n-1}$ as I_m when $t = n - 1$ and as L'_{n-1} when $t = n - 2$. Substituting (4.34) into (4.33) leads to

$$\begin{aligned} \hat{\varepsilon}_t &= H_t(F_t^{-1} v_t - K'_t Z'_{t+1} F_{t+1}^{-1} v_{t+1} - K'_t L'_{t+1} Z'_{t+2} F_{t+2}^{-1} v_{t+2} - \cdots \\ &\quad - K'_t L'_{t+1} \cdots L'_{n-1} Z'_n F_n^{-1} v_n) \\ &= H_t(F_t^{-1} v_t - K'_t r_t) \\ &= H_t u_t, \qquad t = n, \ldots, 1, \end{aligned} \tag{4.36}$$

where r_t is defined in (4.24) and

$$u_t = F_t^{-1} v_t - K'_t r_t. \tag{4.37}$$

We refer to the vector u_t as the *smoothing error*.

The smoothed estimate of η_t is denoted by $\hat{\eta}_t = E(\eta_t|y)$ and analogously to (4.33) we have

$$\hat{\eta}_t = \sum_{j=t}^n E(\eta_t v'_j)F_j^{-1} v_j, \qquad t = 1, \ldots, n. \tag{4.38}$$

The relations (4.18) imply that

$$E(\eta_t v'_j) = \begin{cases} Q_t R'_t Z'_{t+1}, & j = t+1, \\ E(\eta_t x'_j)Z'_j & j = t+2, \ldots, n, \end{cases} \tag{4.39}$$

with

$$E(\eta_t x'_{t+2}) = Q_t R'_t L'_{t+1},$$
$$E(\eta_t x'_{t+3}) = Q_t R'_t L'_{t+1} L'_{t+2},$$
$$\vdots \tag{4.40}$$
$$E(\eta_t x'_n) = Q_t R'_t L'_{t+1} \cdots L'_{n-1},$$

for $t = 1, \ldots, n - 1$. Substituting (4.39) into (4.38) and noting that $E(\eta_t v_t') = 0$ leads to

$$\hat{\eta}_t = Q_t R_t' \left(Z_{t+1}' F_{t+1}^{-1} v_{t+1} + L_{t+1}' Z_{t+2}' F_{t+2}^{-1} v_{t+2} + \cdots + L_{t+1}' \cdots L_{n-1}' Z_n' F_n^{-1} v_n \right)$$

$$= Q_t R_t' r_t, \qquad t = n, \ldots, 1, \tag{4.41}$$

where r_t is obtained from (4.25). This result is useful as we will show in the next section but it also gives the vector r_t the interpretation as the 'scaled' smoothed estimator of η_t. Note that in many practical cases the matrix $Q_t R_t'$ is diagonal or sparse. Equations (4.36) and (4.44) below were first given by de Jong (1988a) and Kohn and Ansley (1989). Equations (4.41) and (4.47) below were given by Koopman (1993).

4.4.2   FAST STATE SMOOTHING

The smoothing recursion for the disturbance vector η_t of the transition equation is particularly useful since it leads to a computationally more efficient method of calculating $\hat{\alpha}_t$ for $t = 1, \ldots, n$ than (4.26). Since the state equation is

$$\alpha_{t+1} = T_t \alpha_t + R_t \eta_t,$$

it follows immediately that

$$\hat{\alpha}_{t+1} = T_t \hat{\alpha}_t + R_t \hat{\eta}_t$$

$$= T_t \hat{\alpha}_t + R_t Q_t R_t' r_t, \qquad t = 1, \ldots, n, \tag{4.42}$$

which is initialised via the relation (4.22) for $t = 1$, that is, $\hat{\alpha}_1 = a_1 + P_1 r_0$ where r_0 is obtained from (4.25). This recursion, due to Koopman (1993), can be used to generate the smoothed states $\hat{\alpha}_1, \ldots, \hat{\alpha}_n$ by an algorithm different from (4.26); it does not require the storage of a_t and P_t and it does not involve multiplications by the full matrix P_t, for $t = 1, \ldots, n$. After the Kalman filter and the storage of v_t, F_t^{-1} and K_t has taken place, the backwards recursion (4.25) is undertaken and the vector r_t is stored for which the storage space of K_t can be used so no additional storage space is required. It should be kept in mind that the matrices T_t and $R_t Q_t R_t'$ are usually sparse matrices containing many zero and unity values which makes the application of (4.42) rapid; this property does not apply to P_t which is a full variance matrix. This approach, however, cannot be used to obtain a recursion for the calculation of $V_t = \mathrm{Var}(\alpha_t | y)$; if V_t is required then (4.26) and (4.31) should be used.

4.4.3   SMOOTHED DISTURBANCE VARIANCE MATRICES

The error variance matrices of the smoothed disturbances are developed by the same approach that we used in §4.3.2 to derive the error variance matrix of the smoothed state vector. Using (2.50) of the regression lemma in §2.13 we have

$$\mathrm{Var}(\varepsilon_t | y) = \mathrm{Var}(\varepsilon_t | Y_{t-1}, v_t, \ldots, v_n)$$

$$= \mathrm{Var}(\varepsilon_t | Y_{t-1}) - \sum_{j=t}^{n} \mathrm{Cov}(\varepsilon_t, v_j) \, \mathrm{Var}(v_j)^{-1} \, \mathrm{Cov}(\varepsilon_t, v_j)'$$

$$= H_t - \sum_{j=1}^{n} \mathrm{Cov}(\varepsilon_t, v_j) F_j^{-1} \mathrm{Cov}(\varepsilon_t, v_j)', \tag{4.43}$$

where $\text{Cov}(\varepsilon_t, v_j) = \text{E}(\varepsilon_t, v_j')$ which is given by (4.34). By substitution we obtain

$$
\begin{aligned}
\text{Var}(\varepsilon_t | y) = {}& H_t - H_t \left(F_t^{-1} + K_t' Z_{t+1}' F_{t+1}^{-1} Z_{t+1} K_t \right. \\
& - K_t' L_{t+1}' Z_{t+2}' F_{t+2}^{-1} Z_{t+2} L_{t+1} K_t - \cdots \\
& \left. - K_t' L_{t+1}' \cdots L_{n-1}' Z_n' F_n^{-1} Z_n L_{n-1} \cdots L_{t+1} K_t \right) H_t' \\
= {}& H_t - H_t \left(F_t^{-1} + K_t' N_t K_t \right) H_t \\
= {}& H_t - H_t D_t H_t, \quad (4.44)
\end{aligned}
$$

with

$$
D_t = F_t^{-1} + K_t' N_t K_t, \quad (4.45)
$$

where N_t is defined in (4.29) and can be obtained from the backwards recursion (4.30).

In a similar way the variance matrix $\text{Var}(\eta_t | y)$ is given by

$$
\text{Var}(\eta_t | y) = \text{Var}(\eta_t) - \sum_{j=t}^{n} \text{Cov}(\eta_t, v_j) F_j^{-1} \text{Cov}(\eta_t, v_j)', \quad (4.46)
$$

where $\text{Cov}(\eta_t, v_j) = \text{E}(\eta_t, v_j')$ which is given by (4.39). Substitution gives

$$
\begin{aligned}
\text{Var}(\eta_t | y) = {}& Q_t - Q_t R_t' \left(Z_{t+1}' F_{t+1}^{-1} Z_{t+1} + L_{t+1}' Z_{t+2}' F_{t+2}^{-1} Z_{t+2} L_{t+1} + \cdots \right. \\
& \left. + L_{t+1}' \cdots L_{n-1}' Z_n' F_n^{-1} Z_n L_{n-1} \cdots L_{t+1} \right) R_t Q_t \\
= {}& Q_t - Q_t R_t' N_t R_t Q_t, \quad (4.47)
\end{aligned}
$$

where N_t is obtained from (4.30).

4.4.4   DISTURBANCE SMOOTHING RECURSION

For convenience we collect together the smoothing equations for the disturbance vectors,

$$
\begin{aligned}
\hat{\varepsilon}_t &= H_t \left(F_t^{-1} v_t - K_t' r_t \right), & \text{Var}(\varepsilon_t | y) &= H_t - H_t \left(F_t^{-1} + K_t' N_t K_t \right) H_t, \\
\hat{\eta}_t &= Q_t R_t' r_t, & \text{Var}(\eta_t | y) &= Q_t - Q_t R_t' N_t R_t Q_t, \\
r_{t-1} &= Z_t' F_t^{-1} v_t + L_t' r_t, & N_{t-1} &= Z_t' F_t^{-1} Z_t + L_t' N_t L_t,
\end{aligned}
$$

$$(4.48)$$

for $t = n, \ldots, 1$ where $r_n = 0$ and $N_n = 0$. These equations can be reformulated as

$$
\begin{aligned}
\hat{\varepsilon}_t &= H_t u_t, & \text{Var}(\varepsilon_t | y) &= H_t - H_t D_t H_t, \\
\hat{\eta}_t &= Q_t R_t' r_t, & \text{Var}(\eta_t | y) &= Q_t - Q_t R_t' N_t R_t Q_t, \\
u_t &= F_t^{-1} v_t - K_t' r_t, & D_t &= F_t^{-1} + K_t' N_t K_t, \\
r_{t-1} &= Z_t' u_t + T_t' r_t, & N_{t-1} &= Z_t' D_t Z_t + T_t' N_t T_t - Z_t' K_t' N_t T_t - T_t' N_t K_t Z_t,
\end{aligned}
$$

for $t = n, \ldots, 1$, which are computationally more efficient since they rely directly on the system matrices Z_t and T_t which have the property that they usually contain many zeros and ones. We refer to these equations collectively as the *disturbance smoothing recursion*. The smoothing error u_t and vector r_t are important in their own rights for a variety of reasons which we will discuss in §7.5. The dimensions of the vectors and matrices of disturbance smoothing are given in Table 4.3.

We see that disturbance smoothing is performed in a similar way to state smoothing: we proceed forwards through the series using (4.13) and backwards through the series using (4.48) to obtain $\hat{\varepsilon}_t$ and $\hat{\eta}_t$ together with the corresponding conditional variances for $t = 1, \ldots, n$. The storage requirement for (4.48) during the forwards pass is less than for the state smoothing recursion (4.32) since here we only need v_t, F_t and K_t of the Kalman filter. Also, the computations are quicker for disturbance smoothing since they do not involve the vector a_t and the matrix P_t which are not sparse.

4.5 Covariance matrices of smoothed estimators

In this section we develop expressions for the covariances between the errors of the smoothed estimators $\hat{\varepsilon}_t$, $\hat{\eta}_t$ and $\hat{\alpha}_t$ contemporaneously and for all leads and lags.

It turns out that the covariances of smoothed estimators rely basically on the cross-expectations $E(\varepsilon_t r'_j)$, $E(\eta_t r'_j)$ and $E(\alpha_t r'_j)$ for $j = t + 1, \ldots, n$. To develop these expressions we collect from equations (4.34), (4.35), (4.39), (4.40), (4.21) and (4.20) the results

$$
\begin{aligned}
&E(\varepsilon_t x'_t) = 0, && E(\varepsilon_t v'_t) = H_t, \\
&E(\varepsilon_t x'_j) = -H_t K'_t L'_{t+1} \cdots L'_{j-1}, && E(\varepsilon_t v'_j) = E(\varepsilon_t x'_j) Z'_j, \\
&E(\eta_t x'_t) = 0, && E(\eta_t v'_t) = 0, \\
&E(\eta_t x'_j) = Q_t R'_t L'_{t+1} \cdots L'_{j-1}, && E(\eta_t v'_j) = E(\eta_t x'_j) Z'_j, && (4.49) \\
&E(\alpha_t x'_t) = P_t, && E(\alpha_t v'_t) = P_t Z'_t, \\
&E(\alpha_t x'_j) = P_t L'_t L'_{t+1} \cdots L'_{j-1}, && E(\alpha_t v'_j) = E(\alpha_t x'_j) Z'_j,
\end{aligned}
$$

for $j = t + 1, \ldots, n$. For the case $j = t + 1$, we replace $L'_{t+1} \cdots L'_t$ by the identity matrix I_m.

We derive the cross-expectations below using the definitions

$$
r_j = \sum_{k=j+1}^{n} L'_{j+1} \cdots L'_{k-1} Z'_k F_k^{-1} v_k,
$$

$$
N_j = \sum_{k=j+1}^{n} L'_{j+1} \cdots L'_{k-1} Z'_k F_k^{-1} Z_k L_{k-1} \cdots L_{j+1},
$$

which are given by (4.23) and (4.28), respectively. It follows that

$$
\begin{aligned}
\mathrm{E}(\varepsilon_t r_j') &= \mathrm{E}(\varepsilon_t v_{j+1}')F_{j+1}^{-1}Z_{j+1} + \mathrm{E}(\varepsilon_t v_{j+2}')F_{j+2}^{-1}Z_{j+2}L_{j+1} + \cdots \\
&\quad + \mathrm{E}(\varepsilon_t v_n')F_n^{-1}Z_n L_{n-1}\cdots L_{j+1} \\
&= -H_t K_t' L_{t+1}' \cdots L_j' Z_{j+1}' F_{j+1}^{-1}Z_{j+1} \\
&\quad - H_t K_t' L_{t+1}' \cdots L_{j+1}' Z_{j+2}' F_{j+2}^{-1}Z_{j+2}L_{j+1} - \cdots \\
&\quad - H_t K_t' L_{t+1}' \cdots L_{n-1}' Z_n' F_n^{-1}Z_n L_{n-1}\cdots L_{j+1} \\
&= -H_t K_t' L_{t+1}' \cdots L_{j-1}' L_j' N_j,
\end{aligned}
\tag{4.50}
$$

$$
\begin{aligned}
\mathrm{E}(\eta_t r_j') &= \mathrm{E}(\eta_t v_{j+1}')F_{j+1}^{-1}Z_{j+1} + \mathrm{E}(\eta_t v_{j+2}')F_{j+2}^{-1}Z_{j+2}L_{j+1} + \cdots \\
&\quad + \mathrm{E}(\eta_t v_n')F_n^{-1}Z_n L_{n-1}\cdots L_{j+1} \\
&= Q_t R_t' L_{t+1}' \cdots L_j' Z_{j+1}' F_{j+1}^{-1}Z_{j+1} \\
&\quad + Q_t R_t' L_{t+1}' \cdots L_{j+1}' Z_{j+2}' F_{j+2}^{-1}Z_{j+2}L_{j+1} + \cdots \\
&\quad + Q_t R_t' L_{t+1}' \cdots L_{n-1}' Z_n' F_n^{-1}Z_n L_{n-1}\cdots L_{j+1} \\
&= Q_t R_t' L_{t+1}' \cdots L_{j-1}' L_j' N_j,
\end{aligned}
\tag{4.51}
$$

$$
\begin{aligned}
\mathrm{E}(\alpha_t r_j') &= \mathrm{E}(\alpha_t v_{j+1}')F_{j+1}^{-1}Z_{j+1} + \mathrm{E}(\alpha_t v_{j+2}')F_{j+2}^{-1}Z_{j+2}L_{j+1} + \cdots \\
&\quad + \mathrm{E}(\alpha_t v_n')F_n^{-1}Z_n L_{n-1}\cdots L_{j+1} \\
&= P_t L_t' L_{t+1}' \cdots L_j' Z_{j+1}' F_{j+1}^{-1}Z_{j+1} \\
&\quad + P_t L_t' L_{t+1}' \cdots L_{j+1}' Z_{j+2}' F_{j+2}^{-1}Z_{j+2}L_{j+1} + \cdots \\
&\quad + P_t L_t' L_{t+1}' \cdots L_{n-1}' Z_n' F_n^{-1}Z_n L_{n-1}\cdots L_{j+1} \\
&= P_t L_t' L_{t+1}' \cdots L_{j-1}' L_j' N_j,
\end{aligned}
\tag{4.52}
$$

for $j = t, \ldots, n$. Hence

$$
\begin{aligned}
\mathrm{E}(\varepsilon_t r_j') &= \mathrm{E}(\varepsilon_t x_{t+1}')N_{t+1,j}^*, \\
\mathrm{E}(\eta_t r_j') &= \mathrm{E}(\eta_t x_{t+1}')N_{t+1,j}^*, \\
\mathrm{E}(\alpha_t r_j') &= \mathrm{E}(\alpha_t x_{t+1}')N_{t+1,j}^*,
\end{aligned}
\tag{4.53}
$$

where $N_{tj}^* = L_t' \cdots L_{j-1}' L_j' N_j$ for $j = t, \ldots, n$.

The cross-expectations of ε_t, η_t and α_t between the smoothed estimators

$$
\hat{\varepsilon}_j = H_j\big(F_j^{-1}v_j - K_j' r_j\big), \qquad \hat{\eta}_j = Q_j R_j' r_j, \qquad \alpha_j - \hat{\alpha}_j = x_j - P_j r_{j-1},
$$

for $j = t+1, \ldots, n$, are given by

$$
\mathrm{E}(\varepsilon_t \hat{\varepsilon}_j') = \mathrm{E}(\varepsilon_t v_j')F_j^{-1}H_j - \mathrm{E}(\varepsilon_t r_j')K_j H_j,
$$

$$
\mathrm{E}(\varepsilon_t \hat{\eta}_j') = \mathrm{E}(\varepsilon_t r_j')R_j Q_j,
$$

$$
\mathrm{E}[\varepsilon_t(\alpha_j - \hat{\alpha}_j)'] = \mathrm{E}(\varepsilon_t x_j') - \mathrm{E}(\varepsilon_t r_{j-1}')P_j,
$$

$$
\mathrm{E}(\eta_t \hat{\varepsilon}_j') = \mathrm{E}(\eta_t v_j')F_j^{-1}H_j - \mathrm{E}(\eta_t r_j')K_j H_j,
$$

$$E(\eta_t \hat{\eta}'_j) = E(\eta_t r'_j) R_j Q_j,$$

$$E[\eta_t (\alpha_j - \hat{\alpha}_j)'] = E(\eta_t x'_j) - E(\eta_t r'_{j-1}) P_j,$$

$$E(\alpha_t \hat{\varepsilon}'_j) = E(\alpha_t v'_j) F_j^{-1} H_j - E(\alpha_t r'_j) K_j H_j,$$

$$E(\alpha_t \hat{\eta}'_j) = E(\alpha_t r'_j) R_j Q_j,$$

$$E[\alpha_t (\alpha_j - \hat{\alpha}_j)'] = E(\alpha_t x'_j) - E(\alpha_t r'_{j-1}) P_j,$$

into which the expressions in equations (4.49), (4.50), (4.51) and (4.52) can be substituted.

The covariance matrices of the smoothed estimators at different times are derived as follows. We first consider the covariance matrix for the smoothed diturbance vector $\hat{\varepsilon}_t$, that is, $\mathrm{Cov}(\varepsilon_t - \hat{\varepsilon}_t, \varepsilon_j - \hat{\varepsilon}_j)$ for $t = 1, \ldots, n$ and $j = t + 1, \ldots, n$. Since

$$E[\hat{\varepsilon}_t (\varepsilon_j - \hat{\varepsilon}_j)'] = E[E\{\hat{\varepsilon}_t (\varepsilon_j - \hat{\varepsilon}_j)' | y\}] = 0,$$

we have

$$\begin{aligned}
\mathrm{Cov}(\varepsilon_t - \hat{\varepsilon}_t, \varepsilon_j - \hat{\varepsilon}_j) &= E[\varepsilon_t (\varepsilon_j - \hat{\varepsilon}_j)'] \\
&= -E(\varepsilon_t \hat{\varepsilon}'_j) \\
&= H_t K'_t L'_{t+1} \cdots L'_{j-1} Z'_j F_j^{-1} H_j \\
&\quad + H_t K'_t L'_{t+1} \cdots L'_{j-1} L'_j N_j K_j H_j \\
&= H_t K'_t L'_{t+1} \cdots L'_{j-1} W'_j,
\end{aligned}$$

where

$$W_j = H_j \left(F_j^{-1} Z_j - K'_j N_j L_j \right), \qquad (4.54)$$

for $j = t + 1, \ldots, n$. In a similar way obtain

$$\begin{aligned}
\mathrm{Cov}(\eta_t - \hat{\eta}_t, \eta_j - \hat{\eta}_j) &= -E(\eta_t \hat{\eta}'_j) \\
&= -Q_t R'_t L'_{t+1} \cdots L'_{j-1} L'_j N_j R_j Q_j, \\
\mathrm{Cov}(\alpha_t - \hat{\alpha}_t, \alpha_j - \hat{\alpha}_j) &= -E[\alpha_t (\alpha_j - \hat{\alpha}_j)'] \\
&= P_t L'_t L'_{t+1} \cdots L'_{j-1} - P_t L'_t L'_{t+1} \cdots L'_{j-1} N_{j-1} P_j \\
&= P_t L'_t L'_{t+1} \cdots L'_{j-1} (I - N_{j-1} P_j),
\end{aligned}$$

for $j = t + 1, \ldots, n$.

The cross-covariance matrices of the smoothed disturbances arc obtained as follows. We have

$$\begin{aligned}
\mathrm{Cov}(\varepsilon_t - \hat{\varepsilon}_t, \eta_j - \hat{\eta}_j) &= E[(\varepsilon_t - \hat{\varepsilon}_t)(\eta_j - \hat{\eta}_j)'] \\
&= E[\varepsilon_t (\eta_j - \hat{\eta}_j)'] \\
&= -E(\varepsilon_t \hat{\eta}'_j) \\
&= H_t K'_t L'_{t+1} \cdots L'_{j-1} L'_j N_j R_j Q_j,
\end{aligned}$$

for $j = t, t + 1, \ldots, n$, and

$$\begin{aligned}
\mathrm{Cov}(\eta_t - \hat{\eta}_t, \varepsilon_j - \hat{\varepsilon}_j) &= -\mathrm{E}(\eta_t \hat{\varepsilon}_j') \\
&= -Q_t R_t' L_{t+1}' \cdots L_{j-1}' Z_j' F_j^{-1} H_j \\
&\quad + Q_t R_t' L_{t+1}' \cdots L_{j-1}' N_j' K_j H_j \\
&= -Q_t R_t' L_{t+1}' \cdots L_{j-1}' W_j',
\end{aligned}$$

for $j = t + 1, \ldots, n$.

The cross-covariances between the smoothed state vector and the smoothed disturbances are obtained in a similar way. We have

$$\begin{aligned}
\mathrm{Cov}(\alpha_t - \hat{\alpha}_t, \varepsilon_j - \hat{\varepsilon}_j) &= -\mathrm{E}(\alpha_t \hat{\varepsilon}_j') \\
&= -P_t L_t' L_{t+1}' \cdots L_{j-1}' Z_j' F_j^{-1} H_j \\
&\quad + P_t L_t' L_{t+1}' \cdots L_{j-1}' N_j' K_j H_j \\
&= -P_t L_t' L_{t+1}' \cdots L_{j-1}' W_j', \\
\mathrm{Cov}(\alpha_t - \hat{\alpha}_t, \eta_j - \hat{\eta}_j) &= -\mathrm{E}(\alpha_t \hat{\eta}_j') \\
&= -P_t L_t' L_{t+1}' \cdots L_{j-1}' L_j' N_j R_j Q_j,
\end{aligned}$$

for $j = t, t + 1, \ldots, n$, and

$$\begin{aligned}
\mathrm{Cov}(\varepsilon_t - \hat{\varepsilon}_t, \alpha_j - \hat{\alpha}_j) &= \mathrm{E}[\varepsilon_t(\alpha_j - \hat{\alpha}_j)'] \\
&= -H_t K_t' L_{t+1}' \cdots L_{j-1}' \\
&\quad + H_t K_t' L_{t+1}' \cdots L_{j-1}' N_{j-1} P_j \\
&= -H_t K_t' L_{t+1}' \cdots L_{j-1}' (I - N_{j-1} P_j), \\
\mathrm{Cov}(\eta_t - \hat{\eta}_t, \alpha_j - \hat{\alpha}_j) &= \mathrm{E}[\eta_t(\alpha_j - \hat{\alpha}_j)'] \\
&= Q_t R_t' L_{t+1}' \cdots L_{j-1}' (I - N_{j-1} P_j),
\end{aligned}$$

for $j = t + 1, \ldots, n$.

Table 4.4. Covariances of smoothed estimators for $t = 1, \ldots, n$.

$\hat{\varepsilon}_t$			
	$\hat{\varepsilon}_j$	$H_t K_t' L_{t+1}' \cdots L_{j-1}' W_j'$	$j > t$
	$\hat{\eta}_j$	$H_t K_t' L_{t+1}' \cdots L_{j-1}' L_j' N_j R_j Q_j$	$j \geq t$
	$\hat{\alpha}_j$	$-H_t K_t' L_{t+1}' \cdots L_{j-1}' (I_m - N_{j-1} P_j)$	$j > t$
$\hat{\eta}_t$			
	$\hat{\varepsilon}_j$	$-Q_t R_t' L_{t+1}' \cdots L_{j-1}' W_j'$	$j > t$
	$\hat{\eta}_j$	$-Q_t R_t' L_{t+1}' \cdots L_{j-1}' L_j' N_j R_j Q_j$	$j > t$
	$\hat{\alpha}_j$	$Q_t R_t' L_{t+1}' \cdots L_{j-1}' (I_m - N_{j-1} P_j)$	$j > t$
$\hat{\alpha}_t$			
	$\hat{\varepsilon}_j$	$-P_t L_t' L_{t+1}' \cdots L_{j-1}' W_j'$	$j \geq t$
	$\hat{\eta}_j$	$-P_t L_t' L_{t+1}' \cdots L_{j-1}' L_j' N_j R_j Q_j$	$j \geq t$
	$\hat{\alpha}_j$	$P_t L_t' L_{t+1}' \cdots L_{j-1}' (I_m - N_{j-1} P_j)$	$j \geq t$

These results here have been developed by de Jong and MacKinnon (1988), who derived the covariances between smoothed state vector estimators, and by Koopman (1993) who derived the covariances between the smoothed disturbance vectors estimators. The results in this section have also been reviewed by de Jong (1998). The auto- and cross-covariance matrices are for convenience collected in Table 4.4.

4.6   Weight functions

4.6.1   INTRODUCTION

Up to this point we have developed recursions for the evaluation of the conditional mean vector and variance matrix of the state vector α_t given the observations $y_1, \ldots, y_{t-1}$ (filtering), given the observations $y_1, \ldots, y_t$ (contemporaneous filtering) and given the observations $y_1, \ldots, y_n$ (smoothing). We also have developed recursions for the conditional mean vectors and variance matrices of the disturbance vectors ε_t and η_t given the observation $y_1, \ldots, y_n$. It follows that these conditional means are weighted sums of past (filtering), of past and present (contemporaneous filtering) and of all (smoothing) observations. It is of interest to study these weights to gain a better understanding of the properties of the estimators as is argued in Koopman and Harvey (1999). For example, the weights for the smoothed estimator of a trend component around time $t = n/2$, that is, at the middle of the series, should be symmetric and centred around t with exponentially declining weights unless specific circumstances require a different pattern. Models which produce weight patterns for the trend components which differ from what is regarded as appropriate should be investigated. In effect, the weights can be regarded as what are known as kernel functions in the field of nonparametric regression; see, for example, Green and Silverman (1994, Chapter 2).

In the case when the state vector contains regression coefficients, the associated weights for the smoothed state vector can be interpreted as leverage statistics as studied in Cook and Weisberg (1982) and Atkinson (1985) in the context of regression models. Such statistics for state space models have been developed with the emphasis on the smoothed signal estimator $Z_t \hat{\alpha}_t$ by, for example, Kohn and Ansley (1989), de Jong (1989), Harrison and West (1991) and de Jong (1998). Since the concept of leverage is more useful in a regression context, we will refer to the expressions below as weights. Given the results of this chapter so far, it is straightforward to develop the weight expressions.

4.6.2   FILTERING WEIGHTS

It follows from the linear properties of the normal distribution that the filtered estimator of the state vector can be expressed as a weighted vector sum of past observations, that is

$$a_t = \sum_{j=1}^{t-1} \omega_{jt} y_j,$$

Table 4.5. Expressions for $\mathrm{E}(s_t \varepsilon'_j)$ with $1 \leq t \leq n$ given (filtering).

s_t	$j < t$	$j = t$	$j > t$
a_t	$L_{t-1} \cdots L_{j+1} K_j H_j$	0	0
$a_{t\mid t}$	$(I - M_t F_t^{-1} Z_t) L_{t-1} \cdots L_{j+1} K_j H_j$	$M_t F_t^{-1} H_t$	0
$Z_t a_t$	$Z_t L_{t-1} \cdots L_{j+1} K_j H_j$	0	0
$Z_t a_{t\mid t}$	$H_t F_t^{-1} Z_t L_{t-1} \cdots L_{j+1} K_j H_j$	$\left(I - H_t F_t^{-1}\right) H_t$	0
v_t	$-Z_t L_{t-1} \cdots L_{j+1} K_j H_j$	H_t	0

where ω_{jt} is an $m \times p$ matrix of weights associated with the estimator a_t and the jth observation. An expression for the weight matrix can be obtained using the fact that

$$\mathrm{E}(a_t \varepsilon'_j) = \omega_{jt} \mathrm{E}(y_j \varepsilon'_j) = \omega_{jt} H_j.$$

Since $x_t = \alpha_t - a_t$ and $\mathrm{E}(\alpha_t \varepsilon'_j) = 0$, we can use (4.17) to obtain

$$\mathrm{E}(a_t \varepsilon'_j) = \mathrm{E}(x_t \varepsilon'_j) = L_{t-1} \mathrm{E}(x_{t-1} \varepsilon'_j)$$
$$= L_{t-1} L_{t-2} \cdots L_{j+1} K_j H_j,$$

which gives

$$\omega_{jt} = L_{t-1} L_{t-2} \cdots L_{j+1} K_j,$$

for $j = t - 1, \ldots, 1$. In a similar way we can obtain the weights associated with other filtering estimators. In Table 4.5 we give a selection of such expressions from which the weights are obtained by disregarding the last matrix H_j. Finally, the expression for weights of $Z_t a_{t\mid t}$ follows since $Z_t M_t = F_t - H_t$ and

$$Z_t \left(I - M_t F_t^{-1} Z_t\right) = \left[I - (F_t - H_t) F_t^{-1}\right] Z_t = H_t F_t^{-1} Z_t.$$

4.6.3   SMOOTHING WEIGHTS

The weighting expressions for smoothing estimators can be obtained in a similar way to that used for filtering. For example, the smoothed estimator of the measurement disturbance vector can be expressed as a weighted vector sum of past, current and future observations, that is

$$\hat{\varepsilon}_t = \sum_{j=1}^{n} \omega^{\varepsilon}_{jt} y_j,$$

where $\omega^{\varepsilon}_{jt}$ is a $p \times p$ matrix of weights associated with the estimator $\hat{\varepsilon}_t$ and the jth observation. An expression for the weight matrix can be obtained using the fact that

$$\mathrm{E}(\hat{\varepsilon}_t \varepsilon'_j) = \omega^{\varepsilon}_{jt} \mathrm{E}(y_j \varepsilon'_j) = \omega^{\varepsilon}_{jt} H_j.$$

Table 4.6. Expressions for $E(s_t \varepsilon_j')$ with $1 \le t \le n$ given (smoothing).

s_t	$j < t$	$j = t$	$j > t$
$\hat{\varepsilon}_t$	$-W_t L_{t-1} \cdots L_{j+1} K_j H_j$	$H_t D_t H_t$	$-H_t K_t' L_{t+1}' \cdots L_{j-1}' W_j'$
$\hat{\eta}_t$	$-Q_t R_t' N_t L_t L_{t-1} \cdots L_{j+1} K_j H_j$	$Q_t R_t' N_t K_t H_t$	$Q_t R_t' L_{t+1}' \cdots L_{j-1}' W_j'$
$\hat{\alpha}_t$	$(I - P_t N_{t-1}) L_{t-1} \cdots L_{j+1} K_j H_j$	$P_t W_t'$	$P_t L_t' L_{t+1}' \cdots L_{j-1}' W_j'$
$Z_t \hat{\alpha}_t$	$W_t L_{t-1} \cdots L_{j+1} K_j H_j$	$(I - H_t D_t) H_t$	$H_t K_t' L_{t+1}' \cdots L_{j-1}' W_j'$

Expressions for the covariance matrices for smoothed disturbances are developed in §4.5 and they are directly related to the expression for $E(\hat{\varepsilon}_t \varepsilon_j')$ because

$$\text{Cov}(\varepsilon_t - \hat{\varepsilon}_t, \varepsilon_j - \hat{\varepsilon}_j) = E[(\varepsilon_t - \hat{\varepsilon}_t)\varepsilon_j'] = -E(\hat{\varepsilon}_t \varepsilon_j'),$$

with $1 \le t \le n$ and $j = 1, \ldots, n$. Therefore, no new derivations need to be given here and we only state the results as presented in Table 4.6.

For example, to obtain the weights for the smoothed estimator of α_t, we require

$$\begin{aligned} E(\hat{\alpha}_t \varepsilon_j') &= -E[(\alpha_t - \hat{\alpha}_t)\varepsilon_j'] = -E[\varepsilon_j(\alpha_t - \hat{\alpha}_t)']' \\ &= \text{Cov}(\varepsilon_j - \hat{\varepsilon}_j, \alpha_t - \hat{\alpha}_t)', \end{aligned}$$

for $j < t$. An expression for this latter quantity can be directly obtained from Table 4.4 but notice that the indices j and t of Table 4.4 need to be reversed here. Further,

$$E(\hat{\alpha}_t \varepsilon_j') = \text{Cov}(\alpha_t - \hat{\alpha}_t, \varepsilon_j - \hat{\varepsilon}_j)$$

for $j \ge t$ can also be obtained from Table 4.4. In the same way we can obtain the weights for the smoothed estimators of ε_t and η_t from Table 4.4 as reported in Table 4.6. Finally, the expression for weights of $Z_t \hat{\alpha}_t$ follows since $Z_t M_t = F_t - H_t$ and $Z_t P_t L_t' = H_t K_t'$. Hence, by using the equations (4.30), (4.45) and (4.54), we have

$$\begin{aligned} Z_t(I - P_t N_{t-1}) &= Z_t - Z_t P_t Z_t' F_t^{-1} Z_t + Z_t P_t L_t' N_t L_t \\ &= H_t F_t^{-1} Z_t + H_t K_t' N_t L_t = W_t, \\ Z_t P_t W_t' &= (Z_t P_t Z_t' F_t^{-1} - Z_t P_t L_t' N_t K_t) H_t \\ &= [(F_t - H_t) F_t^{-1} - H_t K_t' N_t K_t] H_t \\ &= (I - H_t D_t) H_t. \end{aligned}$$

4.7 Simulation smoothing

The drawing of samples of state or disturbance vectors conditional on the observations held fixed is called *simulation smoothing*. Such samples are useful for investigating the performance of techniques of analysis proposed for the linear Gaussian model, and for Bayesian analysis based on this model. The primary purpose of simulation smoothing in this book, however, will be to serve as the basis for the importance sampling techniques we shall develop in Part II for dealing with non-Gaussian and nonlinear models from both classical and Bayesian perspectives.

In this section we will show how to draw random samples of the disturbance vectors ε_t and η_t, and the state vector α_t, for $t = 1, \ldots, n$, generated by the linear Gaussian model (4.1) conditional on the observed vector y. The resulting algorithm is sometimes called a *forwards filtering, backwards sampling* algorithm.

Fruhwirth-Schnatter (1994) and Carter and Kohn (1994) independently developed methods for simulation smoothing of the state vector based on the identity

$$p(\alpha_1, \ldots, \alpha_n | y) = p(\alpha_n | y) p(\alpha_{n-1} | y, \alpha_n) \cdots p(\alpha_1 | y, \alpha_2, \ldots, \alpha_n). \quad (4.55)$$

de Jong and Shephard (1995) made significant progress by first concentrating on sampling the disturbances and subsequently sampling the states. Our treatment to follow is based on the de Jong-Shephard approach.

[*Added in Proof*: Since this book went to the printer we have discovered a new simulation smoother which is simpler than the one presented below. A paper on it with the title 'A simple and efficient simulation smoother for state space time series analysis' can be found on the book's website at http://www.ssfpack.com/dkbook/]

4.7.1 SIMULATING OBSERVATION DISTURBANCES

We begin by considering how to draw a sample of values $\varepsilon_1, \ldots, \varepsilon_n$ from the conditional density $p(\varepsilon_1, \ldots, \varepsilon_n | y)$. We do this by proceeding sequentially, first drawing ε_n from $p(\varepsilon_n | y)$, then drawing ε_{n-1} from $p(\varepsilon_{n-1} | y, \varepsilon_n)$, then drawing ε_{n-2} from $p(\varepsilon_{n-2} | y, \varepsilon_n, \varepsilon_{n-1})$, and so on, making use of the relation

$$p(\varepsilon_1, \ldots, \varepsilon_n | y) = p(\varepsilon_n | y) p(\varepsilon_{n-1} | y, \varepsilon_n) \cdots p(\varepsilon_1 | y, \varepsilon_2, \ldots, \varepsilon_n). \quad (4.56)$$

Since we deal with Gaussian models, we only need expressions for $E(\varepsilon_t | y, \varepsilon_{t+1}, \ldots, \varepsilon_n)$ and $Var(\varepsilon_t | y, \varepsilon_{t+1}, \ldots, \varepsilon_n)$ to draw from $p(\varepsilon_t | y, \varepsilon_{t+1}, \ldots, \varepsilon_n)$. These expressions for the conditional means and variances can be evaluated recursively using backwards algorithms as we will now show.

We first develop a recursion for the calculation of

$$\bar{\varepsilon}_t = E(\varepsilon_t | y_1, \ldots, y_n, \varepsilon_{t+1}, \ldots, \varepsilon_n), \qquad t = n - 1, \ldots, 1, \quad (4.57)$$

with $\bar{\varepsilon}_n = E(\varepsilon_n | y_1, \ldots, y_n)$. Let $v_j = y_j - E(y_j | y_1, \ldots, y_{j-1})$, as in §4.2, for $j = 1, \ldots, n$. Then

$$\bar{\varepsilon}_t = E(\varepsilon_t | v_1, v_2, \ldots, v_n, \varepsilon_{t+1}, \ldots, \varepsilon_n)$$
$$= E(\varepsilon_t | v_t, v_{t+1}, \ldots, v_n, \varepsilon_{t+1}, \ldots, \varepsilon_n), \qquad t = n - 1, \ldots, 1,$$

since ε_t and $\varepsilon_{t+1}, \ldots, \varepsilon_n$ are independent of $v_1, \ldots, v_{t-1}$. Now let

$$d_j = \varepsilon_j - \bar{\varepsilon}_j, \qquad j = t + 1, \ldots, n. \quad (4.58)$$

Since $v_t, \ldots, v_n, d_{t+1}, \ldots, d_n$ are fixed when $v_t, \ldots, v_n, \varepsilon_{t+1}, \ldots, \varepsilon_n$ are fixed and vice versa,

$$\bar{\varepsilon}_t = E(\varepsilon_t | v_t, v_{t+1}, \ldots, v_n, d_{t+1}, \ldots, d_n), \qquad t = n - 1, \ldots, 1,$$

with $\bar{\varepsilon}_n = E(\varepsilon_n | v_1, \ldots, v_n) = \hat{\varepsilon}_n = E(\varepsilon_n | v_n)$.

We know from §4.2.4 that $v_t, \ldots, v_n$ are independent so the joint distribution of

$$v_t, \ldots, v_n, \varepsilon_{t+1}, \ldots, \varepsilon_n,$$

is

$$\prod_{j=t}^{n} p(v_j) \prod_{k=t+1}^{n-1} p(\varepsilon_k | v_k, \ldots, v_n, \varepsilon_{k+1}, \ldots, \varepsilon_n) p(\varepsilon_n | v_n),$$

since

$$p(\varepsilon_k | v_t, \ldots, v_n, \varepsilon_{t+1}, \ldots, \varepsilon_n) = p(\varepsilon_k | v_k, \ldots, v_n, \varepsilon_{k+1}, \ldots, \varepsilon_n),$$

due to the independence of $v_1, \ldots, v_{k-1}$ and ε_k. Transforming from $v_t, \ldots, v_n$, $\varepsilon_{t+1}, \ldots, \varepsilon_n$ to $v_t, \ldots, v_n, d_{t+1}, \ldots, d_n$, noting that the Jacobian is one and that

$$p(\varepsilon_k | v_k, \ldots, v_n, \varepsilon_{k+1}, \ldots, \varepsilon_n) = p(d_k),$$

it follows that $v_t, \ldots, v_n, d_{t+1}, \ldots, d_n$ are independently distributed.

From the results in §4.4.1, we have

$$\hat{\varepsilon}_t = E(\varepsilon_t | y_1, \ldots, y_n) = E(\varepsilon_t | v_t, \ldots, v_n) = H_t \left(F_t^{-1} v_t - K_t' r_t \right), \quad (4.59)$$

where r_t is generated by the backwards recursion

$$r_{t-1} = Z_t' F_t^{-1} v_t + L_t' r_t,$$

for $t = n, \ldots, 1$ with $r_n = 0$. It follows since $\bar{\varepsilon}_n = \hat{\varepsilon}_n = E(\varepsilon_n | v_n)$ that

$$d_n = \varepsilon_n - H_n F_n^{-1} v_n. \qquad (4.60)$$

For the next conditional mean $\bar{\varepsilon}_{n-1}$ we have

$$\begin{aligned}
\bar{\varepsilon}_{n-1} &= E(\varepsilon_{n-1} | v_{n-1}, v_n, d_n) \\
&= E(\varepsilon_{n-1} | v_{n-1}, v_n) + E(\varepsilon_{n-1} | d_n) \\
&= \hat{\varepsilon}_{n-1} + E(\varepsilon_{n-1} d_n') C_n^{-1} d_n, \qquad (4.61)
\end{aligned}$$

where we define

$$C_t = \text{Var}(d_t), \qquad t = 1, \ldots, n.$$

Now

$$\begin{aligned}
E(\varepsilon_{n-1} d_n') &= E[\varepsilon_{n-1} (\varepsilon_n - H_n F_n^{-1} v_n)'] \\
&= -E(\varepsilon_{n-1} v_n') F_n^{-1} H_n \\
&= H_{n-1} K_{n-1}' Z_n' F_n^{-1} H_n,
\end{aligned}$$

using (4.34), so we have from (4.59),

$$\bar{\varepsilon}_{n-1} = H_{n-1}F_{n-1}^{-1}v_{n-1} - H_{n-1}K'_{n-1}r_{n-1} + H_{n-1}K'_{n-1}Z'_n F_n^{-1}H_n C_n^{-1}d_n.$$

We note for comparison with a general result that will be obtained later that this can be written in a form analogous to $\hat{\varepsilon}_{n-1}$ given by (4.59) with $t = n - 1$,

$$\bar{\varepsilon}_{n-1} = H_{n-1}\left(F_{n-1}^{-1}v_{n-1} - K'_{n-1}\tilde{r}_{n-1}\right), \tag{4.62}$$

where $\tilde{r}_{n-1} = r_{n-1} - Z'_n F_n^{-1}H_n C_n^{-1}d_n$.

For calculating

$$\bar{\varepsilon}_{n-2} = E(\varepsilon_{n-2}|v_{n-2}, v_{n-1}, v_n, d_{n-1}, d_n)$$
$$= \hat{\varepsilon}_{n-2} + E(\varepsilon_{n-2}d'_{n-1})C_{n-1}^{-1}d_{n-1} + E(\varepsilon_{n-2}d'_n)C_n^{-1}d_n, \tag{4.63}$$

we need the results from (4.34) and (4.50),

$$E(\varepsilon_t v'_{t+1}) = -H_t K'_t Z'_{t+1}, \qquad E(\varepsilon_t v'_{t+2}) = -H_t K'_t L'_{t+1}Z'_{t+2},$$
$$E(\varepsilon_t r'_{t+1}) = -H_t K'_t N_{t+1}, \qquad E(\varepsilon_t r'_{t+2}) = -H_t K'_t L'_{t+1}N_{t+2},$$

where N_t is generated by the backwards recursion

$$N_{t-1} = Z'_t F_t^{-1}Z_t + L'_t N_t L_t,$$

for $t = n, \ldots, 1$ with $N_n = 0$. This recursion was developed in §4.3.2. Further,

$$d_n = \varepsilon_n - H_n F_n^{-1}v_n,$$
$$d_{n-1} = \varepsilon_{n-1} - H_{n-1}F_{n-1}^{-1}v_{n-1} + H_{n-1}K'_{n-1}r_{n-1}$$
$$\qquad - H_{n-1}K'_{n-1}Z'_n F_n^{-1}H_n C_n^{-1}d_n,$$

so that

$$E(\varepsilon_{n-2}d'_{n-1}) = -E(\varepsilon_{n-2}\bar{\varepsilon}'_{n-1})$$
$$= -E(\varepsilon_{n-2}v'_{n-1})F_{n-1}^{-1}H_{n-1} + E(\varepsilon_{n-2}r'_{n-1})K_{n-1}H_{n-1}$$
$$\qquad - E(\varepsilon_{n-2}d'_n)C_n^{-1}H_n F_n^{-1}Z_n K_{n-1}H_{n-1}$$
$$= H_{n-2}K'_{n-2}Z'_{n-1}F_{n-1}^{-1}H_{n-1} - H_{n-2}K'_{n-2}N_{n-1}K_{n-1}H_{n-1}$$
$$\qquad + H_{n-2}K'_{n-2}L'_{n-1}Z'_n F_n^{-1}H_n C_n^{-1}H_n F_n^{-1}Z_n K_{n-1}H_{n-1}.$$

Also,

$$E(\varepsilon_{n-2}d'_n) = -E(\varepsilon_{n-2}\bar{\varepsilon}'_n)$$
$$= -E(\varepsilon_{n-2}v'_n)F_n^{-1}H_n$$
$$= H_{n-2}K'_{n-2}L'_{n-1}Z'_n F_n^{-1}H_n.$$

To simplify the substitution of these expressions into (4.63), we adopt a notation that will be useful for generalisation later,

$$\tilde{W}_n = H_n F_n^{-1}Z_n,$$
$$\tilde{W}_{n-1} = H_{n-1}\left(F_{n-1}^{-1}Z_{n-1} - K'_{n-1}\tilde{N}_{n-1}L_{n-1}\right),$$
$$\tilde{N}_{n-1} = Z'_n F_n^{-1}Z_n + \tilde{W}'_n C_n^{-1}\tilde{W}_n.$$

Thus using (4.59) we can rewrite (4.63) as

$$\bar{\varepsilon}_{n-2} = H_{n-2}\big(F_{n-2}^{-1}v_{n-2} - K_{n-2}'r_{n-2}\big) - H_{n-2}K_{n-2}'\tilde{W}_{n-1}C_{n-1}^{-1}d_{n-1}$$
$$+ H_{n-2}K_{n-2}'L_{n-1}'\tilde{W}_nC_n^{-1}d_n.$$

It follows that we can put this in the form analogous to (4.62),

$$\bar{\varepsilon}_{n-2} = H_{n-2}\big(F_{n-2}^{-1}v_{n-2} - K_{n-2}'\tilde{r}_{n-2}\big), \tag{4.64}$$

where $\tilde{r}_{n-2}$ is determined by the backwards recursion

$$\tilde{r}_{t-1} = Z_t'F_t^{-1}v_t - \tilde{W}_t'C_t^{-1}d_t + L_t'\tilde{r}_t, \qquad t = n, n-1, \tag{4.65}$$

with $\tilde{r}_n = 0$.

The results of the previous subsection lead us to conjecture the general form

$$\bar{\varepsilon}_t = H_t\big(F_t^{-1}v_t - K_t'\tilde{r}_t\big), \qquad t = n, n-1, \ldots, 1, \tag{4.66}$$

where $\tilde{r}_t$ is determined by the recursion

$$\tilde{r}_{t-1} = Z_t'F_t^{-1}v_t - \tilde{W}_t'C_t^{-1}d_t + L_t'\tilde{r}_t, \qquad t = n, n-1, \ldots, 1 \tag{4.67}$$

with $\tilde{r}_n = 0$ and

$$\tilde{W}_t = H_t\big(F_t^{-1}Z_t - K_t'\tilde{N}_tL_t\big), \tag{4.68}$$

$$\tilde{N}_{t-1} = Z_t'F_t^{-1}Z_t + \tilde{W}_t'C_t^{-1}\tilde{W}_t + L_t'\tilde{N}_tL_t, \tag{4.69}$$

with $\tilde{N}_n = 0$, for $t = n, n-1, \ldots, 1$. All related definitions given earlier are consistent with these general definitions. We shall prove the result (4.66) by backwards induction assuming that it holds when t is replaced by $t + 1, \ldots, n - 1$. Note that throughout the proof we take $j > t$.

We have

$$\bar{\varepsilon}_t = \mathrm{E}(\varepsilon_t|v_t, \ldots, v_n, d_{t+1}, \ldots, d_n)$$

$$= \mathrm{E}(\varepsilon_t|v_t, \ldots, v_n) + \sum_{j=t+1}^{n} \mathrm{E}(\varepsilon_t|d_j)$$

$$= \hat{\varepsilon}_t + \sum_{j=t+1}^{n} \mathrm{E}(\varepsilon_t d_j')C_j^{-1}d_j, \tag{4.70}$$

where $\hat{\varepsilon}_t$ is given by (4.59). Since we are assuming that (4.66) holds when t is replaced by j with $j > t$ we have

$$\mathrm{E}(\varepsilon_t d_j') = -\mathrm{E}(\varepsilon_t \bar{\varepsilon}_j')$$

$$= -\mathrm{E}(\varepsilon_t v_j')F_j^{-1}H_j + \mathrm{E}(\varepsilon_t \tilde{r}_j')K_jH_j$$

$$= H_tK_t'L_{t+1}' \cdots L_{j-1}'Z_j'F_j^{-1}H_j + \mathrm{E}(\varepsilon_t \tilde{r}_j')K_jH_j, \tag{4.71}$$

by (4.34), noting that $L_{t+1}' \cdots L_{j-1}' = I_m$ for $j = t + 1$ and $L_{t+1}' \cdots L_{j-1}' = L_{t+1}'$ for $j = t + 2$.

We now prove by induction that

$$E(\varepsilon_t \tilde{r}_j') = -H_t K_t' L_{t+1}' \cdots L_j' \tilde{N}_j, \qquad j = t+1, \ldots, n, \qquad (4.72)$$

assuming that it holds when j is replaced by $j + 1$. Using (4.67), (4.34) and also (4.71) and (4.72) with j replaced by $j + 1$, we have

$$
\begin{aligned}
E(\varepsilon_t \tilde{r}_j') &= E(\varepsilon_t v_{j+1}') F_{j+1}^{-1} Z_{j+1} - E(\varepsilon_t d_{j+1}') C_{j+1}^{-1} \tilde{W}_{j+1} + E(\varepsilon_t \tilde{r}_{j+1}') L_{j+1} \\
&= -H_t K_t' L_{t+1}' \cdots L_j' \big[Z_{j+1}' F_{j+1}^{-1} Z_{j+1} \\
&\quad + (Z_{j+1}' F_{j+1}^{-1} H_{j+1} - L_{j+1}' \tilde{N}_{j+1} K_{j+1} H_{j+1}) C_{j+1}^{-1} \tilde{W}_{j+1} + L_{j+1}' \tilde{N}_{j+1} L_{j+1} \big] \\
&= -H_t K_t' L_{t+1}' \cdots L_j' (Z_{j+1}' F_{j+1}^{-1} Z_{j+1} + \tilde{W}_{j+1}' C_{j+1}^{-1} \tilde{W}_{j+1} + L_{j+1}' \tilde{N}_{j+1} L_{j+1}) \\
&= -H_t K_t' L_{t+1}' \cdots L_j' \tilde{N}_j.
\end{aligned}
$$

The second last equality follows from (4.68) and the last equality follows from (4.69) with $t = j + 1$. This completes the proof of (4.72). Substituting (4.72) into (4.71) gives

$$
\begin{aligned}
E(\varepsilon_t d_j') &= H_t K_t' L_{t+1}' \cdots L_{j-1}' (Z_j' F_j^{-1} H_j - L_j' \tilde{N}_j K_j H_j) \\
&= H_t K_t' L_{t+1}' \cdots L_{j-1}' \tilde{W}_j'. \qquad (4.73)
\end{aligned}
$$

Substituting (4.73) into (4.70), we obtain

$$\bar{\varepsilon}_t = \hat{\varepsilon}_t + H_t K_t' \sum_{j=t+1}^n L_{t+1}' \cdots L_{j-1}' \tilde{W}_j' C_j^{-1} d_j,$$

with the interpretation that for $j = t + 1$, $L_{t+1}' \cdots L_{j-1}' = I_m$ and for $j = t + 2$, $L_{t+1}' \cdots L_{j-1}' = L_{t+1}'$. It follows from (4.59) that

$$\hat{\varepsilon}_t = H_t F_t^{-1} v_t - H_t K_t' \sum_{j=t+1}^n L_{t+1}' \cdots L_{j-1}' Z_j' F_j^{-1} v_j.$$

Combining the two last expressions gives (4.66), that is,

$$
\begin{aligned}
\bar{\varepsilon}_t &= H_t F_t^{-1} v_t - H_t K_t' \sum_{j=t+1}^n L_{t+1}' \cdots L_{j-1}' (Z_j' F_j^{-1} v_j - \tilde{W}_j' C_j^{-1} d_j) \\
&= H_t (F_t^{-1} v_t - K_t' \tilde{r}_t),
\end{aligned}
$$

since it follows from (4.67) that

$$\tilde{r}_t = \sum_{j=t+1}^n L_{t+1}' \cdots L_{j-1}' (Z_j' F_j^{-1} v_j - \tilde{W}_j' C_j^{-1} d_j).$$

Thus (4.66) is proved.

To perform the simulation we take $\varepsilon_t = \bar{\varepsilon}_t + d_t$ for $t = 1, \ldots, n$ where d_t is a random draw from the normal distribution

$$d_t \sim N(0, C_t).$$

The variance matrix C_t is given by

$$
\begin{aligned}
C_t &= \mathrm{Var}(\varepsilon_t | y, \varepsilon_{t+1}, \ldots, \varepsilon_n) \\
&= \mathrm{Var}(\varepsilon_t | v_t, v_{t+1}, \ldots, v_n, d_{t+1}, \ldots, d_n) \\
&= H_t - H_t D_t H_t - \sum_{j=t+1}^{n} E(\varepsilon_t d_j') C_j^{-1} E(\varepsilon_t d_j')' \\
&= H_t - H_t \left[F_t^{-1} + K_t' \sum_{j=t+1}^{n} L_{t+1}' \cdots L_{j-1}' (Z_j' F_j^{-1} Z_j \right. \\
&\qquad \left. + \tilde{W}_j' C_j^{-1} \tilde{W}_j) L_{j-1} \cdots L_{t+1} K_t \right] H_t,
\end{aligned}
$$

from §4.4.3 and equation (4.73). It follows from (4.69) that

$$\tilde{N}_t = \sum_{j=t+1}^{n} L_{t+1}' \cdots L_{j-1}' (Z_j' F_j^{-1} Z_j + \tilde{W}_j' C_j^{-1} \tilde{W}_j) L_{j-1} \cdots L_{t+1}.$$

Thus we have

$$C_t = H_t - H_t \tilde{D}_t H_t, \qquad t = n, \ldots, 1, \tag{4.74}$$

where

$$\tilde{D}_t = F_t^{-1} + K_t' \tilde{N}_t K_t,$$

and $\tilde{N}_t$ is recursively evaluated by the backwards recursion (4.69).

It is remarkable that in spite of complications caused by the conditioning on $\varepsilon_{t+1}, \ldots, \varepsilon_n$, the structures of the expressions (4.66) for $\bar{\varepsilon}_t$ and (4.74) for $\mathrm{Var}(\varepsilon_t - \bar{\varepsilon}_t)$ are essentially the same as (4.36) for $\hat{\varepsilon}_t$ and (4.44) for $\mathrm{Var}(\varepsilon_t - \hat{\varepsilon}_t)$.

4.7.3 SIMULATION SMOOTHING RECURSION

For convenience we collect together the simulation smoothing recursions for the observation disturbance vectors:

$$
\begin{aligned}
\tilde{r}_{t-1} &= Z_t' F_t^{-1} v_t - \tilde{W}_t' C_t^{-1} d_t + L_t' \tilde{r}_t, \\
\tilde{N}_{t-1} &= Z_t' F_t^{-1} Z_t + \tilde{W}_t' C_t^{-1} \tilde{W}_t + L_t' \tilde{N}_t L_t,
\end{aligned}
\tag{4.75}
$$

for $t = n, \ldots, 1$ and initialised by $\tilde{r}_n = 0$ and $\tilde{N}_n = 0$.

For simulating from the smoothed density of the observation disturbances, we define

$$
\begin{aligned}
\tilde{W}_t &= H_t \left(F_t^{-1} Z_t - K_t' \tilde{N}_t L_t \right), \\
d_t &\sim N(0, C_t), \\
C_t &= H_t - H_t \tilde{D}_t H_t,
\end{aligned}
\tag{4.76}
$$

Table 4.7.   Dimensions of simulation smoothing recursions of §§4.7.3 and 4.7.4.

Observation disturbance		State disturbance	
$\tilde{\varepsilon}_t, d_t$	$p \times 1$	$\tilde{\eta}_t, d_t$	$r \times 1$
$\tilde{W}_t$	$p \times m$	$\tilde{W}_t$	$r \times m$
C_t	$p \times p$	C_t	$r \times r$

where $\tilde{D}_t = F_t^{-1} + K_t' \tilde{N}_t K_t$. A draw $\tilde{\varepsilon}_t$, which is the tth vector of a draw from the smoothed joint density $p(\varepsilon|y)$, is then computed by

$$\tilde{\varepsilon}_t = d_t + H_t \left(F_t^{-1} v_t - K_t' \tilde{r}_t \right). \tag{4.77}$$

Simulation smoothing is performed in a similar way to disturbance smoothing: we proceed forwards through the series using (4.13) and backwards through the series using (4.75), (4.76) and (4.77) to obtain the series of draws $\tilde{\varepsilon}_t$ for $t = 1, \ldots, n$. The quantities v_t, F_t and K_t of the Kalman filter need to be stored for $t = 1, \ldots, n$. In Table 4.7 we present the dimensions of the vectors and matrices occurring in the simulation smoothing recursion.

4.7.4   SIMULATING STATE DISTURBANCES

Similar techniques to those of §4.7.1 can be developed for drawing random samples from the smoothed density $p(\eta_1, \ldots, \eta_n|y)$. We first consider obtaining a recursion for calculating

$$\bar{\eta}_t = \mathrm{E}(\eta_t|y, \eta_{t+1}, \ldots, \eta_n), \qquad t = 1, \ldots, n.$$

For comparison with the treatment of §4.7.1 it is convenient to use some of the same symbols, but with different interpretations. Thus we take

$$\begin{aligned} \tilde{W}_j &= Q_j R_j' \tilde{N}_j L_j, \\ d_j &= \eta_j - \bar{\eta}_j, \\ C_j &= \mathrm{Var}(d_j), \end{aligned} \tag{4.78}$$

where $\tilde{N}_j$ is determined by the backwards recursion (4.69) with values of $\tilde{W}_j$ as given in (4.78); this form of $\tilde{W}_j$ can be found to be appropriate by experimenting with $j = n, n - 1$ as was done in §4.7.1. We shall explain how to prove by induction that

$$\bar{\eta}_t = Q_t R_t' \tilde{r}_t, \qquad t = n, \ldots, 1, \tag{4.79}$$

where $\tilde{r}_t$ is determined by recursion (4.67) with $\tilde{W}_j$ defined by (4.78), leaving some details to the reader.

As in §4.7.1,

$$\bar{\eta}_t = \mathrm{E}(\eta_t|y) + \sum_{j=t+1}^{n} \mathrm{E}(\eta_t|d_j)$$

$$= Q_t R_t' r_t + \sum_{j=t+1}^{n} \mathrm{E}(\eta_t d_j') C_j^{-1} d_j, \tag{4.80}$$

where

$$\mathrm{E}(\eta_t d_j') = -\mathrm{E}(\eta_t \bar{\eta}_j')$$
$$= -\mathrm{E}(\eta_t \tilde{r}_j') R_j Q_j.$$

By methods similar to those leading to (4.72) we can show that

$$\mathrm{E}(\eta_t \tilde{r}_j') = Q_t R_t' L_{t+1}' \cdots L_j' \tilde{N}_j, \qquad j = t+1, \ldots, n. \tag{4.81}$$

Substituting in (4.79) gives

$$\mathrm{E}(\eta_t d_j') = -Q_t R_t' L_{t+1}' \cdots L_j' \tilde{N}_j R_j Q_j,$$
$$= -Q_t R_t' L_{t+1}' \cdots L_{j-1}' \tilde{W}_j', \tag{4.82}$$

whence substituting in (4.80) and expanding r_t proves (4.79) analogously to (4.66).

By similar means we can show that

$$C_t = Q_t - Q_t R_t' \tilde{N}_t R_t Q_t, \qquad t = n, \ldots, 1. \tag{4.83}$$

As with the simulation smoother for the observation disturbances, it is remarkable that $\bar{\eta}_t$ and C_t have the same forms as for $\mathrm{E}(\eta_t|y)$ and $\mathrm{Var}(\eta_t|y)$ in (4.41) and (4.47), apart from the replacement of r_t and N_t by $\tilde{r}_t$ and $\tilde{N}_t$.

The procedure for drawing a vector $\tilde{\eta}_t$ from the smoothed density $p(\eta|y)$ is then: draw a random element d_t from $\mathrm{N}(0, C_t)$ and take

$$\tilde{\eta}_t = d_t + Q_t R_t' \tilde{r}_t, \qquad t = n, \ldots, 1. \tag{4.84}$$

The quantities v_t, F_t and K_t of the Kalman filter need to be stored for $t = 1, \ldots, n$. The storage space of K_t is used to store the $\tilde{\eta}_t$'s during the backwards recursion.

4.7.5 SIMULATING STATE VECTORS

By adopting the same arguments as we used for developing the quick state smoother in §4.4.2, we obtain the following forwards recursion for simulating from the conditional density $p(\alpha|y)$

$$\tilde{\alpha}_{t+1} = T_t \tilde{\alpha}_t + R_t \tilde{\eta}_t, \tag{4.85}$$

for $t = 1, \ldots, n$ with $\tilde{\alpha}_1 = a_1 + P_1 \tilde{r}_0$. For simulation state smoothing we proceed forwards through the series using (4.13) and backwards through the series using (4.75) and (4.84) to obtain the series of draws $\tilde{\eta}_t$ for $t = 1, \ldots, n$. Then we proceed forwards again through the series using (4.85) to obtain $\tilde{\alpha}_t$ for $t = 1, \ldots, n$.

4.7.6 SIMULATING MULTIPLE SAMPLES

In the previous subsections we have derived methods to generate draws from the linear Gaussian model conditional on the observations. When more than one draw is required we can simply repeat the algorithm with different random values. However, we can save a considerable amount of computing after the first draw at the cost of storage. When simulating observation disturbance vectors using the method described in §4.7.3, we can store $\tilde{N}_{t-1}$, $\tilde{W}_t$ and $\tilde{C}_t$ for $t = 1, \ldots, n$. In order to generate a new draw, we only require to draw new values for d_t, apply the recursion for $\tilde{r}_{t-1}$ and compute $\tilde{\varepsilon}_t$ for $t = 1, \ldots, n$. The matrices $\tilde{N}_{t-1}$, $\tilde{W}_t$ and $\tilde{C}_t$ do not change when different values of d_t, $\tilde{r}_{t-1}$ and $\tilde{\varepsilon}_t$ are obtained. This implies huge computational savings when multiple draws are required which is usually the case when the methods and techniques described in Part II of the book are used. The same arguments apply when multiple draws are required for state disturbance vectors or state vectors.

4.8 Missing observations

In §2.7 we discussed the effects of missing observations on filtering and smoothing for the local level model. In case of the general state space model the effects are similar. Therefore, we need not repeat details of the theory for missing observations but merely indicate the extension to the general model of the results of §2.7. Thus when the set of observations y_t for $t = \tau, \ldots, \tau^* - 1$ is missing, the vector v_t and the matrix K_t of the Kalman filter are set to zero for these values, that is, $v_t = 0$ and $K_t = 0$, and the Kalman updates become

$$a_{t+1} = T_t a_t, \qquad P_{t+1} = T_t P_t T_t' + R_t Q_t R_t', \qquad t = \tau, \ldots, \tau^* - 1; \quad (4.86)$$

similarly, the backwards smoothing recursions (4.25) and (4.30) become

$$r_{t-1} = T_t' r_t, \qquad N_{t-1} = T_t' N_t T_t, \qquad t = \tau^* - 1, \ldots, \tau. \quad (4.87)$$

Other relevant equations for smoothing remain the same. The reader can easily verify these results by applying the arguments of §2.7 to the general model. This simple treatment of missing observations is one of the attractions of the state space methods for time series analysis.

Suppose that at time t some but not all of the elements of the observation vector y_t are missing. Let y_t^* be the vector of values actually observed. Then $y_t^* = W_t y_t$ where W_t is a known matrix whose rows are a subset of the rows of I_p. Consequently, at time points where not all elements of y_t are available the first equation of (4.1) is replaced by the equation

$$y_t^* = Z_t^* \alpha_t + \varepsilon_t^*, \qquad \varepsilon_t^* \sim N(0, H_t^*),$$

where $Z_t^* = W_t Z_t, \varepsilon_t^* = W_t \varepsilon_t$ and $H_t^* = W_t H_t W_t'$. The Kalman filter and smoother then proceed exactly as in the standard case, provided that y_t, Z_t and H_t are replaced by y_t^*, Z_t^* and H_t^* at relevant time points. Of course, the dimensionality of the observation vector varies over time, but this does not affect the validity of the formulae. The missing elements can be estimated by appropriate elements of $Z_t \hat{\alpha}_t$ where $\hat{\alpha}_t$ is the smoothed value. A more convenient method for dealing with missing elements for such multivariate models is given in §6.4 which is based on an element by element treatment of the observation vector y_t.

When observations or observational elements are missing, the corresponding values should be omitted from the simulation samples obtained by the methods described in §4.7. When the observation vector at time t is missing, such that the Kalman filter gives $v_t = 0$, $F_t^{-1} = 0$ and $K_t = 0$, it follows that $\tilde{W}_t = 0$, $\tilde{D}_t = 0$ and $C_t = H_t$ in the recursive equations (4.75). The consequence is that a draw from $p(\varepsilon_t | y)$ as given by (4.77) becomes

$$\tilde{\varepsilon}_t \sim N(0, H_t),$$

when y_t is missing. We notice that this draw does not enter the simulation smoothing recursions since $\tilde{W}_t = 0$ when y_t is missing. When only some elements of the observational vector are missing, we only need to adjust the dimensions of the system matrices Z_t and H_t. The equations for simulating state disturbances when vector y_t is missing remain unaltered except that $v_t = 0$, $F_t^{-1} = 0$ and $K_t = 0$.

4.9 Forecasting

Suppose we have vector observations $y_1, \ldots, y_n$ which follow the state space model (4.1) and we wish to forecast y_{n+j} for $j = 1, \ldots, J$. For this purpose let us choose the estimate $\bar{y}_{n+j}$ which has minimum mean square error matrix given Y_n, that is, $\bar{F}_{n+j} = E[(\bar{y}_{n+j} - y_{n+j})(\bar{y}_{n+j} - y_{n+j})' | y]$ is a minimum in the matrix sense for all estimates of y_{n+j}. It is standard knowledge that if x is a random vector with mean μ and finite variance matrix, then the value of a constant vector λ which minimises $E[(\lambda - x)(\lambda - x)']$ is $\lambda = \mu$. It follows that the minimum mean square error forecast of y_{n+j} given Y_n is the conditional mean $\bar{y}_{n+j} = E(y_{n+j} | y)$.

For $j = 1$ the forecast is straightforward. We have $y_{n+1} = Z_{n+1} \alpha_{n+1} + \varepsilon_{n+1}$ so

$$\begin{aligned} \bar{y}_{n+1} &= Z_{n+1} E(\alpha_{n+1} | y) \\ &= Z_{n+1} a_{n+1}, \end{aligned}$$

where a_{n+1} is the estimate (4.8) of α_{n+1} produced by the Kalman filter. The error variance matrix or mean square error matrix

$$\begin{aligned} \bar{F}_{n+1} &= E[(\bar{y}_{n+1} - y_{n+1})(\bar{y}_{n+1} - y_{n+1})'] \\ &= Z_{n+1} P_{n+1} Z_{n+1}' + H_{n+1}, \end{aligned}$$

is produced by the Kalman filter relation (4.7). We now demonstrate that we can generate the forecasts $\bar{y}_{n+j}$ for $j = 2, \ldots, J$ merely by treating $y_{n+1}, \ldots, y_{n+J}$ as missing values as in §4.8. Let $\bar{a}_{n+j} = E(\alpha_{n+j} | y)$ and $\bar{P}_{n+j} = E[(\bar{a}_{n+j} - \alpha_{n+j})$

$(\bar{a}_{n+j} - \alpha_{n+j})'|y]$. Since $y_{n+j} = Z_{n+j}\alpha_{n+j} + \varepsilon_{n+j}$ we have

$$\bar{y}_{n+j} = Z_{n+j}\,\mathrm{E}(\alpha_{n+j}|y)$$
$$= Z_{n+j}\bar{a}_{n+j},$$

with mean square error matrix

$$\bar{F}_{n+j} = \mathrm{E}[\{Z_{n+j}(\bar{a}_{n+j} - \alpha_{n+j}) - \varepsilon_{n+j}\}\{Z_{n+j}(\bar{a}_{n+j} - \alpha_{n+j}) - \varepsilon_{n+j}\}']$$
$$= Z_{n+j}\bar{P}_{n+j}Z'_{n+j} + H_{n+j}.$$

We now derive recursions for calculating $\bar{a}_{n+j}$ and $\bar{P}_{n+j}$. We have $\alpha_{n+j+1} = T_{n+j}\alpha_{n+j} + R_{n+j}\eta_{n+j}$ so

$$\bar{a}_{n+j+1} = T_{n+j}\,\mathrm{E}(\alpha_{n+j}|y)$$
$$= T_{n+j}\bar{a}_{n+j},$$

for $j = 1, \ldots, J-1$ and with $\bar{a}_{n+1} = a_{n+1}$. Also,

$$\bar{P}_{n+j+1} = \mathrm{E}[(\bar{a}_{n+j+1} - \alpha_{n+j+1})(\bar{a}_{n+j+1} - \alpha_{n+j+1})'|y]$$
$$= T_{n+j}\,\mathrm{E}[(\bar{a}_{n+j} - \alpha_{n+j})(\bar{a}_{n+j} - \alpha_{n+j})'|y]T'_{n+j}$$
$$\quad + R_{n+j}\,\mathrm{E}[\eta_{n+j}\eta'_{n+j}]R'_{n+j}$$
$$= T_{n+j}\bar{P}_{n+j}T'_{n+j} + R_{n+j}Q_{n+j}R'_{n+j}.$$

We observe that the recursions for $\bar{a}_{n+j}$ and $\bar{P}_{n+j}$ are the same as the recursions for a_{n+j} and P_{n+j} of the Kalman filter (4.13) provided that we take $v_{n+j} = 0$ and $K_{n+j} = 0$ for $j = 1, \ldots, J-1$. But these are precisely the conditions that in §4.8 enabled us to deal with missing observations by routine application of the Kalman filter. We have therefore demonstrated that forecasts of $y_{n+1}, \ldots, y_{n+J}$ together with their forecast error variance matrices can be obtained merely by treating y_t for $t > n$ as missing observations and using the results of §4.8. In a sense this conclusion could be regarded as intuitively obvious; however, we thought it worth while demonstrating it algebraically. To sum up, forecasts and their associated error variance matrices can be obtained routinely in state space time series analysis based on linear Gaussian models by continuing the Kalman filter beyond $t = n$ with $v_t = 0$ and $K_t = 0$ for $t > n$. These results for forecasting are a particularly elegant feature of state space methods for time series analysis.

4.10   Dimensionality of observational vector

Throughout this chapter we have assumed both for convenience of exposition, and also because this is by far the most common case in practice, that the dimensionality of the observation vector y_t is a fixed value p. It is easy to verify however that none of the basic formulae that we have derived depend on this assumption. For example, the filtering recursion (4.13) and the disturbance smoother (4.48) both remain valid when the dimensionality of y_t is allowed to vary. This convenient generalisation arises because of the recursive nature of the formulae. In fact we made use of this property in the treatment of missing observational elements in §4.8. We do not, however, consider explicitly in this book the situation where the dimensionality of the state vector α_t varies with t, apart from the treatment of missing observations just referred to and the conversion of multivariate series to univariate series in §6.4.

4.11 General matrix form for filtering and smoothing

The state space model can itself be represented in a general matrix form. The observation equation can be formulated as

$$y = Z\alpha + \varepsilon, \qquad \varepsilon \sim N(0, H), \tag{4.88}$$

with

$$y = \begin{pmatrix} y_1 \\ \vdots \\ y_n \end{pmatrix}, \qquad Z = \begin{bmatrix} Z_1 & 0 & 0 \\ & \ddots & \vdots \\ 0 & Z_n & 0 \end{bmatrix}, \qquad \alpha = \begin{pmatrix} \alpha_1 \\ \vdots \\ \alpha_n \\ \alpha_{n+1} \end{pmatrix},$$

$$\varepsilon = \begin{pmatrix} \varepsilon_1 \\ \vdots \\ \varepsilon_n \end{pmatrix}, \qquad H = \begin{bmatrix} H_1 & 0 \\ & \ddots \\ 0 & H_n \end{bmatrix}. \tag{4.89}$$

The state equation takes the form

$$\alpha = T(\alpha_1^* + R\eta), \qquad \eta \sim N(0, Q), \tag{4.90}$$

with

$$T = \begin{bmatrix} I_m & 0 & 0 & 0 & 0 & 0 \\ T_1 & I_m & 0 & 0 & 0 & 0 \\ T_2 T_1 & T_2 & I_m & 0 & 0 & 0 \\ T_3 T_2 T_1 & T_3 T_2 & T_3 & I_m & 0 & 0 \\ & & & & \ddots & \vdots \\ T_{n-1} \cdots T_1 & T_{n-1} \cdots T_2 & T_{n-1} \cdots T_3 & T_{n-1} \cdots T_4 & I_m & 0 \\ T_n \cdots T_1 & T_n \cdots T_2 & T_n \cdots T_3 & T_n \cdots T_4 & \cdots & T_n & I_m \end{bmatrix}, \tag{4.91}$$

$$\alpha_1^* = \begin{pmatrix} \alpha_1 \\ 0 \\ 0 \\ \vdots \\ 0 \end{pmatrix}, \qquad R = \begin{bmatrix} 0 & 0 & \cdots & 0 \\ R_1 & 0 & & 0 \\ 0 & R_2 & & 0 \\ & & \ddots & \vdots \\ 0 & 0 & \cdots & R_n \end{bmatrix}, \tag{4.92}$$

$$\eta = \begin{pmatrix} \eta_1 \\ \vdots \\ \eta_n \end{pmatrix}, \qquad Q = \begin{bmatrix} Q_1 & 0 \\ & \ddots \\ 0 & Q_n \end{bmatrix}. \tag{4.93}$$

This representation of the state space model is useful for getting a better understanding of some of the results in this chapter. For example, it can now be seen immediately that the observation vectors y_t are linear functions of the initial state vector α_1 and the disturbance vectors ε_t and η_t for $t = 1, \ldots, n$ since it follows by substitution that

$$y = ZT\alpha_1^* + ZTR\eta + \varepsilon. \tag{4.94}$$

It also follows that

$$E(y) = ZTa_1^*, \qquad \text{Var}(y) = \Sigma = ZT(P_1^* + RQR')T'Z' + H, \tag{4.95}$$

where

$$a_1^* = \begin{pmatrix} a_1 \\ 0 \\ 0 \\ \vdots \\ 0 \end{pmatrix}, \qquad P_1^* = \begin{bmatrix} P_1 & 0 & 0 & \cdots & 0 \\ 0 & 0 & 0 & & 0 \\ 0 & 0 & 0 & & 0 \\ \vdots & & & \ddots & \\ 0 & 0 & 0 & & 0 \end{bmatrix}.$$

We now show that the vector of innovations can be represented as $v = Cy - Ba_1^*$ where $v = (v_1', \ldots, v_n')'$ and where C and B are matrices of which C is lower block triangular. Firstly we observe that

$$a_{t+1} = L_t a_t + K_t y_t,$$

which follows from (4.8) with $v_t = y_t - Z_t a_t$ and $L_t = T_t - K_t Z_t$. Then by substituting repeatedly we have

$$a_{t+1} = L_t L_{t-1} \cdots L_1 a_1 + \sum_{j=1}^{t-1} L_t L_{t-1} \cdots L_{j+1} K_j y_j + K_t y_t$$

and

$$
\begin{aligned}
v_1 &= y_1 - Z_1 a_1, \\
v_2 &= -Z_2 L_1 a_1 + y_2 - Z_2 K_1 y_1, \\
v_3 &= -Z_3 L_2 L_1 a_1 + y_3 - Z_3 K_2 y_2 - Z_3 L_2 K_1 y_1, \\
v_4 &= -Z_4 L_3 L_2 L_1 a_1 + y_4 - Z_4 K_3 y_3 - Z_4 L_3 K_2 y_2 - Z_4 L_3 L_2 K_1 y_1,
\end{aligned}
$$

and so on. Generally,

$$
\begin{aligned}
v_t = {} &-Z_t L_{t-1} L_{t-2} \cdots L_1 a_1 + y_t - Z_t K_{t-1} y_{t-1} \\
&- Z_t \sum_{j=1}^{t-2} L_{t-1} \cdots L_{j+1} K_j y_j.
\end{aligned}
$$

Note that the matrices K_t and L_t depend on P_1, Z, T, R, H and Q but do not depend on the initial mean vector a_1 or the observations $y_1, \ldots, y_n$, for $t = 1, \ldots, n$. The innovations can thus be represented as

$$v = (I - ZLK)y - ZLa_1^*$$
$$= Cy - ZLa_1^*,$$

where $C = I - ZLK$,

$$L = \begin{bmatrix} I & 0 & 0 & 0 & 0 & 0 \\ L_1 & I & 0 & 0 & 0 & 0 \\ L_2L_1 & L_2 & I & 0 & 0 & 0 \\ L_3L_2L_1 & L_3L_2 & L_3 & I & 0 & 0 \\ & & & & \ddots & \vdots \\ L_{n-1}\cdots L_1 & L_{n-1}\cdots L_2 & L_{n-1}\cdots L_3 & L_{n-1}\cdots L_4 & I & 0 \\ L_n\cdots L_1 & L_n\cdots L_2 & L_n\cdots L_3 & L_n\cdots L_4 & \cdots & L_n & I \end{bmatrix},$$

$$K = \begin{bmatrix} 0 & 0 & \cdots & 0 \\ K_1 & 0 & \cdots & 0 \\ 0 & K_2 & & 0 \\ \vdots & & \ddots & \\ 0 & 0 & & K_n \end{bmatrix},$$

and matrix Z is defined in (4.89). It can be easily verified that matrix C is lower block triangular with identity matrices on the leading diagonal blocks. Since $v = Cy - ZLa_1^*$, $\text{Var}(y) = \Sigma$ and a_1^* is constant, it follows that $\text{Var}(v) = C\Sigma C'$. However, we know from §4.2.4 that the innovations are independent of each other so that $\text{Var}(v)$ is the block diagonal matrix

$$F = \begin{bmatrix} F_1 & 0 & \cdots & 0 \\ 0 & F_2 & & 0 \\ \vdots & & \ddots & \\ 0 & 0 & & F_n \end{bmatrix}.$$

This shows that in effect the Kalman filter is essentially equivalent to a block version of a Cholesky algorithm applied to the observational variance matrix implied by the state space model (4.1).

Given the special structure of $C = I - ZLK$ we can easily reproduce some smoothing results in matrix form. We first notice that $\text{E}(v) = C\text{E}(y) - ZLa_1^* = 0$ and since $\text{E}(y) = ZTa_1^*$, we obtain the identity $CZT = ZL$. It follows that

$$v = C[y - \text{E}(y)], \qquad F = \text{Var}(v) = C\Sigma C',$$

with $\text{E}(y) = ZLa_1^*$. Further we notice that C is nonsingular and

$$\Sigma^{-1} = C'F^{-1}C.$$

The stack of smoothed observation disturbance vectors can be obtained directly by

$$\hat{\varepsilon} = E(\varepsilon|y) = Cov(\varepsilon, y)\Sigma^{-1}[y - E(y)],$$

where $y - E(y) = ZTR\eta + \varepsilon$. Given the results $v = C[y - E(y)]$ and $\Sigma^{-1} = C'F^{-1}C$ it follows straightforwardly that

$$\begin{aligned}
\hat{\varepsilon} &= HC'F^{-1}v \\
&= H(I - K'L'Z')F^{-1}v \\
&= H(F^{-1}v - K'r) \\
&= Hu,
\end{aligned}$$

where $u = F^{-1}v - K'r$ and $r = L'Z'F^{-1}v$. Notice that

$$u = \Sigma^{-1}[y - E(y)] = C'F^{-1}v.$$

It can be verified that the definitions of u and r are consistent with the stack of vectors u_t of (4.37) and r_t of (4.25), respectively.

In a similar way we obtain the stack of smoothed state disturbance vectors directly via

$$\begin{aligned}
\hat{\eta} &= Cov(\eta, y)\Sigma^{-1}[y - E(y)] \\
&= QR'T'Z'u \\
&= QR'r,
\end{aligned}$$

where $r = T'Z'u$. This result is consistent with the stack of $\hat{\eta}_t = Q_t R'_t r_t$ where r_t is evaluated by $r_{t-1} = Z'_t u_t + T'_t r_t$; see §4.4.4. Notice the equality $r = L'Z'F^{-1}v = T'Z'u$.

Finally, we obtain the smoothed estimator of α via

$$\begin{aligned}
\hat{\alpha} &= Cov(\alpha, y)\Sigma^{-1}[y - E(y)] \\
&= Cov(\alpha, y)u \\
&= Cov(\alpha, \eta)R'T'Z'u + Cov(\alpha, \varepsilon)u \\
&= TRQR'T'Z'u \\
&= TR\hat{\eta},
\end{aligned}$$

since $y - E(y) = ZTR\eta + \varepsilon$. This is consistent with the way $\hat{\alpha}_t$ is evaluated using fast state smoothing as described in §4.2.2.

5
Initialisation of filter and smoother

5.1 Introduction

In the previous chapter we have considered the operations of filtering and smoothing for the linear Gaussian state space model

$$y_t = Z_t \alpha_t + \varepsilon_t, \qquad \varepsilon_t \sim N(0, H_t),$$
$$\alpha_{t+1} = T_t \alpha_t + R_t \eta_t, \qquad \eta_t \sim N(0, Q_t), \tag{5.1}$$

under the assumption that $\alpha_1 \sim N(a_1, P_1)$ where a_1 and P_1 are known. In most practical applications, however, at least some of the elements of a_1 and P_1 are unknown. We now develop methods of starting up the series when this is the situation; the process is called *initialisation*. We shall consider the general case where some elements of α_1 have a known joint distribution while about other elements we are completely ignorant.

A general model for the initial state vector α_1 is

$$\alpha_1 = a + A\delta + R_0 \eta_0, \qquad \eta_0 \sim N(0, Q_0), \tag{5.2}$$

where the $m \times 1$ vector a is known, δ is a $q \times 1$ vector of unknown quantities, the $m \times q$ matrix A and the $m \times (m-q)$ matrix R_0 are selection matrices, that is, they consist of columns of the identity matrix I_m; they are defined so that when taken together, their columns constitute a set of g columns of I_m with $g \leq m$ and $A' R_0 = 0$. The matrix Q_0 is assumed to be positive definite and known. In most cases vector a will be treated as a zero vector unless some elements of the initial state vector are known constants. When all elements of the state vector α_t are stationary, the initial means, variances and covariances of these initial state elements can be derived from the model parameters. For example, in the case of a stationary ARMA model it is straightforward to obtain the unconditional variance matrix Q_0 as we will show in §5.6.2. The Kalman filter (4.13) can then be applied routinely with $a_1 = 0$ and $P_1 = Q_0$.

To illustrate the structure and notation of (5.2) for readers unfamiliar with the subject, we present a simple example in which

$$y_t = \mu_t + \rho_t + \varepsilon_t, \qquad \varepsilon_t \sim N(0, \sigma_\varepsilon^2),$$

where

$$\mu_{t+1} = \mu_t + v_t + \xi_t, \qquad \xi_t \sim N(0, \sigma_\xi^2),$$
$$v_{t+1} = v_t + \zeta_t, \qquad \zeta_t \sim N(0, \sigma_\zeta^2),$$
$$\rho_{t+1} = \phi \rho_t + \tau_t, \qquad \tau_t \sim N(0, \sigma_\tau^2),$$

in which $|\phi| < 1$ and the disturbances are all mutually and serially uncorrelated. Thus μ_t is a local linear trend as in (3.2), which is nonstationary, while ρ_t is an unobserved stationary first order AR(1) series with zero mean. In state space form this is

$$y_t = (1 \quad 1 \quad 0) \begin{pmatrix} \rho_t \\ \mu_t \\ v_t \end{pmatrix} + \varepsilon_t,$$

$$\begin{pmatrix} \rho_{t+1} \\ \mu_{t+1} \\ v_{t+1} \end{pmatrix} = \begin{bmatrix} \phi & 0 & 0 \\ 0 & 1 & 1 \\ 0 & 0 & 1 \end{bmatrix} \begin{pmatrix} \rho_t \\ \mu_t \\ v_t \end{pmatrix} + \begin{bmatrix} 1 & 0 & 0 \\ 0 & 1 & 0 \\ 0 & 0 & 1 \end{bmatrix} \begin{pmatrix} \tau_t \\ \xi_t \\ \zeta_t \end{pmatrix}.$$

Thus we have

$$a = \begin{pmatrix} 0 \\ 0 \\ 0 \end{pmatrix}, \qquad A = \begin{bmatrix} 0 & 0 \\ 1 & 0 \\ 0 & 1 \end{bmatrix}, \qquad R_0 = \begin{pmatrix} 1 \\ 0 \\ 0 \end{pmatrix},$$

$\eta_0 = \rho_1$ and $Q_0 = \sigma_\tau^2/(1 - \phi^2)$ where $\sigma_\tau^2/(1 - \phi^2)$ is the variance of the stationary series ρ_t.

Although we treat ϕ as known for the purpose of this section, in practice it is an unknown parameter which in a classical analysis is replaced by its maximum likelihood estimate. We see that the object of the representation (5.2) is to separate out α_1 into a constant part a, a nonstationary part $A\delta$ and a stationary part $R_0\eta_0$. In a Bayesian analysis, α_1 can be treated as having a known or noninformative prior density.

The vector δ can be treated as a fixed vector of unknown parameters or as a vector of random normal variables with infinite variances. For the case where δ is fixed and unknown, we may estimate it by maximum likelihood; a technique for doing this was developed by Rosenberg (1973) and we will discuss this in §5.7. For the case where δ is random we assume that

$$\delta \sim N(0, \kappa I_q), \tag{5.3}$$

where we let $\kappa \to \infty$. We begin by considering the Kalman filter with initial conditions $a_1 = E(\alpha_1) = a$ and $P_1 = \text{Var}(\alpha_1)$ where

$$P_1 = \kappa P_\infty + P_*, \tag{5.4}$$

and we let $\kappa \to \infty$ at a suitable point later. Here $P_\infty = AA'$ and $P_* = R_0 Q_0 R_0'$; since A consists of columns of I_m it follows that P_∞ is an $m \times m$ diagonal matrix with q diagonal elements equal to one and the other elements equal to zero. Also, without loss of generality, when a diagonal element of P_∞ is nonzero we take the corresponding element of a to be zero. A vector δ with distribution $N(0, \kappa I_q)$ as $\kappa \to \infty$ is said to be *diffuse*. Initialisation of the Kalman filter when some elements of α_1 are diffuse is called *diffuse initialisation* of the filter. We now consider the modifications required to the Kalman filter in the diffuse initialisation case.

A simple approximate technique is to replace κ in (5.4) by an arbitrary large number and then use the standard Kalman filter (4.13). This approach was employed by Harvey and Phillips (1979). While the device can be useful for approximate exploratory work, it is not recommended for general use since it can lead to large rounding errors. We therefore develop an exact treatment.

The technique we shall use is to expand matrix products as power series in κ^{-1}, taking only the first two or three terms as required, and then let $\kappa \to \infty$ to obtain the dominant term. The underlying idea was introduced by Ansley and Kohn (1985) in a somewhat inaccessible way. Koopman (1997) presented a more transparent treatment of diffuse filtering and smoothing based on the same idea. Further developments were given by Koopman and Durbin (2001) who obtained the results that form the basis of §5.2 on filtering and §5.3 on state smoothing. This approach gives recursions different from those obtained from the augmentation technique of de Jong (1991) which is based on ideas of Rosenberg (1973); see §5.7. Illustrations of these initialisation methods are given in §§5.6 and 5.7.4.

A direct approach to the initialisation problem for the general multivariate linear Gaussian state space model turns out to be somewhat complicated as can be seen from the treatment of Koopman (1997). The reason for this is that for multivariate series the inverse matrix F_t^{-1} does not have a simple general expansion in powers of κ^{-1} for the first few terms of the series, due to the fact that in very specific situations the part of F_t associated with P_∞ can be singular with varying rank. Rank deficiencies may occur when observations are missing near the beginning of the series, for example. For univariate series, however, the treatment is much simpler since F_t is a scalar so the part associated with P_∞ can only be either zero or positive, both of which are easily dealt with. In complicated cases, it turns out to be simpler in the multivariate case to adopt the filtering and smoothing approach of §6.4 in which the multivariate series is converted to a univariate series by introducing the elements of the observational vector y_t one at a time, rather than deal with the series directly as a multivariate series. We therefore begin by assuming that the part of F_t associated with P_∞ is nonsingular or zero for any t. In this way we can treat most multivariate series, for which this assumption holds directly, and at the same time obtain general results for all univariate time series. We shall use these results in §6.4 for the univariate treatment of multivariate series.

5.2 The exact initial Kalman filter

In this section we use the notation $O(\kappa^{-j})$ to denote a function $f(\kappa)$ of κ such that the limit of $\kappa^j f(\kappa)$ as $\kappa \to \infty$ is finite for $j = 1, 2$.

5.2.1 THE BASIC RECURSIONS

Analogously to the decomposition of the initial matrix P_1 in (5.4) we show that the mean square error matrix P_t has the decomposition

$$P_t = \kappa P_{\infty,t} + P_{*,t} + O(\kappa^{-1}), \qquad t = 2, \dots, n, \qquad (5.5)$$

where $P_{\infty,t}$ and $P_{*,t}$ do not depend on κ. It will be shown that $P_{\infty,t} = 0$ for $t > d$ where d is a positive integer which in normal circumstances is small relative to n. The consequence is that the usual Kalman filter (4.13) applies without change for $t = d+1, \dots, n$ with $P_t = P_{*,t}$. Note that when all initial state elements have a known joint distribution or are fixed and known, matrix $P_\infty = 0$ and therefore $d = 0$.

The decomposition (5.5) leads to the similar decompositions

$$F_t = \kappa F_{\infty,t} + F_{*,t} + O(\kappa^{-1}), \qquad M_t = \kappa M_{\infty,t} + M_{*,t} + O(\kappa^{-1}), \qquad (5.6)$$

and, since $F_t = Z_t P_t Z_t' + H_t$ and $M_t = P_t Z_t'$, we have

$$
\begin{aligned}
F_{\infty,t} &= Z_t P_{\infty,t} Z_t', & F_{*,t} &= Z_t P_{*,t} Z_t' + H_t, \\
M_{\infty,t} &= P_{\infty,t} Z_t', & M_{*,t} &= P_{*,t} Z_t',
\end{aligned}
\qquad (5.7)
$$

for $t = 1, \dots, d$. The Kalman filter that we shall derive as $\kappa \to \infty$ we shall call the *exact initial Kalman filter*. We use the word *exact* here to distinguish the resulting filter from the approximate filter obtained by choosing an arbitrary large value for κ and applying the standard Kalman filter (4.13). In deriving it, it is important to note from (5.7) that a zero matrix $M_{\infty,t}$ (whether $P_{\infty,t}$ is a zero matrix or not) implies that $F_{\infty,t} = 0$. As in the development of the Kalman filter in §4.2.2 we assume that F_t is nonsingular. The derivation of the exact initial Kalman filter is based on the expansion for $F_t^{-1} = [\kappa F_{\infty,t} + F_{*,t} + O(\kappa^{-1})]^{-1}$ as a power series in κ^{-1}, that is

$$F_t^{-1} = F_t^{(0)} + \kappa^{-1} F_t^{(1)} + \kappa^{-2} F_t^{(2)} + O(\kappa^{-3}), \qquad (5.8)$$

for large κ. Since $I_p = F_t F_t^{-1}$ we have

$$
\begin{aligned}
I_p = (\kappa F_{\infty,t} + F_{*,t} + \kappa^{-1} F_{a,t} + \kappa^{-2} F_{b,t} + \cdots) \\
\times \left(F_t^{(0)} + \kappa^{-1} F_t^{(1)} + \kappa^{-2} F_t^{(2)} + \cdots \right).
\end{aligned}
$$

On equating coefficients of κ^j for $j = 0, -1, -2, \dots$ we obtain

$$
\begin{aligned}
F_{\infty,t} F_t^{(0)} &= 0, \\
F_{*,t} F_t^{(0)} + F_{\infty,t} F_t^{(1)} &= I_p, \\
F_{a,t} F_t^{(0)} + F_{*,t} F_t^{(1)} + F_{\infty,t} F_t^{(2)} &= 0, \quad \text{etc.}
\end{aligned}
\qquad (5.9)
$$

We need to solve equations (5.9) for $F_t^{(0)}$, $F_t^{(1)}$ and $F_t^{(2)}$; further terms are not required. We shall consider only the cases where $F_{\infty,t}$ is nonsingular or $F_{\infty,t} = 0$. This limitation of the treatment is justified for three reasons. First, it gives a complete solution for the important special case of univariate series, since if y_t is univariate $F_{\infty,t}$ must obviously be positive or zero. Secondly, if y_t is multivariate the restriction is satisfied in most practical cases. Thirdly, for those rare cases where y_t is multivariate but the restriction is not satisfied, the series can be dealt with as a univariate series by the technique described in §6.4. By limiting the treatment at this point to these two cases, the derivations are essentially no more difficult than those required for treating the univariate case. However, solutions for the general case where no restrictions are placed on $F_{\infty,t}$ are algebraically complicated; see Koopman (1997). We mention that although $F_{\infty,t}$ nonsingular is the most common case, situations can arise in practice where $F_{\infty,t} = 0$ even when $P_{\infty,t} \neq 0$ if $M_{\infty,t} = P_{\infty,t} Z_t' = 0$.

Taking first the case where $F_{\infty,t}$ is nonsingular we have from (5.9),

$$F_t^{(0)} = 0, \qquad F_t^{(1)} = F_{\infty,t}^{-1}, \qquad F_t^{(2)} = -F_{\infty,t}^{-1} F_{*,t} F_{\infty,t}^{-1}. \qquad (5.10)$$

The matrices $K_t = T_t M_t F_t^{-1}$ and $L_t = T_t - K_t Z_t$ depend on the inverse matrix F_t^{-1} so they also can be expressed as power series in κ^{-1}. We have

$$K_t = T_t[\kappa M_{\infty,t} + M_{*,t} + O(\kappa^{-1})](\kappa^{-1} F_t^{(1)} + \kappa^{-2} F_t^{(2)} + \cdots),$$

so

$$K_t = K_t^{(0)} + \kappa^{-1} K_t^{(1)} + O(\kappa^{-2}), \qquad L_t = L_t^{(0)} + \kappa^{-1} L_t^{(1)} + O(\kappa^{-2}), \quad (5.11)$$

where

$$\begin{aligned} K_t^{(0)} &= T_t M_{\infty,t} F_t^{(1)}, & L_t^{(0)} &= T_t - K_t^{(0)} Z_t, \\ K_t^{(1)} &= T_t M_{*,t} F_t^{(1)} + T_t M_{\infty,t} F_t^{(2)}, & L_t^{(1)} &= -K_t^{(1)} Z_t. \end{aligned} \qquad (5.12)$$

By following the recursion (4.8) for a_{t+1} starting with $t = 1$ we find that a_t has the form

$$a_t = a_t^{(0)} + \kappa^{-1} a_t^{(1)} + O(\kappa^{-2}),$$

where $a_1^{(0)} = a$ and $a_1^{(1)} = 0$. As a consequence v_t has the form

$$v_t = v_t^{(0)} + \kappa^{-1} v_t^{(1)} + O(\kappa^{-2}),$$

where $v_t^{(0)} = y_t - Z_t a_t^{(0)}$ and $v_t^{(1)} = -Z_t a_t^{(1)}$. The updating equation (4.8) for a_{t+1} can now be expressed as

$$\begin{aligned} a_{t+1} &= T_t a_t + K_t v_t \\ &= T_t[a_t^{(0)} + \kappa^{-1} a_t^{(1)} + O(\kappa^{-2})] \\ &\quad + [K_t^{(0)} + \kappa^{-1} K_t^{(1)} + O(\kappa^{-2})][v_t^{(0)} + \kappa^{-1} v_t^{(1)} + O(\kappa^{-2})], \end{aligned}$$

which becomes as $\kappa \to \infty$,

$$a_{t+1}^{(0)} = T_t a_t^{(0)} + K_t^{(0)} v_t^{(0)}, \qquad t = 1, \ldots, n. \qquad (5.13)$$

The updating equation (4.11) for P_{t+1} is

$$\begin{aligned} P_{t+1} &= T_t P_t L_t' + R_t Q_t R_t' \\ &= T_t[\kappa P_{\infty,t} + P_{*,t} + O(\kappa^{-1})][L_t^{(0)} + \kappa^{-1} L_t^{(1)} + O(\kappa^{-2})]' + R_t Q_t R_t'. \end{aligned}$$

Consequently, on letting $\kappa \to \infty$, the updates for $P_{\infty,t+1}$ and $P_{*,t+1}$ are given by

$$
\begin{aligned}
P_{\infty,t+1} &= T_t P_{\infty,t} L_t^{(0)\prime}, \\
P_{*,t+1} &= T_t P_{\infty,t} L_t^{(1)\prime} + T_t P_{*,t} L_t^{(0)\prime} + R_t Q_t R_t',
\end{aligned}
\tag{5.14}
$$

for $t = 1, \ldots, n$. The matrix P_{t+1} also depends on terms in κ^{-1}, κ^{-2}, etc. but these terms will not be multiplied by κ or higher powers of κ within the Kalman filter recursions. Thus the updating equations for P_{t+1} do not need to take account of these terms. Recursions (5.13) and (5.14) constitute the exact Kalman filter.

In the case where $F_{\infty,t} = 0$, we have

$$
F_t = F_{*,t} + O(\kappa^{-1}), \qquad M_t = M_{*,t} + O(\kappa^{-1}),
$$

and the inverse matrix F_t^{-1} is given by

$$
F_t^{-1} = F_{*,t}^{-1} + O(\kappa^{-1}).
$$

Therefore,

$$
\begin{aligned}
K_t &= T_t[M_{*,t} + O(\kappa^{-1})][F_{*,t}^{-1} + O(\kappa^{-1})] \\
&= T_t M_{*,t} F_{*,t}^{-1} + O(\kappa^{-1}).
\end{aligned}
$$

The updating equation for $a_{t+1}^{(0)}$ (5.13) has

$$
K_t^{(0)} = T_t M_{*,t} F_{*,t}^{-1},
\tag{5.15}
$$

and the updating equation for P_{t+1} becomes

$$
\begin{aligned}
P_{t+1} &= T_t P_t L_t' + R_t Q_t R_t' \\
&= T_t[\kappa P_{\infty,t} + P_{*,t} + O(\kappa^{-1})][L_t^{(0)} + \kappa^{-1} L_t^{(1)} + O(\kappa^{-2})]' + R_t Q_t R_t',
\end{aligned}
$$

where $L_t^{(0)} = T_t - K_t^{(0)} Z_t$ and $L_t^{(1)} = -K_t^{(1)} Z_t$. The updates for $P_{\infty,t+1}$ and $P_{*,t}$ can be simplified considerably since $M_{\infty,t} = P_{\infty,t} Z_t' = 0$. By letting $\kappa \to \infty$ we have

$$
\begin{aligned}
P_{\infty,t+1} &= T_t P_{\infty,t} L_t^{(0)\prime} \\
&= T_t P_{\infty,t} T_t' - T_t P_{\infty,t} Z_t' K_t^{(0)\prime} \\
&= T_t P_{\infty,t} T_t', \\
P_{*,t+1} &= T_t P_{\infty,t} L_t^{(1)\prime} + T_t P_{*,t} L_t^{(0)\prime} + R_t Q R_t' \\
&= -T_t P_{\infty,t} Z_t' K_t^{(1)\prime} + T_t P_{*,t} L_t^{(0)\prime} + R_t Q R_t' \\
&= T_t P_{*,t} L_t^{(0)\prime} + R_t Q R_t',
\end{aligned}
\tag{5.16}
$$

$$
\tag{5.17}
$$

for $t = 1, \ldots, d$, with $P_{\infty,1} = P_\infty = AA'$ and $P_{*,1} = P_* = R_0 Q_0 R_0'$. It might be thought that an expression of the form $F_{*,t} + \kappa^{-1} F_{**,t} + O(\kappa^{-2})$ should be used for F_t here so that two-term expansions could be carried out throughout. It can be shown however that when $M_{\infty,t} = P_{\infty,t} Z_t' = 0$, so that $F_{\infty,t} = 0$, the contribution of the term $\kappa^{-1} F_{**,t}$ is zero; we have therefore omitted it to simplify the presentation.

5.2.2   TRANSITION TO THE USUAL KALMAN FILTER

We now show that for non-degenerate models there is a value of d of t such that $P_{\infty,t} \neq 0$ for $t \leq d$ and $P_{\infty,t} = 0$ for $t > d$. From (5.2) the vector of diffuse elements of α_1 is δ and its dimensionality is q. For finite κ the log density of δ is

$$\log p(\delta) = -\frac{q}{2} \log 2\pi - \frac{q}{2} \log \kappa - \frac{1}{2\kappa} \delta' \delta,$$

since $E(\delta) = 0$ and $\operatorname{Var}(\delta) = \kappa I_q$. Now consider the joint density of δ and Y_t. In an obvious notation the log conditional density of δ given Y_t is

$$\log p(\delta|Y_t) = \log p(\delta, Y_t) - \log p(Y_t),$$

for $t = 1, \ldots, n$. Differentiating with respect to δ, letting $\kappa \to \infty$, equating to zero and solving for δ, gives a solution $\delta = \tilde{\delta}$ which is the conditional mode, and hence the conditional mean, of δ given Y_t. Since $p(\delta, Y_t)$ is Gaussian, $\log p(\delta, Y_t)$ is quadratic in δ so its second derivative does not depend on δ. The reciprocal of minus the second derivative is the variance matrix of δ given Y_t. Let d be the first value of t for which this variance matrix exists. In practical cases d will usually be small relative to n. If there is no value of t for which the variance matrix exists we say that the model is degenerate, since a series of observations which does not even contain enough information to estimate the initial conditions is clearly useless.

By repeated substitution from the state equation $\alpha_{t+1} = T_t \alpha_t + R_t \eta_t$ we can express α_{t+1} as a linear function of α_1 and $\eta_1, \ldots, \eta_t$. Elements of α_1 other than those in δ and also elements of $\eta_1, \ldots, \eta_t$ have finite unconditional variances and hence have finite conditional variances given Y_t. We have also ensured that elements of δ have finite conditional variances given Y_t for $t \geq d$ by definition of d. It follows that $\operatorname{Var}(\alpha_{t+1}|Y_t) = P_{t+1}$ is finite and hence $P_{\infty,t+1} = 0$ for $t \geq d$. On the other hand, for $t < d$, elements of $\operatorname{Var}(\delta|Y_t)$ become infinite as $\kappa \to \infty$ from which it follows that elements of $\operatorname{Var}(\alpha_{t+1}|Y_t)$ become infinite, so $P_{\infty,t+1} \neq 0$ for $t < d$. This establishes that for non-degenerate models there is a value d of t such that $P_{\infty,t} \neq 0$ for $t \leq d$ and $P_{\infty,t} = 0$ for $t > d$. Thus when $t > d$ we have $P_t = P_{*,t} + O(\kappa^{-1})$ so on letting $\kappa \to \infty$ we can use the usual Kalman filter (4.13) starting with $a_{d+1} = a_{d+1}^{(0)}$ and $P_{d+1} = P_{*,d+1}$. A similar discussion of this point is given by de Jong (1991).

5.2.3   A CONVENIENT REPRESENTATION

The updating equations for $P_{*,t+1}$ and $P_{\infty,t+1}$, for $t = 1, \ldots, d$, can be combined to obtain a very convenient representation. Let

$$P_t^\dagger = [P_{*,t} \quad P_{\infty,t}], \qquad L_t^\dagger = \begin{bmatrix} L_t^{(0)} & L_t^{(1)} \\ 0 & L_t^{(0)} \end{bmatrix}. \tag{5.18}$$

From (5.14), the limiting initial state filtering equations as $\kappa \to \infty$ can be written as

$$P_{t+1}^\dagger = T_t P_t^\dagger L_t^{\dagger\prime} + [R_t Q_t R_t' \quad 0], \qquad t = 1, \ldots, d, \tag{5.19}$$

with the initialisation $P_1^\dagger = P^\dagger = [P_* \quad P_\infty]$. For the case $F_{\infty,t} = 0$, the equations in (5.19) with the definitions in (5.18) are still valid but with

$$K_t^{(0)} = T_t M_{*,t} F_{*,t}^{-1}, \qquad L_t^{(0)} = T_t - K_t^{(0)} Z_t, \qquad L_t^{(1)} = 0.$$

This follows directly from the argument used to derive (5.15), (5.16) and (5.17). The recursion (5.19) for diffuse state filtering is due to Koopman and Durbin (2001). It is similar in form to the standard Kalman filtering (4.13) recursion, leading to simplifications in implementing the computations.

5.3   Exact initial state smoothing

5.3.1   SMOOTHED MEAN OF STATE VECTOR

To obtain the limiting recursions for the smoothing equation $\hat{\alpha}_t = a_t + P_t r_{t-1}$ given in (4.26) for $t = d, \ldots, 1$, we return to the recursion (4.25) for r_{t-1}, that is,

$$r_{t-1} = Z_t' F_t^{-1} v_t + L_t' r_t, \qquad t = n, \ldots, 1,$$

with $r_n = 0$. Since r_{t-1} depends on F_t^{-1} and L_t which can both be expressed as power series in κ^{-1} we write

$$r_{t-1} = r_{t-1}^{(0)} + \kappa^{-1} r_{t-1}^{(1)} + O(\kappa^{-2}), \qquad t = d, \ldots, 1. \tag{5.20}$$

Substituting the relevant expansions into the recursion for r_{t-1} we have for the case $F_{\infty,t}$ nonsingular,

$$r_{t-1}^{(0)} + \kappa^{-1} r_{t-1}^{(1)} + \cdots = Z_t' \left(\kappa^{-1} F_t^{(1)} + \kappa^{-2} F_t^{(2)} + \cdots \right) \left(v_t^{(0)} + \kappa^{-1} v_t^{(1)} + \cdots \right)$$
$$+ \left(L_t^{(0)} + \kappa^{-1} L_t^{(1)} + \cdots \right)' \left(r_t^{(0)} + \kappa^{-1} r_t^{(1)} + \cdots \right),$$

leading to recursions for $r_t^{(0)}$ and $r_t^{(1)}$,

$$r_{t-1}^{(0)} = L_t^{(0)'} r_t^{(0)},$$
$$r_{t-1}^{(1)} = Z_t' F_t^{(1)} v_t^{(0)} + L_t^{(0)'} r_t^{(1)} + L_t^{(1)'} r_t^{(0)}, \tag{5.21}$$

for $t = d, \ldots, 1$ with $r_d^{(0)} = r_d$ and $r_d^{(1)} = 0$.

The smoothed state vector is

$$\hat{\alpha}_t = a_t + P_t r_{t-1}$$
$$= a_t + [\kappa P_{\infty,t} + P_{*,t} + O(\kappa^{-1})] \left[r_{t-1}^{(0)} + \kappa^{-1} r_{t-1}^{(1)} + O(\kappa^{-2}) \right]$$
$$= a_t + \kappa P_{\infty,t} \left(r_{t-1}^{(0)} + \kappa^{-1} r_{t-1}^{(1)} \right) + P_{*,t} \left(r_{t-1}^{(0)} + \kappa^{-1} r_{t-1}^{(1)} \right) + O(\kappa^{-1})$$
$$= a_t + \kappa P_{\infty,t} r_{t-1}^{(0)} + P_{*,t} r_{t-1}^{(0)} + P_{\infty,t} r_{t-1}^{(1)} + O(\kappa^{-1}), \tag{5.22}$$

where $a_t = a_t^{(0)} + \kappa^{-1} a_t^{(1)} + \cdots$. It is immediately obvious that for this expression to make sense we must have $P_{\infty,t} r_{t-1}^{(0)} = 0$ for all t. This will be the case if we can show that $\text{Var}(\alpha_t | Y_n)$ is finite for all t as $\kappa \to \infty$. Analogously to the argument in §5.2.2 we can express α_t as a linear function of $\delta, \eta_0, \eta_1, \ldots, \eta_{t-1}$. But $\text{Var}(\delta | Y_d)$ is finite by definition of d so $\text{Var}(\delta | Y_n)$ must be finite as $\kappa \to \infty$ since $d < n$. Also,

$Q_j = \text{Var}(\eta_j)$ is finite so $\text{Var}(\eta_j|Y_n)$ is finite for $j = 0, \ldots, t-1$. It follows that $\text{Var}(\alpha_t|Y_n)$ is finite for all t as $\kappa \to \infty$ so from (5.22) $P_{\infty,t}r_{t-1}^{(0)} = 0$.

Letting $\kappa \to \infty$ we obtain

$$\hat{\alpha}_t = a_t^{(0)} + P_{*,t}r_{t-1}^{(0)} + P_{\infty,t}r_{t-1}^{(1)}, \qquad t = d, \ldots, 1, \qquad (5.23)$$

with $r_d^{(0)} = r_d$ and $r_d^{(1)} = 0$. The equations (5.21) and (5.23) can be re-formulated to obtain

$$r_{t-1}^{\dagger} = \begin{pmatrix} 0 \\ Z_t'F_t^{(1)}v_t^{(0)} \end{pmatrix} + L_t^{\dagger\prime}r_t^{\dagger}, \qquad \hat{\alpha}_t = a_t^{(0)} + P_t^{\dagger}r_{t-1}^{\dagger}, \qquad t = d, \ldots, 1,$$

$$(5.24)$$

where

$$r_{t-1}^{\dagger} = \begin{pmatrix} r_{t-1}^{(0)} \\ r_{t-1}^{(1)} \end{pmatrix}, \qquad \text{with} \quad r_d^{\dagger} = \begin{pmatrix} r_d \\ 0 \end{pmatrix},$$

and the partioned matrices $P_t^{\dagger}$ and $L_t^{\dagger}$ are defined in (5.18). This formulation is convenient since it has the same form as the standard smoothing recursion (4.26). Considering the complexity introduced into the model by the presence of the diffuse elements in α_1, it is very interesting that the state smoothing equations in (5.24) have the same basic structure as the corresponding equations (4.26). This is a useful property in constructing software for implementation of the algorithms.

To avoid extending the treatment further, and since the case $F_{\infty,t} = 0$ is rare in practice, we omit consideration of it here and in §5.3.2 and refer the reader to the discussion in Koopman and Durbin (2001).

5.3.2 SMOOTHED VARIANCE OF STATE VECTOR

We now consider the evaluation of the variance matrix of the estimation error $\hat{\alpha}_t - \alpha_t$ for $t = d, \ldots, 1$ in the diffuse case. We shall not derive the cross-covariances between the estimation errors at different time points in the diffuse case because in practice there is little interest in these quantities.

From §4.3.2, the error variance matrix of the smoothed state vector is given by $V_t = P_t - P_t N_{t-1} P_t$ with the recursion $N_{t-1} = Z_t'F_t^{-1}Z_t + L_t'N_tL_t$, for $t = n, \ldots, 1$, and $N_n = 0$. To obtain exact finite expressions for V_t and N_{t-1} where $F_{\infty,t}$ is nonsingular and $\kappa \to \infty$, for $t = d, \ldots, 1$, we find that we need to take three-term expansions instead of the two-term expressions previously employed. Thus we write

$$N_t = N_t^{(0)} + \kappa^{-1}N_t^{(1)} + \kappa^{-2}N_t^{(2)} + O(\kappa^{-3}). \qquad (5.25)$$

Ignoring residual terms and on substituting in the expression for N_{t-1}, we obtain the recursion for N_{t-1} as

$$N_{t-1}^{(0)} + \kappa^{-1}N_{t-1}^{(1)} + \kappa^{-2}N_{t-1}^{(2)} + \cdots$$
$$= Z_t'\left(\kappa^{-1}F_t^{(1)} + \kappa^{-2}F_t^{(2)} + \cdots\right)Z_t + \left(L_t^{(0)} + \kappa^{-1}L_t^{(1)} + \kappa^{-2}L_t^{(2)} + \cdots\right)'$$
$$\times \left(N_t^{(0)} + \kappa^{-1}N_t^{(1)} + \kappa^{-2}N_t^{(2)} + \cdots\right)\left(L_t^{(0)} + \kappa^{-1}L_t^{(1)} + \kappa^{-2}L_t^{(2)} + \cdots\right),$$

which leads to the set of recursions

$$N_{t-1}^{(0)} = L_t^{(0)\prime} N_t^{(0)} L_t^{(0)},$$

$$N_{t-1}^{(1)} = Z_t' F_t^{(1)} Z_t + L_t^{(0)\prime} N_t^{(1)} L_t^{(0)} + L_t^{(1)\prime} N_t^{(0)} L_t^{(0)} + L_t^{(0)\prime} N_t^{(0)} L_t^{(1)},$$

$$N_{t-1}^{(2)} = Z_t' F_t^{(2)} Z_t + L_t^{(0)\prime} N_t^{(2)} L_t^{(0)} + L_t^{(0)\prime} N_t^{(1)} L_t^{(1)} + L_t^{(1)\prime} N_t^{(1)} L_t^{(0)}$$

$$+ L_t^{(0)\prime} N_t^{(0)} L_t^{(2)} + L_t^{(2)\prime} N_t^{(0)} L_t^{(0)} + L_t^{(1)\prime} N_t^{(0)} L_t^{(1)}, \qquad (5.26)$$

with $N_d^{(0)} = N_d$ and $N_d^{(1)} = N_d^{(2)} = 0$.

Substituting the power series in κ^{-1}, κ^{-2}, etc. and the expression $P_t = \kappa P_{\infty,t} + P_{*,t}$ into the relation $V_t = P_t - P_t N_{t-1} P_t$ we obtain

$$V_t = \kappa P_{\infty,t} + P_{*,t}$$
$$- (\kappa P_{\infty,t} + P_{*,t})\big(N_{t-1}^{(0)} + \kappa^{-1} N_{t-1}^{(1)} + \kappa^{-2} N_{t-1}^{(2)} + \cdots\big)(\kappa P_{\infty,t} + P_{*,t})$$
$$= -\kappa^2 P_{\infty,t} N_{t-1}^{(0)} P_{\infty,t}$$
$$+ \kappa\big(P_{\infty,t} - P_{\infty,t} N_{t-1}^{(0)} P_{*,t} - P_{*,t} N_{t-1}^{(0)} P_{\infty,t} - P_{\infty,t} N_{t-1}^{(1)} P_{\infty,t}\big)$$
$$+ P_{*,t} - P_{*,t} N_{t-1}^{(0)} P_{*,t} - P_{*,t} N_{t-1}^{(1)} P_{\infty,t} - P_{\infty,t} N_{t-1}^{(1)} P_{*,t}$$
$$- P_{\infty,t} N_{t-1}^{(2)} P_{\infty,t} + O(\kappa^{-1}). \qquad (5.27)$$

It was shown in the previous section that $V_t = \mathrm{Var}(\alpha_t|Y_n)$ is finite for $t = 1, \dots, n$. Thus the two matrix terms associated with κ and κ^2 in (5.27) must be zero. Letting $\kappa \to \infty$, the smoothed state variance matrix is given by

$$V_t = P_{*,t} - P_{*,t} N_{t-1}^{(0)} P_{*,t} - P_{*,t} N_{t-1}^{(1)} P_{\infty,t} - P_{\infty,t} N_{t-1}^{(1)} P_{*,t} - P_{\infty,t} N_{t-1}^{(2)} P_{\infty,t}. \qquad (5.28)$$

Note that $P_{*,t} N_{t-1}^{(1)} P_{\infty,t} = (P_{\infty,t} N_{t-1}^{(1)} P_{*,t})'$.

Koopman and Durbin (2001) have shown by exploiting the equality $P_{\infty,t} L_t^{(0)} N_t^{(0)} = 0$, that when the recursions for $N_t^{(1)}$ and $N_t^{(2)}$ in (5.26) are used to calculate the terms in $N_{t-1}^{(1)}$ and $N_{t-1}^{(2)}$, respectively, in (5.28), various items vanish and that the effect is that we can proceed in effect as if the recursions are

$$N_{t-1}^{(0)} = L_t^{(0)\prime} N_t^{(0)} L_t^{(0)},$$

$$N_{t-1}^{(1)} = Z_t' F_t^{(1)} Z_t + L_t^{(0)\prime} N_t^{(1)} L_t^{(0)} + L_t^{(1)\prime} N_t^{(0)} L_t^{(0)}, \qquad (5.29)$$

$$N_{t-1}^{(2)} = Z_t' F_t^{(2)} Z_t + L_t^{(0)\prime} N_t^{(2)} L_t^{(0)} + L_t^{(0)\prime} N_t^{(1)} L_t^{(1)} + L_t^{(1)\prime} N_t^{(1)} L_t^{(0)} + L_t^{(1)\prime} N_t^{(0)} L_t^{(1)},$$

and that we can compute V_t by

$$V_t = P_{*,t} - P_{*,t} N_{t-1}^{(0)} P_{*,t} - \big(P_{\infty,t} N_{t-1}^{(1)} P_{*,t}\big)' - P_{\infty,t} N_{t-1}^{(1)} P_{*,t} - P_{\infty,t} N_{t-1}^{(2)} P_{\infty,t}. \qquad (5.30)$$

Thus the matrices calculated from (5.26) can be replaced by the ones computed by

(5.29) to obtain the correct value for V_t. The new recursions in (5.29) are convenient since the matrix $L_t^{(2)}$ drops out from our calculations for $N_t^{(2)}$.

Indeed the matrix recursions for $N_t^{(1)}$ and $N_t^{(2)}$ in (5.29) are not the same as the recursions for $N_t^{(1)}$ and $N_t^{(2)}$ in (5.26). Also, it can be noticed that matrix $N_t^{(1)}$ in (5.26) is symmetric while $N_t^{(1)}$ in (5.29) is not symmetric. However, the same notation is employed because $N_t^{(1)}$ is only relevant for computing V_t and matrix $P_{\infty,t} N_{t-1}^{(1)}$ is the same when $N_{t-1}^{(1)}$ is computed by (5.26) as when it is computed by (5.29). The same argument applies to matrix $N_t^{(2)}$.

It can now be easily verified that equations (5.30) and the modified recursions (5.29) can be re-formulated as

$$N_{t-1}^{\dagger} = \begin{bmatrix} 0 & Z_t' F_t^{(1)} Z_t \\ Z_t' F_t^{(1)} Z_t & Z_t' F_t^{(2)} Z_t \end{bmatrix} + L_t^{\dagger\prime} N_t^{\dagger} L_t^{\dagger}, \qquad V_t = P_{*,t} - P_t^{\dagger} N_{t-1}^{\dagger} P_t^{\dagger\prime},$$

(5.31)

for $t = d, \ldots, 1$, where

$$N_{t-1}^{\dagger} = \begin{bmatrix} N_{t-1}^{(0)} & N_{t-1}^{(1)\prime} \\ N_{t-1}^{(1)} & N_{t-1}^{(2)} \end{bmatrix}, \qquad \text{with} \quad N_d^{\dagger} = \begin{bmatrix} N_d & 0 \\ 0 & 0 \end{bmatrix},$$

and the partioned matrices $P_t^{\dagger}$ and $L_t^{\dagger}$ are defined in (5.18). Again, this formulation has the same form as the standard smoothing recursion (4.31) which is a useful property when writing software. The formulations (5.24) and (5.31) are given by Koopman and Durbin (2001).

5.4 Exact initial disturbance smoothing

Calculating the smoothed disturbances does not require as much computing as calculating the smoothed state vector when the initial state vector is diffuse. This is because the smoothed disturbance equations do not involve matrix multiplications which depend on terms in κ or higher order terms. From (4.36) the smoothed estimator is $\hat{\varepsilon}_t = H_t(F_t^{-1} v_t - K_t' r_t)$ where $F_t^{-1} = O(\kappa^{-1})$ for $F_{\infty,t}$ positive definite, $F_t^{-1} = F_{*,t}^{-1} + O(\kappa^{-1})$ for $F_{\infty,t} = 0$, $K_t = K_t^{(0)} + O(\kappa^{-1})$ and $r_t = r_t^{(0)} + O(\kappa^{-1})$ so, as $\kappa \to \infty$, we have

$$\hat{\varepsilon}_t = \begin{cases} -H_t K_t^{(0)\prime} r_t^{(0)} & \text{if } F_{\infty,t} \text{ is nonsingular,} \\ H_t\big(F_{*,t}^{-1} v_t - K_t^{(0)\prime} r_t^{(0)}\big) & \text{if } F_{\infty,t} = 0, \end{cases}$$

for $t = d, \ldots, 1$. Other results for disturbance smoothing are obtained in a similar way and for convenience we collect them together in the form

$$\hat{\varepsilon}_t = -H_t K_t^{(0)\prime} r_t^{(0)},$$
$$\hat{\eta}_t = Q_t R_t' r_t^{(0)},$$
$$\text{Var}(\varepsilon_t | y) = H_t - H_t K_t^{(0)\prime} N_t^{(0)} K_t^{(0)} H_t,$$
$$\text{Var}(\eta_t | y) = Q_t - Q_t R_t' N_t^{(0)} R_t Q_t,$$

for the case where $F_{\infty,t} \neq 0$ and

$$\hat{\varepsilon}_t = H_t\left(F_{*,t}^{-1}v_t - K_t^{(0)\prime}r_t^{(0)}\right),$$

$$\hat{\eta}_t = Q_t R_t' r_t^{(0)},$$

$$\mathrm{Var}(\varepsilon_t|y) = H_t - H_t\left(F_{*,t}^{-1} + K_t^{(0)\prime}N_t^{(0)}K_t^{(0)}\right)H_t,$$

$$\mathrm{Var}(\eta_t|y) = Q_t - Q_t R_t' N_t^{(0)} R_t Q_t,$$

for the case $F_{\infty,t} = 0$ and for $t = d, \ldots, 1$. It is fortunate that disturbance smoothing in the diffuse case does not require as much extra computing as for state smoothing. This is particularly convenient when the score vector is computed repeatedly within the process of parameter estimation as we will discuss in §7.3.3.

5.5   Exact initial simulation smoothing

When the initial state vector is diffuse it turns out that the simulation smoother of §4.7 can be used without the complexities of §5.3 required for diffuse state smoothing. We first consider how to obtain a simulated sample of α given y.

Taking $\alpha = (\alpha_1, \ldots, \alpha_n, \alpha_{n+1})'$ as before, define $\alpha_{/1}$ as α but without α_1. It follows that

$$p(\alpha|y) = p(\alpha_1|y)p(\alpha_{/1}|y, \alpha_1). \tag{5.32}$$

Obtain a simulated value $\tilde{\alpha}_1$ of α_1 by drawing a sample value from $p(\alpha_1|y) = N(\hat{\alpha}_1, V_1)$ for which $\hat{\alpha}_1$ and V_1 are computed by the exact initial state smoother as developed in §5.3. Next initialise the Kalman filter with $a_1 = \tilde{\alpha}_1$ and $P_1 = 0$, since we now treat $\tilde{\alpha}_1$ as given, and apply the Kalman filter and simulation smoother as decribed in §4.7. This procedure for obtaining a sample value of α given y is justified by equation (5.32).

To obtain multiple samples, we repeat this procedure. This requires computing a new value of $\tilde{\alpha}_1$, and new values of v_t from the Kalman filter for each new draw. The Kalman filter quantities F_t, K_t and P_{t+1} do not need to be re-computed. Note that for a model with non-diffuse initial conditions, we do not re-compute the v_t's for generating multiple samples; see §4.7.6.

A similar procedure can be followed for simulating disturbance vectors given y: we initialise the Kalman filter with $a_1 = \tilde{\alpha}_1$ and $P_1 = 0$ as above and then use the simulation smoothing recursions of §4.7 to generate samples for the disturbances.

5.6   Examples of initial conditions for some models

In this section we give some examples of the exact initial Kalman filter for $t = 1, \ldots, d$ for a range of state space models.

5.6.1   STRUCTURAL TIME SERIES MODELS

Structural time series models are usually set up in terms of nonstationary components. Therefore, most of the models in this class have the initial state vector equals δ, that is, $\alpha_1 = \delta$ so that $a_1 = 0$, $P_* = 0$ and $P_\infty = I_m$. We then proceed with the algorithms provided by §§5.2, 5.3 and 5.4.

To illustrate the exact initial Kalman filter in detail we consider the local linear trend model (3.2) with system matrices

$$Z_t = (1 \quad 0), \qquad T_t = \begin{bmatrix} 1 & 1 \\ 0 & 1 \end{bmatrix}, \qquad Q_t = \sigma_\varepsilon^2 \begin{bmatrix} q_\xi & 0 \\ 0 & q_\zeta \end{bmatrix},$$

and with $H_t = \sigma_\varepsilon^2$, $R_t = I_2$, where $q_\xi = \sigma_\xi^2 / \sigma_\varepsilon^2$ and $q_\zeta = \sigma_\zeta^2 / \sigma_\varepsilon^2$. The exact initial Kalman filter is started with

$$a_1 = 0, \qquad P_{*,1} = 0, \qquad P_{\infty,1} = I_2,$$

and the first update is based on

$$K_1^{(0)} = \begin{pmatrix} 1 \\ 0 \end{pmatrix}, \qquad L_1^{(0)} = \begin{bmatrix} 0 & 1 \\ 0 & 1 \end{bmatrix}, \qquad K_1^{(1)} = -\sigma_\varepsilon^2 \begin{pmatrix} 1 \\ 0 \end{pmatrix},$$

$$L_1^{(1)} = \sigma_\varepsilon^2 \begin{bmatrix} 1 & 0 \\ 0 & 0 \end{bmatrix},$$

such that

$$a_2 = \begin{pmatrix} y_1 \\ 0 \end{pmatrix}, \qquad P_{*,2} = \sigma_\varepsilon^2 \begin{bmatrix} 1 + q_\xi & 0 \\ 0 & q_\zeta \end{bmatrix}, \qquad P_{\infty,2} = \begin{bmatrix} 1 & 1 \\ 1 & 1 \end{bmatrix}.$$

The second update gives the quantities

$$K_2^{(0)} = \begin{pmatrix} 2 \\ 1 \end{pmatrix}, \qquad L_2^{(0)} = \begin{bmatrix} -1 & 1 \\ -1 & 1 \end{bmatrix},$$

and

$$K_2^{(1)} = -\sigma_\varepsilon^2 \begin{pmatrix} 3 + q_\xi \\ 2 + q_\xi \end{pmatrix}, \qquad L_2^{(1)} = \sigma_\varepsilon^2 \begin{bmatrix} 3 + q_\xi & 0 \\ 2 + q_\xi & 0 \end{bmatrix},$$

together with the state update results

$$a_3 = \begin{pmatrix} 2y_2 - y_1 \\ y_2 - y_1 \end{pmatrix}, \qquad P_{*,3} = \sigma_\varepsilon^2 \begin{bmatrix} 5 + 2q_\xi + q_\zeta & 3 + q_\xi + q_\zeta \\ 3 + q_\xi + q_\zeta & 2 + q_\xi + 2q_\zeta \end{bmatrix},$$

$$P_{\infty,3} = \begin{bmatrix} 0 & 0 \\ 0 & 0 \end{bmatrix}.$$

It follows that the usual Kalman filter (4.13) can be used for $t = 3, \ldots, n$.

5.6.2 STATIONARY ARMA MODELS

The univariate stationary ARMA model with zero mean of order p and q is given by

$$y_t = \phi_1 y_{t-1} + \cdots + \phi_p y_{t-p} + \zeta_t + \theta_1 \zeta_{t-1} + \cdots + \theta_q \zeta_{t-q}, \qquad \zeta_t \sim N(0, \sigma^2).$$

The state space form is

$$y_t = (1, 0, \ldots, 0)\alpha_t,$$
$$\alpha_{t+1} = T\alpha_t + R\zeta_{t+1},$$

where the system matrices T and R are given by (3.17) with $r = \max(p, q + 1)$. All elements of the state vector are stationary so that the part $a + A\delta$ in (5.2) is zero and $R_0 = I_m$. The unconditional distribution of the initial state vector α_1 is therefore given by

$$\alpha_1 \sim N(0, \sigma^2 Q_0),$$

where, since $Var(\alpha_{t+1}) = Var(T\alpha_t + R\zeta_{t+1})$, $\sigma^2 Q_0 = \sigma^2 T Q_0 T' + \sigma^2 R R'$. This equation needs to be solved for Q_0. It can be shown that a solution can be obtained by solving the linear equation $(I_{m^2} - T \otimes T) \text{vec}(Q_0) = \text{vec}(RR')$ for Q_0, where $\text{vec}(Q_0)$ and $\text{vec}(RR')$ are the stacked columns of Q_0 and RR' and where

$$T \otimes T = \begin{bmatrix} t_{11}T & \cdots & t_{1m}T \\ t_{21}T & \cdots & t_{2m}T \\ \vdots & & \\ t_{m1}T & \cdots & t_{mm}T \end{bmatrix},$$

with t_{ij} denoting the (i, j) element of matrix T; see, for example, Magnus and Neudecker (1988, Theorem 2, p. 30) who give a general treatment of problems of this type. The Kalman filter is initialised by $a_1 = 0$ and $P_1 = Q_0$.

As an example, consider the ARMA(1, 1) model

$$y_t = \phi y_{t-1} + \zeta_t + \theta \zeta_{t-1}, \qquad \zeta_t \sim N(0, \sigma^2).$$

Then

$$T = \begin{bmatrix} \phi & 1 \\ 0 & 0 \end{bmatrix} \quad \text{and} \quad R = \begin{pmatrix} 1 \\ \theta \end{pmatrix},$$

so the solution is

$$Q_0 = \begin{bmatrix} (1 - \phi^2)^{-1}(1 + \theta^2 + 2\phi\theta) & \theta \\ \theta & \theta^2 \end{bmatrix}.$$

5.6.3 NONSTATIONARY ARIMA MODELS

The univariate nonstationary ARIMA model of order p, d and q can be put in the form

$$y_t^* = \phi_1 y_{t-1}^* + \cdots + \phi_p y_{t-p}^* + \zeta_t + \theta_1 \zeta_{t-1} + \cdots + \theta_q \zeta_{t-q}, \qquad \zeta_t \sim N(0, \sigma^2).$$

where $y_t^* = \Delta^d y_t$. The state space form of the ARIMA model with $p = 2$, $d = 1$ and $q = 1$ is given in §3.3 with the state vector given by

$$\alpha_t = \begin{pmatrix} y_{t-1} \\ y_t^* \\ \phi_2 y_{t-1}^* + \theta_1 \zeta_t \end{pmatrix},$$

where $y_t^* = \Delta y_t = y_t - y_{t-1}$. The first element of the initial state vector α_1, that is y_0, is nonstationary while the other elements are stationary. Therefore, the initial vector $\alpha_1 = a + A\delta + R_0 \eta_0$ is given by

$$\alpha_1 = \begin{pmatrix} 0 \\ 0 \\ 0 \end{pmatrix} + \begin{pmatrix} 1 \\ 0 \\ 0 \end{pmatrix} \delta + \begin{bmatrix} 0 & 0 \\ 1 & 0 \\ 0 & 1 \end{bmatrix} \eta_0, \qquad \eta_0 \sim N(0, Q_0),$$

where Q_0 is the 2×2 unconditional variance matrix for an ARMA model with $p = 2$ and $q = 1$ which we obtain from §5.6.2. When δ is diffuse, the mean vector and variance matrix are

$$a_1 = 0, \qquad P_1 = \kappa P_\infty + P_*,$$

where

$$P_\infty = \begin{bmatrix} 1 & 0 & 0 \\ 0 & 0 & 0 \\ 0 & 0 & 0 \end{bmatrix}, \qquad P_* = \begin{bmatrix} 0 & 0 \\ 0 & Q_0 \end{bmatrix}.$$

Analysis then proceeds using the exact initial Kalman filter and smoother of §§5.2, 5.3 and 5.4. The initial state specification for ARIMA models with $d = 1$ but with other values for p and q is obtained in a similar way.

From §3.3, the initial state vector for the ARIMA model with $p = 2$, $d = 2$ and $q = 1$ is given by

$$\alpha_1 = \begin{pmatrix} y_0 \\ \Delta y_0 \\ y_1^* \\ \phi_2 y_0^* + \theta_1 \zeta_1 \end{pmatrix}.$$

The first two elements of α_1, that is, y_0 and Δy_0, are nonstationary and we therefore treat them as diffuse. Thus we write

$$\alpha_1 = \begin{pmatrix} 0 \\ 0 \\ 0 \\ 0 \end{pmatrix} + \begin{bmatrix} 1 & 0 \\ 0 & 1 \\ 0 & 0 \\ 0 & 0 \end{bmatrix} \delta + \begin{bmatrix} 0 & 0 \\ 0 & 0 \\ 1 & 0 \\ 0 & 1 \end{bmatrix} \eta_0, \qquad \eta_0 \sim N(0, Q_0),$$

where Q_0 is as for the previous case. It follows that the mean vector and variance matrix of α_1 are

$$a_1 = 0, \qquad P_1 = \kappa P_\infty + P_*,$$

where

$$P_\infty = \begin{bmatrix} I_2 & 0 \\ 0 & 0 \end{bmatrix}, \qquad P_* = \begin{bmatrix} 0 & 0 \\ 0 & Q_0 \end{bmatrix}.$$

We then proceed with the methods of §§5.2, 5.3 and 5.4. The initial conditions for non-seasonal ARIMA models with other values for p, d and q and seasonal models are derived in similar ways.

5.6.4   REGRESSION MODEL WITH ARMA ERRORS

The regression model with k explanatory variables and ARMA(p, q) errors (3.27) can be written in state space form as in §3.7. The initial state vector is

$$\alpha_1 = \begin{pmatrix} 0 \\ 0 \end{pmatrix} + \begin{bmatrix} I_k \\ 0 \end{bmatrix} \delta + \begin{bmatrix} 0 \\ I_r \end{bmatrix} \eta_0, \qquad \eta_0 \sim N(0, Q_0),$$

where Q_0 is obtained as in §5.6.2 and $r = \max(p, q + 1)$. When δ is treated as diffuse we have $\alpha_1 \sim N(a_1, P_1)$ where $a_1 = 0$ and $P_1 = \kappa P_\infty + P_*$ with

$$P_\infty = \begin{bmatrix} I_k & 0 \\ 0 & 0 \end{bmatrix}, \qquad P_* = \begin{bmatrix} 0 & 0 \\ 0 & Q_0 \end{bmatrix}.$$

We then proceed as described in the last section.

To illustrate the use of the exact initial Kalman filter we consider the simple case of an AR(1) model with a constant, that is

$$y_t = \mu + \xi_t,$$
$$\xi_t = \phi \xi_{t-1} + \zeta_t, \qquad \zeta_t \sim N(0, \sigma^2).$$

In state space form we have

$$\alpha_t = \begin{pmatrix} \mu \\ \xi_t \end{pmatrix}$$

and the system matrices are given by

$$Z_t = (1 \quad 1), \qquad T_t = \begin{bmatrix} 1 & 0 \\ 0 & \phi \end{bmatrix}, \qquad R_t = \begin{pmatrix} 0 \\ 1 \end{pmatrix},$$

with $H_t = 0$ and $Q_t = \sigma^2$. The exact initial Kalman filter is started with

$$a_1 = \begin{pmatrix} 0 \\ 0 \end{pmatrix}, \qquad P_{*,1} = c \begin{bmatrix} 0 & 0 \\ 0 & 1 \end{bmatrix}, \qquad P_{\infty,1} = \begin{bmatrix} 1 & 0 \\ 0 & 0 \end{bmatrix},$$

where $c = \sigma^2/(1 - \phi^2)$. The first update is based on

$$K_1^{(0)} = \begin{pmatrix} 1 \\ 0 \end{pmatrix}, \qquad L_1^{(0)} = \begin{bmatrix} 0 & -1 \\ 0 & \phi \end{bmatrix}, \qquad K_1^{(1)} = c \begin{pmatrix} -1 \\ \phi \end{pmatrix},$$

$$L_1^{(1)} = c \begin{bmatrix} 1 & 1 \\ -\phi & -\phi \end{bmatrix},$$

such that

$$a_2 = \begin{pmatrix} y_1 \\ 0 \end{pmatrix}, \qquad P_{*,2} = \frac{\sigma^2}{1 - \phi^2} \begin{bmatrix} 1 & -\phi \\ -\phi & 1 \end{bmatrix}, \qquad P_{\infty,2} = \begin{bmatrix} 0 & 0 \\ 0 & 0 \end{bmatrix}.$$

It follows that the usual Kalman filter (4.13) can be used for $t = 2, \ldots, n$.

5.6.5 SPLINE SMOOTHING

The initial state vector for the spline model (3.41) is simply $\alpha_1 = \delta$, implying that $a_1 = 0$, $P_* = 0$ and $P_\infty = I_2$.

5.7 Augmented Kalman filter and smoother

5.7.1 INTRODUCTION

An alternative approach for dealing with the initialisation problem is due to Rosenberg (1973), de Jong (1988b) and de Jong (1991). As in (5.2), the initial state vector is defined as

$$\alpha_1 = a + A\delta + R_0\eta_0, \qquad \eta_0 \sim N(0, Q_0). \qquad (5.33)$$

Rosenberg (1973) treats δ as a fixed unknown vector and he employs maximum likelihood to estimate δ while de Jong (1991) treats δ as a diffuse vector. Since the treatments of Rosenberg and de Jong are both based on the idea of augmenting the observed vector, we will refer to their procedures collectively as the *augmentation approach*. The approach of Rosenberg offers relief to analysts who feel uncomfortable about using diffuse initialising densities on the ground that infinite variances have no counterpart in real data. In fact, as we shall show, the two approaches give effectively the same answer so these analysts could regard the diffuse assumption as a device for achieving initialisation based on maximum likelihood estimates of the unknown initial state elements.

5.7.2 AUGMENTED KALMAN FILTER

In this subsection we establish the groundwork for both the Rosenberg and the de Jong techniques. For given δ, apply the Kalman filter, (4.13) with $a_1 = E(\alpha_1) = a + A\delta$, $P_1 = Var(\alpha_1) = P_* = R_0 Q_0 R_0'$ and denote the resulting value of a_t from the filter output by $a_{\delta,t}$. Since $a_{\delta,t}$ is a linear function of the observations and $a_1 = a + A\delta$ we can write

$$a_{\delta,t} = a_{a,t} + A_{A,t}\delta, \qquad (5.34)$$

where $a_{a,t}$ is the value of a_t in the filter output obtained by taking $a_1 = a$, $P_1 = P_*$ and where the jth column of $A_{A,t}$ is the value of a_t in the filter output obtained from an observational vector of zeros with $a_1 = A_j$, $P_1 = P_*$, where A_j is the jth column of A. Denote the value of v_t in the filter output obtained by taking $a_1 = a + A\delta$, $P_1 = P_*$ by $v_{\delta,t}$. Analogously to (5.34) we can write

$$v_{\delta,t} = v_{a,t} + V_{A,t}\delta, \qquad (5.35)$$

where $v_{a,t}$ and $V_{A,t}$ are given by the same Kalman filters that produced $a_{a,t}$ and $A_{A,t}$.

The matrices $(a_{a,t}, A_{A,t})$ and $(v_{a,t}, V_{A,t})$ can be computed in one pass through a Kalman filter which inputs the observation vector y_t augmented by zero values. This is possible because for each Kalman filter producing a particular column of $(a_{a,t}, A_{A,t})$, the same variance initialisation $P_1 = P_*$ applies, so the variance output, which we denote by $F_{\delta,t}$, $K_{\delta,t}$ and $P_{\delta,t+1}$, is the same for each Kalman filter. Replacing the Kalman filter equations for the vectors v_t and a_t by the corresponding equations for the matrices $(a_{a,t}, A_{A,t})$ and $(v_{a,t}, V_{A,t})$ leads to the equations

$$
\begin{aligned}
(v_{a,t}, V_{A,t}) &= (y_t, 0) - Z_t(a_{a,t}, A_{A,t}), \\
(a_{a,t+1}, A_{A,t+1}) &= T_t(a_{a,t}, A_{A,t}) + K_{\delta,t}(v_{a,t}, V_{A,t}),
\end{aligned}
\tag{5.36}
$$

where $(a_{a,1}, A_{A,1}) = (a, A)$; the recursions corresponding to F_t, K_t and P_{t+1} remain as for the standard Kalman filter, that is,

$$
\begin{aligned}
F_{\delta,t} &= Z_t P_{\delta,t} Z_t' + H_t, \\
K_{\delta,t} &= T_t P_{\delta,t} Z_t' F_{\delta,t}^{-1}, \qquad L_{\delta,t} = T_t - K_{\delta,t} Z_t, \\
P_{\delta,t+1} &= T_t P_{\delta,t} L_{\delta,t}' + R_t Q_t R_t',
\end{aligned}
\tag{5.37}
$$

for $t = 1, \ldots, n$ with $P_{\delta,1} = P_*$. We have included the suffix δ in these expressions not because they depend mathematically on δ but because they have been calculated on the assumption that δ is fixed. The modified Kalman filter (5.36) and (5.37) will be referred to as the *augmented Kalman filter* in this book.

5.7.3 FILTERING BASED ON THE AUGMENTED KALMAN FILTER

With these preliminaries, let us first consider the diffuse case (5.2) with $\delta \sim N(0, \kappa I_q)$ where $\kappa \to \infty$; we will consider later the case where δ is fixed and is estimated by maximum likelihood. From (5.34) we obtain for given κ,

$$
a_{t+1} = \mathrm{E}(\alpha_{t+1}|Y_t) = a_{a,t+1} + A_{A,t+1}\bar{\delta}_t,
\tag{5.38}
$$

where $\bar{\delta}_t = \mathrm{E}(\delta|Y_t)$. Now

$$
\begin{aligned}
\log p(\delta|Y_t) &= \log p(\delta) + \log p(Y_t|\delta) - \log p(Y_t) \\
&= \log p(\delta) + \sum_{j=1}^{t} \log p(v_{\delta,j}) - \log p(Y_t) \\
&= -\frac{1}{2\kappa}\delta'\delta - b_t'\delta - \frac{1}{2}\delta' S_{A,t}\delta + \text{terms independent of } \delta,
\end{aligned}
\tag{5.39}
$$

where

$$
b_t = \sum_{j=1}^{t} V_{A,j}' F_{\delta,j}^{-1} v_{a,j}, \qquad S_{A,t} = \sum_{j=1}^{t} V_{A,j}' F_{\delta,j}^{-1} V_{A,j}.
\tag{5.40}
$$

Since densities are normal, the mean of $p(\delta|Y_t)$ is equal to the mode, and this is the value of δ which maximises $\log p(\delta|Y_t)$, so on differentiating (5.39) with respect to δ and equating to zero, we have

$$\bar{\delta}_t = -\left(S_{A,t} + \frac{1}{\kappa}I_q\right)^{-1}b_t. \tag{5.41}$$

Also,

$$
\begin{aligned}
P_{t+1} &= \mathrm{E}[(a_{t+1} - \alpha_{t+1})(a_{t+1} - \alpha_{t+1})'] \\
&= \mathrm{E}[\{a_{\delta,t+1} - \alpha_{t+1} - A_{A,t+1}(\delta - \bar{\delta}_t)\}\{a_{\delta,t+1} - \alpha_{t+1} - A_{A,t+1}(\delta - \bar{\delta}_t)\}'] \\
&= P_{\delta,t+1} + A_{A,t+1}\,\mathrm{Var}(\delta|Y_t)A'_{A,t+1} \\
&= P_{\delta,t+1} + A_{A,t+1}\left(S_{A,t} + \frac{1}{\kappa}I_q\right)^{-1}A'_{A,t+1}, \tag{5.42}
\end{aligned}
$$

since $\mathrm{Var}(\delta|Y_t) = (S_{A,t} + \frac{1}{\kappa}I_q)^{-1}$.

Letting $\kappa \to \infty$ we have

$$\bar{\delta}_t = -S_{A,t}^{-1}b_t, \tag{5.43}$$

$$\mathrm{Var}(\delta|Y_t) = S_{A,t}^{-1}, \tag{5.44}$$

when $S_{A,t}$ is nonsingular. The calculations of b_t and $S_{A,t}$ are easily incorporated into the augmented Kalman filter (5.36) and (5.37). It follows that

$$a_{t+1} = a_{a,t+1} - A_{A,t+1}S_{A,t}^{-1}b_t, \tag{5.45}$$

$$P_{t+1} = P_{\delta,t+1} + A_{A,t+1}S_{A,t}^{-1}A'_{A,t+1}, \tag{5.46}$$

as $\kappa \to \infty$. For $t < d$, $S_{A,t}$ is singular so values of a_{t+1} and P_{t+1} given by (5.45) and (5.46) do not exist. However, when $t = d$, a_{d+1} and P_{d+1} exist and consequently when $t > d$ the values a_{t+1} and P_{t+1} can be calculated by the standard Kalman filter for $t = d + 1, \ldots, n$. Thus we do not need to use the augmented Kalman filter (5.36) for $t = d + 1, \ldots, n$. These results are due to de Jong (1991) but our derivation here is more transparent.

We now consider a variant of the maximum likelihood method for initialising the filter due to Rosenberg (1973). In this technique, δ is regarded as fixed and unknown and we employ maximum likelihood given Y_t to obtain estimate $\hat{\delta}_t$. The loglikelihood of Y_t given δ is

$$\log p(Y_t|\delta) = \sum_{j=1}^{t} \log p(v_{\delta,j}) = -b_t'\delta - \frac{1}{2}\delta'S_{A,t}\delta + \text{terms independent of } \delta,$$

which is the same as (5.39) apart from the term $-\delta'\delta/(2\kappa)$. Differentiating with respect to δ, equating to zero and taking the second derivative gives

$$\hat{\delta}_t = -S_{A,t}^{-1}b_t, \qquad \mathrm{Var}(\hat{\delta}_t) = S_{A,t}^{-1},$$

when $S_{A,t}$ is nonsingular, that is for $t = d, \ldots, n$. These values are the same as $\bar{\delta}_t$ and $\mathrm{Var}(\delta_t | Y_t)$ when $\kappa \to \infty$. In practice we choose t to be the smallest value for which $\hat{\delta}_t$ exists, which is d. It follows that the values of a_{t+1} and P_{t+1} for $t \geq d$ given by this approach are the same as those obtained in the diffuse case. Thus the solution of the initialisation problem given in §5.2 also applies to the case where δ is fixed and unknown. From a computational point of view the calculations of §5.2 are more efficient than those for the augmented device described in this section when the model is reasonably large. A comparison of the computational efficiency is given in §5.7.5. Rosenberg (1973) used a procedure which differed slightly from this. Although he employed essentially the same augmentation technique, in the notation above he estimated δ by the value $\hat{\delta}_n$ based on all the data.

5.7.4   ILLUSTRATION: THE LOCAL LINEAR TREND MODEL

To illustrate the augmented Kalman filter we consider the same local linear trend model as in §5.6.1. The system matrices of the local linear trend model (3.2) are given by

$$
Z = (1 \quad 0), \qquad T = \begin{bmatrix} 1 & 1 \\ 0 & 1 \end{bmatrix}, \qquad Q = \sigma_\varepsilon^2 \begin{bmatrix} q_\xi & 0 \\ 0 & q_\zeta \end{bmatrix},
$$

with $H = \sigma_\varepsilon^2$ and $R = I_2$ and where $q_\xi = \sigma_\xi^2 / \sigma_\varepsilon^2$ and $q_\zeta = \sigma_\zeta^2 / \sigma_\varepsilon^2$. The augmented Kalman filter is started with

$$
(a_{a,1}, A_{A,1}) = \begin{bmatrix} 0 & 1 & 0 \\ 0 & 0 & 1 \end{bmatrix}, \qquad P_{\delta,1} = \sigma_\varepsilon^2 \begin{bmatrix} 0 & 0 \\ 0 & 0 \end{bmatrix},
$$

and the first update is based on

$$
(v_{a,1}, V_{A,1}) = (y_1 \quad -1 \quad 0), \qquad F_{\delta,1} = \sigma_\varepsilon^2, \qquad K_{\delta,1} = \begin{pmatrix} 0 \\ 0 \end{pmatrix},
$$

$$
L_{\delta,1} = \begin{bmatrix} 1 & 1 \\ 0 & 1 \end{bmatrix},
$$

so

$$
b_1 = -\frac{1}{\sigma_\varepsilon^2} \begin{pmatrix} y_1 \\ 0 \end{pmatrix}, \qquad S_{A,1} = \frac{1}{\sigma_\varepsilon^2} \begin{bmatrix} 1 & 0 \\ 0 & 0 \end{bmatrix},
$$

and

$$
(a_{a,2}, A_{A,2}) = \begin{bmatrix} 0 & 1 & 1 \\ 0 & 0 & 1 \end{bmatrix}, \qquad P_{\delta,2} = \sigma_\varepsilon^2 \begin{bmatrix} q_\xi & 0 \\ 0 & q_\zeta \end{bmatrix}.
$$

The second update gives the quantities

$$
(v_{a,2}, V_{A,2}) = (y_2 \quad -1 \quad -1), \qquad F_{\delta,2} = \sigma_\varepsilon^2 (1 + q_\xi),
$$

$$
K_{\delta,2} = \frac{q_\xi}{1 + q_\xi} \begin{pmatrix} 1 \\ 0 \end{pmatrix}, \qquad L_{\delta,2} = \begin{bmatrix} \frac{1}{1+q_\xi} & 1 \\ 0 & 1 \end{bmatrix},
$$

with

$$b_2 = \frac{-1}{1 + q_\xi} \begin{pmatrix} (1 + q_\xi)y_1 + y_2 \\ y_2 \end{pmatrix}, \qquad S_{A,2} = \frac{1}{\sigma_\varepsilon^2(1 + q_\xi)} \begin{bmatrix} 2 + q_\xi & 1 \\ 1 & 1 \end{bmatrix},$$

and the state update results

$$(a_{a,3}, A_{A,3}) = \frac{1}{1 + q_\xi} \begin{bmatrix} q_\xi y_2 & 1 & 2 + q_\xi \\ 0 & 0 & 1 + q_\xi \end{bmatrix}, \qquad P_{\delta,3} = \sigma_\varepsilon^2 \begin{bmatrix} q_\xi + \frac{q_\xi}{1+q_\xi} + q_\zeta & q_\zeta \\ q_\zeta & 2q_\zeta \end{bmatrix}.$$

The augmented part can be collapsed since $S_{A,2}$ is nonsingular, giving

$$S_{A,2}^{-1} = \sigma_\varepsilon^2 \begin{bmatrix} 1 & -1 \\ -1 & 2 + q_\xi \end{bmatrix}, \qquad \bar\delta_2 = -S_{A,2}^{-1}b_2 = \begin{pmatrix} y_1 \\ y_2 - y_1 \end{pmatrix}.$$

It follows that

$$a_3 = a_{a,3} + A_{A,3}\bar\delta_2 = \begin{pmatrix} 2y_2 - y_1 \\ y_2 - y_1 \end{pmatrix},$$

$$P_3 = P_{\delta,3} + A_{A,3}S_{A,2}^{-1}A'_{A,3} = \sigma_\varepsilon^2 \begin{bmatrix} 5 + 2q_\xi + q_\zeta & 3 + q_\xi + q_\zeta \\ 3 + q_\xi + q_\zeta & 2 + q_\xi + 2q_\zeta \end{bmatrix}.$$

and the usual Kalman filter (4.13) can be used for $t = 3, \ldots, n$. These results are exactly the same as obtained in §5.6.1, though the computations take longer as we will now show.

5.7.5 COMPARISONS OF COMPUTATIONAL EFFICIENCY

The adjusted Kalman filters of §§5.2 and 5.7.2 both require more computations than the Kalman filter (4.13) with known initial conditions. Of course the adjustments are only required for a limited number of updates. The additional computations for the exact initial Kalman filter are due to updating the matrix $P_{\infty,t+1}$ and computing the matrices $K_t^{(1)}$ and $L_t^{(1)}$ when $F_{\infty,t} \neq 0$, for $t = 1, \ldots, d$. For many practical state space models the system matrices Z_t and T_t are sparse selection matrices containing many zeros and ones; this is the case for the models discussed in Chapter 3. Therefore, calculations involving Z_t and T_t are particularly cheap for most models. Table 5.1 compares the number of additional multiplications (compared to the Kalman filter with known initial conditions) required for filtering using the devices of §§5.2 and 5.7.2 applied to several structural time series models which are discussed in §3.2. The results in Table 5.1 show that the additional

Table 5.1. Number of additional multiplications for filtering.

Model	Exact initial	Augmenting	Difference (%)
Local level	3	7	57
Local linear trend	18	46	61
Basic seasonal (s = 4)	225	600	63
Basic seasonal (s = 12)	3549	9464	63

number of computations for the exact initial Kalman filter of §5.2 is less than half the extra computations required for the augmentation device of §5.7.2. Such computational efficiency gains are important when the Kalman filter is used many times as is the case for parameter estimation; a detailed discussion of estimation is given in Chapter 7. It will also be argued in §7.3.5 that many computations for the exact initial Kalman filter only need to be done once for a specific model since the computed values remain the same when the parameters of the model change. This argument does not apply to the augmentation device and this is another important reason why our approach in §5.2 is more efficient than the augmentation approach.

5.7.6   SMOOTHING BASED ON THE AUGMENTED KALMAN FILTER

The smoothing algorithms can also be developed using the augmented approach. The smoothing recursion for r_{t-1} in (4.48) needs to be augmented in the same way as is done for v_t and a_t of the Kalman filter. When the augmented Kalman filter is applied for $t = 1, \ldots, n$, the modifications for smoothing are straightforward after computing $\hat{\delta}_n$ and $\text{Var}(\hat{\delta}_n)$ and then applying similar expressions to those of (5.45) and (5.46). The collapse of the augmented Kalman filter to the standard Kalman filter is computationally efficient for filtering but, as a result, the estimates $\hat{\delta}_n$ and $\text{Var}(\hat{\delta}_n)$ are not available for calculating the smoothed estimates of the state vector. It is not therefore straightforward to do smoothing when the collapsing device is used in the augmentation approach. A solution for this problem has been given by Chu-Chun-Lin and de Jong (1993).

6
Further computational aspects

6.1 Introduction

In this chapter we will discuss a number of remaining computational aspects of
the Kalman filter and smoother. Two different ways of incorporating regression
effects within the Kalman filter are described in §6.2. The standard Kalman
filter recursion for the variance matrix of the filtered state vector does not rule
out the possibility that this matrix becomes negative definite; this is obviously an
undesirable outcome since it indicates the presence of rounding errors. The square
root Kalman filter eliminates this problem at the expense of slowing down the
filtering and smoothing processes; details are given in §6.3. The computational
costs of implementing the filtering and smoothing procedures of Chapters 4 and 5
can become high for high-dimensional multivariate models, particularly in dealing
with the initialisation problem. It turns out that by bringing the elements of the
observation vectors in multivariate models into the filtering and smoothing
computing operations one at a time, substantial gains in computational efficiency
are achieved and the initialisation problem is simplified considerably. Methods
based on this idea are developed in §6.4. The various algorithms presented in
the Chapters 4 and 5 and this chapter need to be implemented efficiently on a
computer. Recently, the computer package *SsfPack* has been developed which
has implemented all the algorithms considered in this book in a computationally
efficient way. Section 6.6 describes the main features of the package together with
an example.

6.2 Regression estimation

6.2.1 INTRODUCTION

As for the structural time series model considered in §3.2, the general state space
model can be extended to allow for the incorporation of explanatory variables and
intervention variables into the model. To accomplish this generalisation we replace
the measurement equation of the state space model (3.1) by

$$y_t = Z_t \alpha_t + X_t \beta + \varepsilon_t, \tag{6.1}$$

where $X_t = (x_{1,t}, \ldots, x_{k,t})$ is a $p \times k$ matrix of explanatory variables and β is a
$k \times 1$ vector of unknown regression coefficients which we assume are constant

over time and which we wish to estimate. We will not discuss here time-varying
regression coefficients because they can be included as part of the state vector α_t
in an obvious way as in §3.6 and are then dealt with by the standard Kalman filter
and smoother. There are two ways in which the inclusion of regression effects
with fixed coefficients can be handled. First, we may include the coefficient vector
β in the state vector. Alternatively, and particularly on occasions where we wish
to keep the dimensionality of the state vector as low as possible, we can use the
augmented Kalman filter and smoother. Both solutions will be discussed in the
next two sections. Different types of residuals exist when regression variables are
included in the model. We show in §6.2.4 how to calculate them within the two
different solutions.

6.2.2   INCLUSION OF COEFFICIENT VECTOR IN STATE VECTOR

The state space model in which the constant coefficient vector β in (6.1) is included
in the state vector has the form

$$y_t = [Z_t \quad X_t] \begin{pmatrix} \alpha_t \\ \beta_t \end{pmatrix} + \varepsilon_t,$$

$$\begin{pmatrix} \alpha_{t+1} \\ \beta_{t+1} \end{pmatrix} = \begin{bmatrix} T_t & 0 \\ 0 & I_k \end{bmatrix} \begin{pmatrix} \alpha_t \\ \beta_t \end{pmatrix} + \begin{bmatrix} R_t \\ 0 \end{bmatrix} \eta_t,$$

for $t = 1, \ldots, n$. In the initial state vector, β_1 can be taken as diffuse or fixed. In
the diffuse case the model for the initial state vector is

$$\begin{pmatrix} \alpha_1 \\ \beta_1 \end{pmatrix} \sim N \left\{ \begin{pmatrix} a \\ 0 \end{pmatrix}, \kappa \begin{bmatrix} P_\infty & 0 \\ 0 & I_k \end{bmatrix} + \begin{bmatrix} P_* & 0 \\ 0 & 0 \end{bmatrix} \right\},$$

where $\kappa \to \infty$; see also §5.6.4 where we give the initial state vector for the
regression model with ARMA errors. We attach suffixes to the β's purely for
convenience in the state space formulation since $\beta_{t+1} = \beta_t = \beta$. The exact initial
Kalman filter (5.19) and the Kalman filter (4.13) can be applied straightforwardly
to this enlarged state space model to obtain the estimate of β. The enlargement of
the state space model will not cause much extra computing because of the sparse
nature of the system matrices.

6.2.3   REGRESSION ESTIMATION BY AUGMENTATION

Another method of estimating β is by augmentation of the Kalman filter. This
technique is essentially the same as was used in the augmented Kalman filter of
§5.7. We will give details of this approach on the assumption that the initial state
vector does not contain diffuse elements. The likelihood function in terms of β is
constructed by applying the Kalman filter to the variables $y_t, x_{1,t}, \ldots, x_{k,t}$ in turn.
Each of the variables $x_{1,t}, \ldots, x_{k,t}$ is treated in the Kalman filter as the observation
vector with the same variance elements as used for y_t. Denote the resulting one-
step forecast errors by $v_t^*, x_{1,t}^*, \ldots, x_{k,t}^*$, respectively. Since the filtering operations

are linear, the one-step forecast errors for the series $y_t - X_t\beta$ are given by $v_t = v_t^* - X_t^*\beta$ where $X_t^* = (x_{1,t}^* \ldots x_{k,t}^*)$. It should be noted that the $k + 1$ Kalman filters are the same except that the values for v_t and a_t in (4.13) are different. We can therefore combine these filters into an augmented Kalman filter where we replace vector y_t by matrix (y_t, X_t) to obtain the 'innovations' matrix (v_t^*, X_t^*); this is analogous to the augmented Kalman filter described in §5.7. The likelihood may be written as

$$\log L(y|\beta) = \text{constant} - \frac{1}{2}\sum_{t=1}^{n} v_t' F_t^{-1} v_t$$

$$= \text{constant} - \frac{1}{2}\sum_{t=1}^{n} (v_t^* - X_t^*\beta)' F_t^{-1} (v_t^* - X_t^*\beta).$$

Maximising this with respect to β gives the generalised least squares estimates $\hat{\beta}$ and $\text{Var}(\hat{\beta})$ where

$$\hat{\beta} = \left(\sum_{t=1}^{n} X_t^{*\prime} F_t^{-1} X_t^*\right)^{-1} \sum_{t=1}^{n} X_t^{*\prime} F_t^{-1} v_t^*, \qquad \text{Var}(\hat{\beta}) = \left(\sum_{t=1}^{n} X_t^{*\prime} F_t^{-1} X_t^*\right)^{-1}.$$

$$(6.2)$$

In the case where the initial state vector contains diffuse elements we can extend the augmented Kalman filter for δ as shown in §5.7. However, we ourselves prefer to use the exact initial Kalman filter for dealing with δ. The equations for $P_{*,t}$ and $P_{\infty,t}$ of the exact initial Kalman filter are not affected since they do not depend on the data. The update for the augmented state vector is given by the equations

$$(v_t^*, x_{1,t}^*, \ldots, x_{k,t}^*) = (y_t, x_{1,t}, \ldots, x_{k,t}) - Z_t(a_{a,t}, A_{x,t}),$$
$$(a_{a,t+1}, A_{x,t+1}) = T_t(a_{a,t}, A_{x,t}) + K_t^{(0)}(v_t^*, x_{1,t}^*, \ldots, x_{k,t}^*),$$
$$(6.3)$$

for $t = 1, \ldots, d$ with $(a_{a,1}, A_{x,1}) = (a, 0, \ldots, 0)$ and where $K_t^{(0)}$ is defined in §5.2. Note that F_t^{-1} in (6.2) must be replaced by zero or $F_{*,t}^{-1}$ depending on the value for $F_{\infty,t}$ in the exact initial Kalman filter. Overall, the treatment given in the previous section where we include β in the state vector, treat δ and β as diffuse and then apply the exact initial Kalman filter is conceptually simpler, though it may not be as efficient computationally for large models.

6.2.4 LEAST SQUARES AND RECURSIVE RESIDUALS

By considering the measurement equation (6.1) we define two different types of residuals following Harvey (1989, §7.4.1): recursive residuals and least squares residuals. The first type are defined as

$$v_t = y_t - Z_t a_t - X_t \hat{\beta}_{t-1}, \qquad t = d + 1, \ldots, n,$$

where $\hat{\beta}_{t-1}$ is the maximum likelihood estimate of β given Y_{t-1}. The residuals v_t are computed easily by including β in the state vector with α_1 diffuse since the filtered state vector of the enlarged model in §6.2.2 contains $\hat{\beta}_{t-1}$. The augmentation method can of course also evaluate v_t but it needs to compute $\hat{\beta}_{t-1}$ at each time point which is not computationally efficient. Note that the residuals v_t are serially uncorrelated. The least squares residuals are given by

$$v_t^+ = y_t - Z_t a_t - X_t \hat{\beta}, \qquad t = d + 1, \ldots, n,$$

where $\hat{\beta}$ is the maximum likelihood estimate of β based on the entire series, so $\hat{\beta} = \hat{\beta}_n$. For the case where the method of §6.2.2 is used to compute $\hat{\beta}$, we require two Kalman filters: one for the enlarged model to compute $\hat{\beta}$ and a Kalman filter for the constructed measurement equation $y_t - X_t \hat{\beta} = Z_t \alpha_t + \varepsilon_t$ whose 'innovations' v_t are in fact v_t^+. The same applies to the method of §6.2.3, except that $\hat{\beta}$ is computed using (6.2). The least squares residuals are correlated due to the presence in them of $\hat{\beta}$, which is calculated from the whole sample.

Both sets of residuals can be used for diagnostic purposes. For these purposes the residuals v_t have the advantage of being serially uncorrelated whereas the residuals v_t^+ have the advantage of being based on the estimate $\hat{\beta}$ calculated from the whole sample. For further discussion we refer to Harvey (1989, §7.4.1).

6.3   Square root filter and smoother

6.3.1   INTRODUCTION

In this section we deal with the situation where, because of rounding errors and matrices being close to singularity, the possibility arises that the calculated value of P_t is negative definite, or close to this, giving rise to unacceptable rounding errors. From (4.13), the state variance matrix P_t is updated by the Kalman filter equations

$$\begin{aligned}
F_t &= Z_t P_t Z_t' + H_t, \\
K_t &= T_t P_t Z_t' F_t^{-1}, \\
P_{t+1} &= T_t P_t L_t' + R_t Q_t R_t' \\
&= T_t P_t T_t' + R_t Q_t R_t' - K_t F_t K_t'.
\end{aligned} \qquad (6.4)$$

It can happen that the calculated value of P_t becomes negative definite when, for example, erratic changes occur in the system matrices over time. The problem can be avoided by using a transformed version of the Kalman filter called the *square root filter*. However, the amount of computation required is substantially larger than that required for the standard Kalman filter. The square root filter is based on orthogonal lower triangular transformations for which we can use Givens rotation techniques. The standard reference to square root filtering is Morf and Kailath (1975).

6.3.2 SQUARE ROOT FORM OF VARIANCE UPDATING

Define the partitioned matrix U_t by

$$U_t = \begin{bmatrix} Z_t \tilde{P}_t & \tilde{H}_t & 0 \\ T_t \tilde{P}_t & 0 & R_t \tilde{Q}_t \end{bmatrix}, \tag{6.5}$$

where

$$P_t = \tilde{P}_t \tilde{P}'_t, \qquad H_t = \tilde{H}_t \tilde{H}'_t, \qquad Q_t = \tilde{Q}_t \tilde{Q}'_t,$$

in which the matrices $\tilde{P}_t$, $\tilde{H}_t$ and $\tilde{Q}_t$ are lower triangular matrices. It follows that

$$U_t U'_t = \begin{bmatrix} F_t & Z_t P_t T'_t \\ T_t P_t Z'_t & T_t P_t T'_t + R_t Q_t R'_t \end{bmatrix}. \tag{6.6}$$

The matrix U_t can be transformed to a lower triangular matrix using the orthogonal matrix G such that $GG' = I_{m+p+r}$. Note that a lower triangular matrix for a rectangular matrix such as U_t, where the number of columns exceeds the number of rows, is defined as a matrix of the form $[A \quad 0]$ where A is a square and lower triangular matrix. Postmultiplying by G we have

$$U_t G = U_t^*, \tag{6.7}$$

and $U_t^* U_t^{*\prime} = U_t U'_t$ as given by (6.6). The lower triangular rectangular matrix U_t^* has the same dimensions as U_t and can be represented as the partitioned matrix

$$U_t^* = \begin{bmatrix} U_{1,t}^* & 0 & 0 \\ U_{2,t}^* & U_{3,t}^* & 0 \end{bmatrix},$$

where $U_{1,t}^*$ and $U_{3,t}^*$ are lower triangular square matrices. It follows that

$$\begin{aligned} U_t^* U_t^{*\prime} &= \begin{bmatrix} U_{1,t}^* U_{1,t}^{*\prime} & U_{1,t}^* U_{2,t}^{*\prime} \\ U_{2,t}^* U_{1,t}^{*\prime} & U_{2,t}^* U_{2,t}^{*\prime} + U_{3,t}^* U_{3,t}^{*\prime} \end{bmatrix} \\ &= \begin{bmatrix} F_t & Z_t P_t T'_t \\ T_t P_t Z'_t & T_t P_t T'_t + R_t Q_t R'_t \end{bmatrix}, \end{aligned}$$

from which we deduce that

$$\begin{aligned} U_{1,t}^* &= \tilde{F}_t, \\ U_{2,t}^* &= T_t P_t Z'_t \tilde{F}_t^{\prime -1} = K_t \tilde{F}_t, \end{aligned}$$

where $F_t = \tilde{F}_t \tilde{F}'_t$ and $\tilde{F}_t$ is lower triangular. It is remarkable to find that $U_{3,t}^* = \tilde{P}_{t+1}$ since

$$\begin{aligned} U_{3,t}^* U_{3,t}^{*\prime} &= T_t P_t T'_t + R_t Q_t R'_t - U_{2,t}^* U_{2,t}^{*\prime} \\ &= T_t P_t T'_t + R_t Q_t R'_t - K_t F_t K'_t \\ &= P_{t+1}. \end{aligned}$$

which follows from (6.4). Thus by transforming U_t in (6.5) to a lower triangular matrix we obtain $\tilde{P}_{t+1}$; this operation can thus be regarded as a square root recursion for P_t. The update for the state vector a_t can be easily incorporated using

$$
\begin{aligned}
a_{t+1} &= T_t a_t + K_t v_t, \\
&= T_t a_t + T_t P_t Z_t' \tilde{F}_t'^{-1} \tilde{F}_t^{-1} v_t \\
&= T_t a_t + U_{2,t}^* U_{1,t}^{*-1} v_t,
\end{aligned}
$$

where $v_t = y_t - Z_t a_t$. Note that the inverse of $U_{1,t}^*$ is easy to calculate since it is a lower triangular matrix.

6.3.3   GIVENS ROTATIONS

Matrix G can be any orthogonal matrix which transforms U_t to a lower triangular matrix. Many different techniques can be used to achieve this objective; for example, Golub and Van Loan (1997) give a detailed treatment of the Householder and Givens matrices for the purpose. We will give here a short description of the latter. The orthogonal 2×2 matrix

$$
G_2 = \begin{bmatrix} c & s \\ -s & c \end{bmatrix},
\tag{6.8}
$$

with $c^2 + s^2 = 1$ is the key to Givens transformations. It is used to transform the vector

$$
x = (x_1 \quad x_2),
$$

into a vector in which the second element is zero, that is

$$
y = x G_2 = (y_1 \quad 0),
$$

by taking

$$
c = \frac{x_1}{\sqrt{x_1^2 + x_2^2}}, \qquad s = -\frac{x_2}{\sqrt{x_1^2 + x_2^2}},
\tag{6.9}
$$

for which $c^2 + s^2 = 1$ and $s x_1 + c x_2 = 0$. Note that $y_1 = c x_1 - s x_2$ and $y G_2' = x G_2 G_2' = x$.

The general Givens matrix G is defined as the identity matrix I_q but with four elements $I_{ii}, I_{jj}, I_{ij}, I_{ji}$ replaced by

$$
\begin{aligned}
G_{ii} = G_{jj} &= c, \\
G_{ij} &= s, \\
G_{ji} &= -s,
\end{aligned}
$$

for $1 \le i < j \le q$ and with c and s given by (6.9) but now enforcing element (i, j) of matrix XG to be zero for all $1 \le i < j \le q$ and for any matrix X. It follows that $GG' = I$ so when Givens rotations are applied repeatedly to create zero blocks in a matrix, the overall transformation matrix is also orthogonal. These properties of the Givens rotations, their computational efficiency and their numerical stability makes them a popular tool to transform nonzero matrices into sparse matrices such as a lower triangular matrix. More details and efficient algorithms for Givens rotations are given by Golub and Van Loan (1997).

6.3.4 SQUARE ROOT SMOOTHING

The backwards recursion (4.30) for N_{t-1} of the basic smoothing equations can also be given in a square root form. These equations use the output of the square root Kalman filter as given by:

$$U_{1,t}^* = \tilde{F}_t,$$
$$U_{2,t}^* = K_t\tilde{F}_t,$$
$$U_{3,t}^* = \tilde{P}_{t+1}.$$

The recursion for N_{t-1} is given by

$$N_{t-1} = Z_t'F_t^{-1}Z_t + L_t'N_tL_t,$$

where

$$F_t^{-1} = (U_{1,t}^*U_{1,t}^{*\prime})^{-1},$$
$$L_t = T_t - U_{2,t}^*U_{1,t}^{*-1}Z_t.$$

We introduce the lower triangular square matrix $\tilde{N}_t$ such that

$$N_t = \tilde{N}_t\tilde{N}_t',$$

and the $m \times (m + p)$ matrix

$$\tilde{N}_{t-1}^* = \begin{bmatrix} Z_t'U_{1,t}^{*-1\prime} & L_t'\tilde{N}_t \end{bmatrix},$$

from which it follows that $N_{t-1} = \tilde{N}_{t-1}^*\tilde{N}_{t-1}^{*\prime}$.

The matrix N_{t-1}^* can be transformed to a lower triangular matrix using some orthogonal matrix G such that $GG' = I_{m+p}$; compare §6.3.2. We have

$$\tilde{N}_{t-1}^*G = \begin{bmatrix} \tilde{N}_{t-1} & 0 \end{bmatrix},$$

such that $N_{t-1} = \tilde{N}_{t-1}^*\tilde{N}_{t-1}^{*\prime} = \tilde{N}_{t-1}\tilde{N}_{t-1}'$. Thus by transforming the matrix N_{t-1}^*, depending on matrices indexed by time t, to a lower triangular matrix we obtain the square root matrix of N_{t-1}. Consequently we have developed a backwards recursion for N_{t-1} in square root form. The backwards recursion for r_{t-1} is not affected apart from the way in which F_t^{-1} and L_t are computed.

6.3.5 SQUARE ROOT FILTERING AND INITIALISATION

The square root formulation could be developed for the exact initial version of the Kalman filter of §5.2. However, the motivation for developing square root versions for filtering and smoothing is to avoid computational numerical instabilities due to rounding errors which are built up during the recursive computations. Since initialisation usually only requires a limited number of d updates, the numerical problems are not substantial during this process. Thus, although use of the square root filter may be important for $t = d, \ldots, n$, it will normally be adequate to employ the standard exact initial Kalman filter as described in §5.2 for $t = 1, \ldots, d$.

The square root formulation of the augmented Kalman filter of §5.7 is more or less the same as the usual Kalman filter because updating equations for F_t, K_t and P_t are unaltered. Some adjustments are required for the updating of the augmented quantities but these can be derived straightforwardly; some details are given by de Jong (1991). Recently, Snyder and Saligari (1996) have proposed a Kalman filter based on Givens rotations, such as the ones developed in §6.3.3, with the fortunate property that diffuse priors $\kappa \to \infty$ can be dealt with explicitly within the Givens operations. Their application of this solution however was limited to filtering only and it does not seem to provide an adequate solution for initial diffuse smoothing.

6.3.6   ILLUSTRATION: LOCAL LINEAR TREND MODEL

For the local linear trend model (3.2) we take

$$
U_t = \begin{bmatrix} \tilde{P}_{11,t} & 0 & \sigma_\varepsilon & 0 & 0 \\ \tilde{P}_{11,t} + \tilde{P}_{21,t} & \tilde{P}_{22,t} & 0 & \sigma_\xi & 0 \\ \tilde{P}_{21,t} & \tilde{P}_{22,t} & 0 & 0 & \sigma_\zeta \end{bmatrix},
$$

which is transformed to the lower triangular matrix

$$
U_t^* = \begin{bmatrix} \tilde{F}_t & 0 & 0 & 0 & 0 \\ K_{11,t}\tilde{F}_t & \tilde{P}_{11,t+1} & 0 & 0 & 0 \\ K_{21,t}\tilde{F}_t & \tilde{P}_{21,t+1} & \tilde{P}_{22,t+1} & 0 & 0 \end{bmatrix}.
$$

The zero elements of U_t^* are created row-wise by a sequence of Givens rotations applied to the matrix U_t. Some zero elements in U_t^* are already zero in U_t and they mostly remain zero within the overall Givens transformation so the number of computations can be limited somewhat.

6.4   Univariate treatment of multivariate series

6.4.1   INTRODUCTION

In Chapters 4 and 5 and in this chapter we have treated the filtering and smoothing of multivariate series in the traditional way by taking the entire observational vectors y_t as the items for analysis. In this section we present an alternative approach in which the elements of y_t are brought into the analysis one at a time, thus in effect converting the multivariate series into a univariate time series. This device not only offers significant computational gains for the filtering and smoothing of the bulk of the series but it also provides substantial simplification of the initialisation process when the initial state vector α_1 is partially or wholly diffuse.

This univariate approach to vector observations was suggested for filtering by Anderson and Moore (1979, §6.4) and for filtering and smoothing longitudinal models by Fahrmeir and Tutz (1994, §8.4). The treatment given by these authors was, however, incomplete and in particular did not deal with the initialisation problem, where the most substantial gains are made. The following discussion of the univariate approach is based on Koopman and Durbin (2000) who gave a complete treatment including a discussion of the initialisation problem.

6.4.2 DETAILS OF UNIVARIATE TREATMENT

Our analysis will be based on the standard model

$$y_t = Z_t \alpha_t + \varepsilon_t, \qquad \alpha_{t+1} = T_t \alpha_t + R_t \eta_t,$$

with $\varepsilon_t \sim N(0, H_t)$ and $\eta_t \sim N(0, Q_t)$ for $t = 1, \ldots, n$. To begin with, let us assume that $\alpha_1 \sim N(a_1, P_1)$ and H_t is diagonal; this latter restriction will be removed later. On the other hand, we introduce two slight generalisations of the basic model: first, we permit the dimensionality of y_t to vary over time by taking the dimension of vector y_t to be $p_t \times 1$ for $t = 1, \ldots, n$; secondly, we do not require the prediction error variance matrix F_t to be non-singular.

Write the observation and disturbance vectors as

$$y_t = \begin{pmatrix} y_{t,1} \\ \vdots \\ y_{t,p_t} \end{pmatrix}, \qquad \varepsilon_t = \begin{pmatrix} \varepsilon_{t,1} \\ \vdots \\ \varepsilon_{t,p_t} \end{pmatrix},$$

and the observation equation matrices as

$$Z_t = \begin{pmatrix} Z_{t,1} \\ \vdots \\ Z_{t,p_t} \end{pmatrix}, \qquad H_t = \begin{pmatrix} \sigma_{t,1}^2 & 0 & 0 \\ 0 & \ddots & 0 \\ 0 & 0 & \sigma_{t,p_t}^2 \end{pmatrix},$$

where $y_{t,i}, \varepsilon_{t,i}$ and $\sigma_{t,i}^2$ are scalars and $Z_{t,i}$ is a $(1 \times m)$ row vector, for $i = 1, \ldots, p_t$. The observation equation for the univariate representation of the model is

$$y_{t,i} = Z_{t,i} \alpha_{t,i} + \varepsilon_{t,i}, \qquad i = 1, \ldots, p_t, \qquad t = 1, \ldots, n, \qquad (6.10)$$

where $\alpha_{t,i} = \alpha_t$. The state equation corresponding to (6.10) is

$$\begin{aligned} \alpha_{t,i+1} &= \alpha_{t,i}, & i &= 1, \ldots, p_t - 1, \\ \alpha_{t+1,1} &= T_t \alpha_{t,p_t} + R_t \eta_t, & t &= 1, \ldots, n, \end{aligned} \qquad (6.11)$$

where the initial state vector $\alpha_{1,1} = \alpha_1 \sim N(a_1, P_1)$.

Define

$$a_{t,1} = E(\alpha_{t,1} | Y_{t-1}), \qquad P_{t,1} = \text{Var}(\alpha_{t,1} | Y_{t-1}),$$

and

$$\begin{aligned} a_{t,i} &= E(\alpha_{t,i} | Y_{t-1}, y_{t,1}, \ldots, y_{t,i-1}), \\ P_{t,i} &= \text{Var}(\alpha_{t,i} | Y_{t-1}, y_{t,1}, \ldots, y_{t,i-1}), \end{aligned}$$

for $i = 2, \ldots, p_t$. By treating the vector series $y_1, \ldots, y_n$ as the scalar series

$$y_{1,1}, \ldots, y_{1,p_1}, y_{2,1}, \ldots, y_{n,p_n},$$

the filtering equations (4.13) can be written as

$$a_{t,i+1} = a_{t,i} + K_{t,i} v_{t,i}, \qquad P_{t,i+1} = P_{t,i} - K_{t,i} F_{t,i} K'_{t,i}, \qquad (6.12)$$

where

$$v_{t,i} = y_{t,i} - Z_{t,i} a_{t,i}, \qquad F_{t,i} = Z_{t,i} P_{t,i} Z'_{t,i} + \sigma_{t,i}^2, \qquad K_{t,i} = P_{t,i} Z'_{t,i} F_{t,i}^{-1}, \qquad (6.13)$$

for $i = 1, \ldots, p_t$ and $t = 1, \ldots, n$. This formulation has $v_{t,i}$ and $F_{t,i}$ as scalars and $K_{t,i}$ as a column vector. The transition from time t to time $t + 1$ is achieved by the relations

$$a_{t+1,1} = T_t a_{t,p_t+1}, \qquad P_{t+1,1} = T_t P_{t,p_t+1} T'_t + R_t Q_t R'_t. \qquad (6.14)$$

The values $a_{t+1,1}$ and $P_{t+1,1}$ are the same as the values a_{t+1} and P_{t+1} computed by the standard Kalman filter.

It is important to note that the elements of the innovation vector v_t are not the same as $v_{t,i}$, for $i = 1, \ldots, p_t$; only the first element of v_t is equal to $v_{t,1}$. The same applies to the diagonal elements of the variance matrix F_t and the variances $F_{t,i}$, for $i = 1, \ldots, p_t$; only the first diagonal element of F_t is equal to $F_{t,1}$. It should be emphasised that there are models for which $F_{t,i}$ can be zero, for example the case where y_t is a multinomial observation with all cell counts included in y_t. This indicates that $y_{t,i}$ is linearly dependent on previous observations for some i. In this case,

$$\begin{aligned} a_{t,i+1} &= \mathrm{E}(\alpha_{t,i+1} | Y_{t-1}, y_{t,1}, \ldots, y_{t,i}) \\ &= \mathrm{E}(\alpha_{t,i+1} | Y_{t-1}, y_{t,1}, \ldots, y_{t,i-1}) = a_{t,i}, \end{aligned}$$

and similarly $P_{t,i+1} = P_{t,i}$. The contingency is therefore easily dealt with.

The basic smoothing recursions (4.48) for the standard state space model can be reformulated for the univariate series

$$y_{1,1}, \ldots, y_{1,p_t}, y_{2,1}, \ldots, y_{n,p_n},$$

as

$$\begin{aligned} r_{t,i-1} &= Z'_{t,i} F_{t,i}^{-1} v_{t,i} + L'_{t,i} r_{t,i}, & N_{t,i-1} &= Z'_{t,i} F_{t,i}^{-1} Z_{t,i} + L'_{t,i} N_{t,i} L_{t,i}, \\ r_{t-1,p_t} &= T'_{t-1} r_{t,0}, & N_{t-1,p_t} &= T'_{t-1} N_{t,0} T_{t-1}, \end{aligned}$$
$$(6.15)$$

where $L_{t,i} = I_m - K_{t,i} Z_{t,i} F_{t,i}^{-1}$, for $i = p_t, \ldots, 1$ and $t = n, \ldots, 1$. The initialisations are $r_{n,p_n} = 0$ and $N_{n,p_n} = 0$. The equations for r_{t-1,p_t} and N_{t-1,p_t} do not apply for $t = 1$. The values for $r_{t,0}$ and $N_{t,0}$ are the same as the values for the smoothing quantities r_{t-1} and N_{t-1} of the standard smoothing equations, respectively.

The smoothed state vector $\hat{\alpha}_t = \mathrm{E}(\alpha_t | y)$ and the variance error matrix $V_t = \mathrm{Var}(\alpha_t | y)$, together with other related smoothing results for the transition equation, are computed by using the standard equations (4.26) and (4.31) with

$$a_t = a_{t,1}, \qquad P_t = P_{t,1}, \qquad r_{t-1} = r_{t,0}, \qquad N_{t-1} = N_{t,0}.$$

Finally, the smoothed estimators for the observation disturbances $\varepsilon_{t,i}$ of (6.10) follow directly from our approach and are given by

$$\widehat{\varepsilon}_{t,i} = \sigma_{t,i}^2 F_{t,i}^{-1}(v_{t,i} - K_{t,i}'r_{t,i}),$$
$$\mathrm{Var}(\widehat{\varepsilon}_{t,i}) = \sigma_{t,i}^4 F_{t,i}^{-2}(F_{t,i} + K_{t,i}'N_{t,i}K_{t,i}).$$

Similar formulae can also be developed for the simulation smoother of §4.7 after conversion to univariate models.

6.4.3 CORRELATION BETWEEN OBSERVATION EQUATIONS

For the case where H_t is not diagonal, the univariate representation of the state space model (6.10) does not apply due to the correlations between the $\varepsilon_{t,i}$'s. In this situation we can pursue two different approaches. Firstly, we can put the disturbance vector ε_t into the state vector. For the observation equation of (3.1) define

$$\bar{\alpha}_t = \begin{pmatrix} \alpha_t \\ \varepsilon_t \end{pmatrix}, \qquad \bar{Z}_t = \begin{pmatrix} Z_t & I_{P_t} \end{pmatrix},$$

and for the state equation define

$$\bar{\eta}_t = \begin{pmatrix} \eta_t \\ \varepsilon_t \end{pmatrix}, \qquad \bar{T}_t = \begin{pmatrix} T_t & 0 \\ 0 & 0 \end{pmatrix},$$

$$\bar{R}_t = \begin{pmatrix} R_t & 0 \\ 0 & I_{P_t} \end{pmatrix}, \qquad \bar{Q}_t = \begin{pmatrix} Q_t & 0 \\ 0 & H_t \end{pmatrix},$$

leading to

$$y_t = \bar{Z}_t\bar{\alpha}_t, \qquad \bar{\alpha}_{t+1} = \bar{T}_t\bar{\alpha}_t + \bar{R}_t\bar{\eta}_t, \qquad \bar{\eta}_t \sim N(0, \bar{Q}_t),$$

for $t = 1, \ldots, n$. We then proceed with the same technique as for the case where H_t is diagonal by treating each element of the observation vector individually. The second approach is to transform the observations. In the case where H_t is not diagonal, we diagonalise it by the Cholesky decomposition

$$H_t = C_t H_t^* C_t',$$

where H_t^* is diagonal and C_t is lower triangular with ones on the diagonal. By transforming the observations, we obtain the observation equation

$$y_t^* = Z_t^* \alpha_t + \varepsilon_t^*, \qquad \varepsilon_t^* \sim N(0, H_t^*),$$

where $y_t^* = C_t^{-1}y_t$, $Z_t^* = C_t^{-1}Z_t$ and $\varepsilon_t^* = C_t^{-1}\varepsilon_t$. Since C_t is a lower triangular matrix, it is easy to compute its inverse. The state vector is not affected by the transformation. Since the elements of ε_t^* are independent we can treat the series y_t^* as a univariate series in the above way.

Table 6.1. Percentage savings for filtering using univariate approach.

State dim.	Obs. dim.	$p = 1$	$p = 2$	$p = 3$	$p = 5$	$p = 10$	$p = 20$
$m = 1$		0	39	61	81	94	98
$m = 2$		0	27	47	69	89	97
$m = 3$		0	21	38	60	83	95
$m = 5$		0	15	27	47	73	90
$m = 10$		0	8	16	30	54	78
$m = 20$		0	5	9	17	35	58

These two approaches for correlated observation disturbances are complementary. The first method has the drawback that the state vector can become large. The second method is illustrated in §6.4.5 where we show that simultaneously transforming the state vector can also be convenient.

6.4.4   COMPUTATIONAL EFFICIENCY

The main motivation for this 'univariate' approach to filtering and smoothing for multivariate state space models is computational efficiency. This approach avoids the inversion of matrix F_t and two matrix multiplications. Also, the implementation of the recursions is more straightforward. Table 6.1 shows that the percentage savings in the number multiplications for filtering using the univariate approach compared to the standard approach are considerable. The calculations concerning the transition are not taken into account in the calculations for this table because matrix T_t is usually sparse with most elements equal to zero or unity.

Table 6.2 presents the considerable percentage savings in the number of multiplications for state smoothing compared to the standard multivariate approach. Again, the computations involving the transition matrix T_t are not taken into account in compiling these figures.

Table 6.2. Percentage savings for smoothing using univariate approach.

State dim.	Obs. dim.	$p = 1$	$p = 2$	$p = 3$	$p = 5$	$p = 10$	$p = 20$
$m = 1$		0	27	43	60	77	87
$m = 2$		0	22	36	53	72	84
$m = 3$		0	19	32	48	68	81
$m = 5$		0	14	25	40	60	76
$m = 10$		0	9	16	28	47	65
$m = 20$		0	5	10	18	33	51

6.4.5 ILLUSTRATION: VECTOR SPLINES

We now consider the application of the univariate approach to vector splines. The generalisation of smoothing splines of Hastie and Tibshirani (1990) to the multivariate case is considered by Fessler (1991) and Yee and Wild (1996). The vector spline model is given by

$$y_i = \theta(t_i) + \varepsilon_i, \qquad \mathrm{E}(\varepsilon_i) = 0, \qquad \mathrm{Var}(\varepsilon_i) = \Sigma_i, \qquad i = 1, \ldots, n,$$

where y_i is a $p \times 1$ vector response at scalar t_i, $\theta(\cdot)$ is an arbitrary smooth vector function and errors ε_i are mutually uncorrelated. The variance matrix Σ_i is assumed to be known and is usually constant for varying i. This is a generalisation of the univariate problem considered in §3.11.2. The standard method of estimating the smooth vector function is by minimising the generalized least squares criterion

$$\sum_{i=1}^{n} \{y_i - \theta(t_i)\}' \, \Sigma_i^{-1} \, \{y_i - \theta(t_i)\} + \sum_{j=1}^{p} \lambda_j \int \theta_j''(t)^2 \, dt,$$

where the non-negative smoothing parameter λ_j determines the smoothness of the jth smooth function $\theta_j(\cdot)$ of vector $\theta(\cdot)$ for $j = 1, \ldots, p$. Note that $t_{i+1} > t_i$ for $i = 1, \ldots, n - 1$ and $\theta_j''(t)$ denotes the second derivative of $\theta_j(t)$ with respect to t. In the same way as for the univariate case in (3.38), we use the discrete model

$$y_i = \mu_i + \varepsilon_i,$$

$$\mu_{i+1} = \mu_i + \delta_i v_i + \eta_i, \qquad \mathrm{Var}(\eta_i) = \frac{\delta_i^3}{3} \Lambda,$$

$$v_{i+1} = v_i + \zeta_i, \qquad \mathrm{Var}(\zeta_i) = \delta_i \Lambda, \qquad \mathrm{Cov}(\eta_i, \zeta_i) = \frac{\delta_i^2}{2} \Lambda,$$

with vector $\mu_i = \theta(t_i)$, scalar $\delta_i = t_{i+1} - t_i$ and diagonal matrix $\Lambda = \mathrm{diag}(\lambda_1, \ldots, \lambda_p)$. This model is equivalent to the continuous-time representation of the multivariate local linear trend model with no disturbance vector for the level equation; see Harvey (1989, Chapter 8). In the case of $\Sigma_i = \Sigma$ and with the Cholesky decomposition $\Sigma = CDC'$ where matrix C is lower triangular and matrix D is diagonal, we obtain the transformed model

$$y_i^* = \mu_i^* + \varepsilon_i^*,$$

$$\mu_{i+1}^* = \mu_i^* + \delta_i v_i^* + \eta_i^*, \qquad \mathrm{Var}(\eta_i) = \frac{\delta_i^3}{3} Q,$$

$$v_{i+1}^* = v_i^* + \zeta_i^*, \qquad \mathrm{Var}(\zeta_i) = \delta_i Q, \qquad \mathrm{Cov}(\eta_i, \zeta_i) = \frac{\delta_i^2}{2} Q,$$

with $y_i^* = C^{-1} y_i$ and $\mathrm{Var}(\varepsilon_i^*) = D$ where we have used (3.39). Furthermore, we have $\mu_i^* = C^{-1} \mu_i$, $v_i^* = C^{-1} v_i$ and $Q = C^{-1} \Lambda C'^{-1}$. The Kalman filter smoother algorithm provides the fitted smoothing spline. The untransformed model and the transformed model can both be handled by the univariate strategy of filtering and smoothing discussed in this section. The advantage of the transformed model is that ε_i^* can be excluded from the state vector which is not possible for the untransformed model because $\text{Var}(\varepsilon_i) = \Sigma_i$ is not necessarily diagonal; see the discussion in §6.4.3.

The percentage computational saving of the univariate approach for spline smoothing depends on the size p. The state vector dimension for the transformed model is $m = 2p$ so the percentage saving in computing for filtering is 30 if $p = 5$ and it is 35 if $p = 10$; see Table 6.1. The percentages for smoothing are 28 and 33, respectively; see Table 6.2.

6.5   Filtering and smoothing under linear restrictions

We now consider how to carry out filtering and smoothing subject to a set of time-varying linear restrictions on the state vector of the form

$$R_t^* \alpha_t = r_t^*, \qquad t = 1, \ldots, n, \tag{6.16}$$

where the matrix R_t^* and the vector r_t^* are known and where the number of rows in R_t^* can vary with t. Although linear restrictions on the state vector can often be easily dealt with by re-specifying the elements of the state vector, an alternative is to proceed as follows. To impose the restrictions (6.16) we augment the obervation equation as

$$\begin{pmatrix} y_t \\ r_t^* \end{pmatrix} = \begin{bmatrix} Z_t \\ R_t^* \end{bmatrix} \alpha_t + \begin{pmatrix} \varepsilon_t \\ 0 \end{pmatrix}, \qquad t = 1, \ldots, n. \tag{6.17}$$

For this augmented model, filtering and smoothing will produce estimates a_t and $\hat{\alpha}_t$ which are subject to the restrictions $R_t^* a_t = r_t^*$ and $R_t^* \hat{\alpha}_t = r_t^*$; for a discussion of this procedure, see Doran (1992). Equation (6.17) represents a multivariate model whether y_t is univariate or not. This can, however, be converted into a univariate model by the device of §6.4.

6.6   The algorithms of Ssf Pack

6.6.1   INTRODUCTION

SsfPack is a suite of C routines for carrying out computations involving the statistical analysis of univariate and multivariate models in state space form. *SsfPack* allows for a range of different state space forms from simple time-invariant models to complicated time-varying models. Functions are specified which put standard models such as ARIMA and spline models in state space form. Routines are available for filtering, smoothing and simulation smoothing. Ready-to-use functions are provided for standard tasks such as likelihood evaluation, forecasting and signal extraction. The headers of these routines are documented by Koopman, Shephard and Doornik (1999). *SsfPack* can be easily used for implementing, fitting and analysing Gaussian and non-Gaussian models relevant to many areas of time series analysis. In future versions, the exact initial Kalman filter and smoother and the square root filter will become part of *SsfPack*. Routines for dealing with multivariate models after conversion to univariate models will also be provided. A Gaussian illustration is given in §6.6.3. A further discussion of SsfPack in the context of non-Gaussian models is given in §14.6.

6.6.2 THE *SSFPACK* FUNCTIONS

A list of *SsfPack* functions is given below; they are grouped in functions which put specific univariate models into state space form, functions which perform the basic filtering and smoothing operations and functions which execute specific important tasks for state space analysis such as likelihood evaluation. The first column contains the function names, the second column gives reference to the section number of the *SsfPack* documentation of Koopman *et al.* (1999) and the third column describes the function with references to equation or section numbers in this book.

Models in state space form

AddSsfReg	§3.3	adds regression effects (3.27) to state space.
GetSsfArma	§3.1	puts ARMA model (3.15) in state space.
GetSsfReg	§3.3	puts regression model (3.27) in state space.
GetSsfSpline	§3.4	puts cubic spline model (3.43) in state space.
GetSsfStsm	§3.2	puts structural time series model of §3.2.1 in state space.
SsfCombine	§6.2	combines system matrices of two models.
SsfCombineSym	§6.2	combines symmetric system matrices of two models.

General state space algorithms

KalmanFil	§4.3	provides output of the Kalman filter in §4.2.2.
KalmanSmo	§4.4	provides output of the basic smoothing algorithm in §4.3.3.
SimSmoDraw	§4.5	provides a simulated sample such as (4.77).
SimSmoWgt	§4.5	provides output of the simulation smoother (4.75).

Ready-to-use functions

SsfCondDens	§4.6	provides mean or a draw from the conditional density (4.56).
SsfLik	§5.1	provides log-likelihood function (7.3).
SsfLikConc	§5.1	provides concentrated log-likelihood function (2.46).
SsfLikSco	§5.1	provides score vector information (7.15).
SsfMomentEst	§5.2 §5.3	provides output from prediction, forecasting and smoothing.
SsfRecursion	§4.2	provides output of the state space recursion (3.1).

These *SsfPack* functions are documented in Koopman *et al.* (1999). We will
not discuss the functions further but an example of *Ox* code, which utilises the link
with the *SsfPack* library, is given below.

6.6.3   ILLUSTRATION: SPLINE SMOOTHING

In the *Ox* code below we consider the continuous spline smoothing problem which
aims at minimising (3.43) for a given value of λ. The aim of the program is to fit a
spline through the Nile time series of Chapter 2 (see §2.2.2 for details). To illustrate
that the *SsfPack* functions can deal with missing observations we have treated two
parts of the data set as missing. The continuous spline model is easily put in
state space form using the function GetSsfSpline. The smoothing parameter λ
is chosen to take the value 2500 (the function requires the input of $\lambda^{-1} = 0.004$).
We need to compute an estimator for the unknown scalar value of σ_ζ^2 in (3.44)
which can be obtained using the function SsfLik. After rescaling, the estimated
spline function using filtering (ST_FIL) and smoothing (ST_SMO) is computed using
the function SsfMomentEst. The output is presented in Figure 6.1 and shows
the filtered and smoothed estimates of the spline function. The two diagrams
illustrate the point that filtering can be interpreted as extrapolation and that when
observations are missing, smoothing is in fact interpolation.

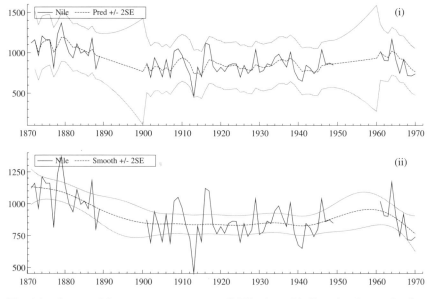

Fig. 6.1.   Output of *Ox* program spline.ox: (i) Nile data with filtered estimate of spline
function with 95% confidence interval; (ii) Nile data with smoothed estimate of spline
function with 95% confidence interval.

spline.ox

```
#include <oxstd.h>
#include <oxdraw.h>
#include <oxfloat.h>
#include <packages/ssfpack/ssfpack.h>

main()
{
  decl mdelta, mphi, momega, msigma, myt, mfil, msmo, cm, dlik, dvar;

  myt = loadmat("Nile.dat")';
  myt[][1890-1871:1900-1871] = M_NAN;     // set 1890..1900 to missing
  myt[][1950-1871:1960-1871] = M_NAN;     // set 1950..1960 to missing

  GetSsfSpline(0.004, <>, &mphi, &momega, &msigma); // SSF for spline
  SsfLik(&dlik, &dvar, myt, mphi, momega);                 // need dVar
  cm = columns(mphi);                             // dimension of state
  momega *= dvar;                           // set correct scale of Omega
  SsfMomentEst(ST_FIL, &mfil, myt, mphi, momega);
  SsfMomentEst(ST_SMO, &msmo, myt, mphi, momega);

  // NB: first filtered estimator doesn't exist
  DrawTMatrix(0, myt, {"Nile"}, 1871, 1, 1);
  DrawTMatrix(0, mfil[cm][1:], {"Pred +/- 2SE"}, 1872, 1, 1, 0, 3);
  DrawZ(sqrt(mfil[2*cm+1][1:]), "", ZMODE_BAND, 2.0, 14);
  DrawTMatrix(1, myt, {"Nile"}, 1871, 1, 1);
  DrawTMatrix(1, msmo[cm][], {"Smooth +/- 2SE"}, 1871, 1, 1, 0, 3);
  DrawZ(sqrt(msmo[2*cm+1][]), "", ZMODE_BAND, 2.0, 14);
  ShowDrawWindow();
}
```

7
Maximum likelihood estimation

7.1 Introduction

In virtually all applications in practical work the models depend on unknown parameters. In this chapter we consider the estimation of these parameters by maximum likelihood. For the linear Gaussian model we shall show that the likelihood can be calculated by a routine application of the Kalman filter, even when the initial state vector is fully or partially diffuse. We go on to consider how the loglikelihood can be maximised by means of iterative numerical procedures. An important part in this process is played by the score vector and we show how this is calculated, both for the case where the initial state vector has a known distribution and for the diffuse case. A useful device for maximisation of the loglikelihood in some cases, particularly in the early stages of maximisation, is the EM algorithm; we give details of this for the linear Gaussian model. We go on to consider biases in estimates due to errors in parameter estimation. The chapter ends with a discussion of some questions of goodness-of-fit and diagnostic checks.

7.2 Likelihood evaluation

7.2.1 LOGLIKELIHOOD WHEN INITIAL CONDITIONS ARE KNOWN

We first assume that the initial state vector has density $N(a_1, P_1)$ where a_1 and P_1 are known. The likelihood is

$$L(y) = p(y_1, \ldots, y_n) = p(y_1) \prod_{t=2}^{n} p(y_t|Y_{t-1}),$$

where $Y_t = \{y_1, \ldots, y_t\}$. In practice we generally work with the loglikelihood

$$\log L(y) = \sum_{t=1}^{n} \log p(y_t|Y_{t-1}), \qquad (7.1)$$

where $p(y_1|Y_0) = p(y_1)$. For model (3.1), $E(y_t|Y_{t-1}) = Z_t a_t$. Putting $v_t = y_t - Z_t a_t$, $F_t = \text{Var}(y_t|Y_{t-1})$ and substituting $N(Z_t a_t, F_t)$ for $p(y_t|Y_{t-1})$ in (7.1), we obtain

$$\log L(y) = -\frac{np}{2} \log 2\pi - \frac{1}{2} \sum_{t=1}^{n} \left(\log|F_t| + v_t' F_t^{-1} v_t \right). \qquad (7.2)$$

The quantities v_t and F_t are calculated routinely by the Kalman filter (4.13) so $\log L(y)$ is easily computed from the Kalman filter output. We assume that F_t is nonsingular for $t = 1, \ldots, n$. If this condition is not satisfied initially it is usually possible to redefine the model so that it is satisfied. The representation (7.2) of the loglikelihood was first given by Schweppe (1965). Harvey (1989, §3.4) refers to it as the *prediction error decomposition*.

7.2.2 DIFFUSE LOGLIKELIHOOD

We now consider the case where some elements of α_1 are diffuse. As in §5.1, we assume that $\alpha_1 = a + A\delta + R_0\eta_0$ where a is a known constant vector, $\delta \sim N(0, \kappa I_q)$, $\eta_0 \sim N(0, Q_0)$ and $A'R_0 = 0$, giving $\alpha_1 \sim N(a_1, P_1)$ where $P_1 = \kappa P_\infty + P_*$ and $\kappa \to \infty$. From (5.6) and (5.7),

$$F_t = \kappa F_{\infty,t} + F_{*,t} + O(\kappa^{-1}) \quad \text{with} \quad F_{\infty,t} = Z_t P_{\infty,t} Z_t',$$

where by definition of d, $P_{\infty,t} \neq 0$ for $t = 1, \ldots, d$. The number of diffuse elements in α_1 is q which is the dimensionality of vector δ. Thus the loglikelihood (7.2) will contain a term $-\frac{q}{2} \log 2\pi\kappa$ so $\log L(y)$ will not converge as $\kappa \to \infty$. Following de Jong (1991), we therefore define the *diffuse loglikelihood* as

$$\log L_d(y) = \lim_{\kappa \to \infty} \left[\log L(y) + \frac{q}{2} \log \kappa \right]$$

and we work with $\log L_d(y)$ in place of $\log L(y)$ for estimation of unknown parameters in the diffuse case. Similar definitions for the diffuse loglikelihood function have been adopted by Harvey and Phillips (1979) and Ansley and Kohn (1986). As in §5.2, and for the same reasons, we assume that $F_{\infty,t}$ is positive definite or is a zero matrix. We also assume that q is a multiple of p. This covers the important special case of univariate series and is generally satisfied in practice for multivariate series; if not, the series can be dealt with as if it were univariate as in §6.4.

Suppose first that $F_{\infty,t}$ is positive definite and therefore has rank p. From (5.8) and (5.10) we have for $t = 1, \ldots, d$,

$$F_t^{-1} = \kappa^{-1} F_{\infty,t}^{-1} + O(\kappa^{-2}).$$

It follows that

$$
\begin{aligned}
-\log|F_t| = \log|F_t^{-1}| &= \log|\kappa^{-1} F_{\infty,t}^{-1} + O(\kappa^{-2})| \\
&= -p \log \kappa + \log|F_{\infty,t}^{-1} + O(\kappa^{-1})|,
\end{aligned}
$$

and

$$\lim_{\kappa \to \infty} (-\log|F_t| + p \log \kappa) = \log|F_{\infty,t}^{-1}| = -\log|F_{\infty,t}|.$$

Moreover,

$$
\begin{aligned}
\lim_{\kappa \to \infty} v_t' F_t^{-1} v_t &= \lim_{\kappa \to \infty} \left[v_t^{(0)} + \kappa^{-1} v_t^{(1)} + O(\kappa^{-2}) \right]' \left[\kappa^{-1} F_{\infty,t}^{-1} + O(\kappa^{-2}) \right] \\
&\quad \times \left[v_t^{(0)} + \kappa^{-1} v_t^{(1)} + O(\kappa^{-2}) \right] \\
&= 0
\end{aligned}
$$

for $t = 1, \ldots, d$, where $v_t^{(0)}$ and $v_t^{(1)}$ are defined in §5.2.1.

When $F_{\infty,t} = 0$, it follows from §5.2.1 that $F_t = F_{*,t} + O(\kappa^{-1})$ and $F_t^{-1} = F_{*,t}^{-1} + O(\kappa^{-1})$. Consequently,

$$
\lim_{\kappa \to \infty} (-\log|F_t|) = -\log|F_{*,t}| \quad \text{and} \quad \lim_{\kappa \to \infty} v_t' F_t^{-1} v_t = v_t^{(0)'} F_{*,t}^{-1} v_t^{(0)}.
$$

Putting these results together, we obtain the diffuse loglikelihood as

$$
\log L_d(y) = -\frac{np}{2} \log 2\pi - \frac{1}{2} \sum_{t=1}^{d} w_t - \frac{1}{2} \sum_{t=d+1}^{n} \left(\log|F_t| + v_t' F_t^{-1} v_t \right), \quad (7.3)
$$

where

$$
w_t = \begin{cases} \log|F_{\infty,t}|, & \text{if } F_{\infty,t} \text{ is positive definite}, \\ \log|F_{*,t}| + v_t^{(0)'} F_{*,t}^{-1} v_t^{(0)}, & \text{if } F_{\infty,t} = 0, \end{cases}
$$

for $t = 1, \ldots, d$. The expression (7.3) for the diffuse loglikelihood is given by Koopman (1997).

7.2.3   DIFFUSE LOGLIKELIHOOD EVALUATED VIA AUGMENTED KALMAN FILTER

In the notation of §5.7.3, the joint density of δ and y for given κ is

$$
\begin{aligned}
p(\delta, y) &= p(\delta) p(y|\delta) \\
&= p(\delta) \sum_{t=1}^{n} p(v_{\delta,t}) \\
&= (2\pi)^{-(np+q)/2} \kappa^{-q/2} \prod_{t=1}^{n} |F_{\delta,t}|^{-1/2} \\
&\quad \times \exp\left[-\frac{1}{2} \left(\frac{\delta'\delta}{\kappa} + S_{a,n} + 2b_n'\delta + \delta' S_{A,n}\delta \right) \right], \quad (7.4)
\end{aligned}
$$

where $v_{\delta,t}$ is defined in (5.35), b_n and $S_{A,n}$ are defined in (5.40), and $S_{a,n} = \sum_{t=1}^{n} v_{a,t}' F_{\delta,t}^{-1} v_{a,t}$. From (5.41) we have $\bar{\delta}_n = \mathrm{E}(\delta|Y_n) = -(S_{A,n} + \kappa^{-1} I_q)^{-1} b_n$. The exponent of (7.4) can now be rewritten as

$$
-\frac{1}{2} [S_{a,n} + (\delta - \bar{\delta}_n)'(S_{A,n} + \kappa^{-1} I_q)(\delta - \bar{\delta}_n) - \bar{\delta}_n'(S_{A,n} + \kappa^{-1} I_q)\bar{\delta}_n],
$$

as is easily verified. Integrating out δ from $p(\delta, y)$ we obtain the marginal density of y. After taking logs, the loglikelihood appears as

$$\log L(y) = -\frac{np}{2} \log 2\pi - \frac{q}{2} \log \kappa - \frac{1}{2} \log|S_{A,n} + \kappa^{-1} I_q|$$
$$- \frac{1}{2} \sum_{t=1}^{n} \log |F_{\delta,t}| - \frac{1}{2}[S_{a,n} - \bar{\delta}'_n(S_{A,n} + \kappa^{-1} I_q)\bar{\delta}_n]. \quad (7.5)$$

Adding $\frac{q}{2} \log \kappa$ and letting $\kappa \to \infty$ we obtain the diffuse loglikelihood

$$\log L_d(y) = -\frac{np}{2} \log 2\pi - \frac{1}{2} \log|S_{A,n}| - \frac{1}{2} \sum_{t=1}^{n} \log|F_{\delta,t}|$$
$$- \frac{1}{2}(S_{a,n} - b'_n S_{A,n}^{-1} b_n), \quad (7.6)$$

which is due to de Jong (1991). In spite of its very different structure (7.6) necessarily has the same numerical value as (7.3).

It is shown in §5.7.3 that the augmented Kalman filter can be collapsed at time point $t = d$. We could therefore form a partial likelihood based on Y_d for fixed κ, integrate out δ and let $\kappa \to \infty$ as in (7.6). Subsequently we could add the contribution from innovations $v_{d+1}, \ldots, v_n$ obtained from the collapsed Kalman filter. However, we will not give detailed formulae here.

These results were originally derived by de Jong (1988b) and de Jong (1991). The calculations required to compute (7.6) are more complicated than those required to compute (7.3). This is another reason why we ourselves prefer the initialisation technique of §5.2 to the augmentation device of §5.7. A further reason for prefering our computation of (7.3) is given in §7.3.5.

7.2.4 LIKELIHOOD WHEN ELEMENTS OF INITIAL STATE VECTOR ARE FIXED BUT UNKNOWN

Now let us consider the case where δ is treated as fixed. The density of y given δ is, as in the previous section,

$$p(y|\delta) = (2\pi)^{-np/2} \prod_{t=1}^{n} |F_{\delta,t}|^{-1/2} \exp\left[-\frac{1}{2}(S_{a,n} + 2b'_n \delta_n + \delta'_n S_{A,n} \delta_n)\right]. \quad (7.7)$$

The usual way to remove the influence of an unknown parameter vector such as δ from the likelihood is to estimate it by its maximum likelihood estimate, $\hat{\delta}_n$ in this case, and to employ the concentrated loglikelihood $\log L_c(y)$ obtained by substituting $\hat{\delta}_n = -S_{A,n}^{-1} b_n$ for δ in $p(y|\delta)$. This gives

$$\log L_c(y) = -\frac{np}{2} \log(2\pi) - \frac{1}{2} \sum_{t=1}^{n} |F_{\delta,t}| - \frac{1}{2}(S_{a,n} - b'_n S_{A,n}^{-1} b_n). \quad (7.8)$$

Comparing (7.6) and (7.8) we see that the only difference between them is the presence in (7.6) of the term $-\frac{1}{2}\log|S_{A,n}|$. The relation between (7.6) and (7.8) was demonstrated by de Jong (1988b) using a different argument.

Harvey and Shephard (1990) argue that parameter estimation should preferably be based on the loglikelihood function (7.3) for which the initial vector δ is treated as diffuse not fixed. They have shown for the local level model of Chapter 2 that maximising (7.8) with respect to the signal to noise ratio q leads to a much higher probability of estimating q to be zero compared to maximising (7.3). This is undesirable from a forecasting point of view since this results in no discounting of past observations.

7.3 Parameter estimation

7.3.1 INTRODUCTION

So far in this book we have assumed that the system matrices Z_t, H_t, T_t, R_t and Q_t in model (3.1) are known for $t = 1, \ldots, n$. We now consider the more usual situation in which at least some of the elements of these matrices depend on a vector ψ of unknown parameters. We shall estimate ψ by maximum likelihood. To make explicit the dependence of the loglikelihood on ψ we write $\log L(y|\psi)$, $\log L_d(y|\psi)$ and $\log L_c(y|\psi)$. In the diffuse case we shall take it for granted that for models of interest, estimates of ψ obtained by maximising $\log L(y|\psi)$ for fixed κ converge to the estimates obtained by maximising the diffuse loglikelihood $\log L_d(y|\psi)$ as $\kappa \to \infty$.

7.3.2 NUMERICAL MAXIMISATION ALGORITHMS

A wide range of numerical search algorithms are available for maximising the loglikelihood. Many of these are based on Newton's method which solves the equation

$$\partial_1(\psi) = \frac{\partial \log L(y|\psi)}{\partial \psi} = 0, \tag{7.9}$$

using the first-order Taylor series

$$\partial_1(\psi) \simeq \tilde{\partial}_1(\psi) + \tilde{\partial}_2(\psi)(\psi - \tilde{\psi}), \tag{7.10}$$

for some trial value $\tilde{\psi}$, where

$$\tilde{\partial}_1(\psi) = \partial_1(\psi)|_{\psi=\tilde{\psi}}, \qquad \tilde{\partial}_2(\psi) = \partial_2(\psi)|_{\psi=\tilde{\psi}},$$

with

$$\partial_2(\psi) = \frac{\partial^2 \log L(y|\psi)}{\partial \psi \, \partial \psi'}. \tag{7.11}$$

By equating (7.10) to zero we obtain a revised value $\bar{\psi}$ from the expression

$$\bar{\psi} = \tilde{\psi} - \tilde{\partial}_2(\psi)^{-1}\tilde{\partial}_1(\psi).$$

This process is repeated until it converges or until a switch is made to another optimisation method. If the Hessian matrix $\ddot{\partial}_2(\psi)$ is negative definite for all ψ the loglikelihood is said to be concave and a unique maximum exists for the likelihood. The *gradient* $\ddot{\partial}_1(\psi)$ determines the direction of the step taken to the optimum and the Hessian modifies the size of the step. It is possible to overstep the maximum in the direction determined by the vector

$$\tilde{\pi}(\psi) = -\ddot{\partial}_2(\psi)^{-1}\ddot{\partial}_1(\psi),$$

and therefore it is common practice to include a line search along the gradient vector within the optimisation process. We obtain the algorithm

$$\bar{\psi} = \tilde{\psi} + s\tilde{\pi}(\psi),$$

where various methods are available to find the optimum value for s which is usually found to be between 0 and 1. In practice it is often computationally demanding or impossible to compute $\partial_1(\psi)$ and $\partial_2(\psi)$ analytically. Numerical evaluation of $\partial_1(\psi)$ is usually feasible. A variety of computational devices are available to approximate $\partial_2(\psi)$ in order to avoid computing it analytically or numerically. For example, the *STAMP* package of Koopman *et al.* (2000) and the *Ox* matrix programming system of Doornik (1998) both use the so-called BFGS (Broyden-Fletcher-Goldfarb-Shannon) method which approximates the Hessian matrix using a device in which at each new value for ψ a new approximate inverse Hessian matrix is obtained via the recursion

$$\ddot{\partial}_2(\psi)^{-1} = \tilde{\partial}_2(\psi)^{-1} + \left(s + \frac{g'g^*}{\tilde{\pi}(\psi)'g}\right)\frac{\tilde{\pi}(\psi)\tilde{\pi}(\psi)'}{\tilde{\pi}(\psi)'g} - \frac{\tilde{\pi}(\psi)g^{*\prime} + g^*\tilde{\pi}(\psi)'}{\tilde{\pi}(\psi)'g},$$

where g is the difference between the gradient $\tilde{\partial}_1(\psi)$ and the gradient for a trial value of ψ prior to $\tilde{\psi}$ and

$$g^* = \tilde{\partial}_2(\psi)^{-1}g.$$

The BFGS method ensures that the approximate Hessian matrix remains nagative definite. The details and derivations of the Newton's method of optimisation and the BFGS method in particular can be found, for example, in Fletcher (1987).

Model parameters are sometimes constrained. For example, the parameters in the local level model (2.3) must satisfy the constraints $\sigma_\varepsilon^2 \geq 0$ and $\sigma_\eta^2 \geq 0$ with $\sigma_\varepsilon^2 + \sigma_\eta^2 > 0$. However, the introduction of constraints such as these within the numerical procedure is inconvenient and it is preferable that the maximisation is performed with respect to quantities which are unconstrained. For this example we therefore make the transformations $\psi_\varepsilon = \frac{1}{2}\log\sigma_\varepsilon^2$ and $\psi_\eta = \frac{1}{2}\log\sigma_\eta^2$ where $-\infty < \psi_\varepsilon, \psi_\eta < \infty$, thus converting the problem to one in unconstrained maximisation. The parameter vector is $\psi = [\psi_\varepsilon, \psi_\eta]'$. Similarly, if we have a parameter χ which is restricted to the range $[-a, a]$ where a is positive we can make a transformation to ψ_χ for which

$$\chi = \frac{a\psi_\chi}{\sqrt{1+\psi_\chi^2}}, \qquad -\infty < \psi_\chi < \infty.$$

7.3.3 THE SCORE VECTOR

We now consider details of the calculation of the gradient or *score vector*

$$\partial_1(\psi) = \frac{\partial \log L(y|\psi)}{\partial \psi}.$$

As indicated in the last section, this vector is important in numerical maximisation since it specifies the direction in the parameter space along which a search should be made.

We begin with the case where the initial vector α_1 has the distribution $\alpha_1 \sim N(a_1, P_1)$ where a_1 and P_1 are known. Let $p(\alpha, y|\psi)$ be the joint density of α and y, let $p(\alpha|y, \psi)$ be the conditional density of α given y and let $p(y|\psi)$ be the marginal density of y for given ψ. We now evaluate the score vector $\partial \log L(y|\psi)/\partial \psi = \partial \log p(y|\psi)/\partial \psi$ at the trial value $\tilde{\psi}$. We have

$$\log p(y|\psi) = \log p(\alpha, y|\psi) - \log p(\alpha|y, \psi).$$

Let $\tilde{\mathrm{E}}$ denote expectation with respect to density $p(\alpha|y, \tilde{\psi})$. Since $p(y|\psi)$ does not depend on α, taking $\tilde{\mathrm{E}}$ of both sides gives

$$\log p(y|\psi) = \tilde{\mathrm{E}}[\log p(\alpha, y|\psi)] - \tilde{\mathrm{E}}[\log p(\alpha|y, \psi)].$$

To obtain the score vector at $\tilde{\psi}$, we differentiate both sides with respect to ψ and put $\psi = \tilde{\psi}$. Assuming that differentiating under integral signs is legitimate,

$$\tilde{\mathrm{E}}\left[\frac{\partial \log p(\alpha|y, \psi)}{\partial \psi}\bigg|_{\psi=\tilde{\psi}} \right] = \int \frac{1}{p(\alpha|y, \tilde{\psi})} \frac{\partial p(\alpha|y, \psi)}{\partial \psi}\bigg|_{\psi=\tilde{\psi}} p(\alpha|y, \tilde{\psi})\, d\alpha$$

$$= \frac{\partial}{\partial \psi} \int p(\alpha|y, \psi)\, d\alpha\bigg|_{\psi=\tilde{\psi}} = 0.$$

Thus

$$\frac{\partial \log p(y|\psi)}{\partial \psi}\bigg|_{\psi=\tilde{\psi}} = \tilde{\mathrm{E}}\left[\frac{\partial \log p(\alpha, y|\psi)}{\partial \psi} \right]\bigg|_{\psi=\tilde{\psi}}.$$

With substitutions $\eta_t = R_t'(\alpha_{t+1} - T_t\alpha_t)$ and $\varepsilon_t = y_t - Z_t\alpha_t$ and putting $\alpha_1 - a_1 = \eta_0$ and $P_1 = Q_0$, we obtain

$\log p(\alpha, y|\psi) = \text{constant}$

$$-\frac{1}{2}\sum_{t=1}^{n}\left(\log|H_t| + \log|Q_{t-1}| + \varepsilon_t' H_t^{-1}\varepsilon_t + \eta_{t-1}' Q_{t-1}^{-1}\eta_{t-1}\right).$$

$$(7.12)$$

On taking the expectation $\tilde{\mathrm{E}}$ and differentiating with respect to ψ, this gives the score vector at $\psi = \tilde{\psi}$,

$$\frac{\partial \log L(y|\psi)}{\partial \psi}\bigg|_{\psi=\tilde{\psi}} = -\frac{1}{2}\frac{\partial}{\partial \psi}\sum_{t=1}^{n}\Big[\big(\log|H_t| + \log|Q_{t-1}|$$

$$+ \mathrm{tr}\big[\{\hat{\varepsilon}_t\hat{\varepsilon}_t' + \mathrm{Var}(\varepsilon_t|y)\}H_t^{-1}\big]$$

$$+ \mathrm{tr}\big[\{\hat{\eta}_{t-1}\hat{\eta}_{t-1}' + \mathrm{Var}(\eta_{t-1}|y)\}Q_{t-1}^{-1}\big]\big|\psi\big)\big|_{\psi=\tilde{\psi}}\Big], \quad (7.13)$$

where $\hat{\varepsilon}_t$, $\hat{\eta}_{t-1}$, $\mathrm{Var}(\varepsilon_t|y)$ and $\mathrm{Var}(\eta_{t-1}|y)$ are obtained for $\psi = \tilde{\psi}$ as in §4.4.

Only the terms in H_t and Q_t in (7.13) require differentiation with respect to ψ. Since in practice H_t and Q_t are often simple functions of ψ, this means that the score vector is often easy to calculate, which can be a considerable advantage in numerical maximisation of the loglikelihood. A similar technique can be developed for the system matrices Z_t and T_t but this requires more computations which involve the state smoothing recursions. Koopman and Shephard (1992), to whom the result (7.13) is due, therefore conclude that the score values for ψ associated with system matrices Z_t and T_t can be evaluated better numerically than analytically.

We now consider the diffuse case. In §5.1 we have specified the initial state vector α_1 as

$$\alpha_1 = a + A\delta + R_0\eta_0, \qquad \delta \sim \mathrm{N}(0, \kappa I_q), \qquad \eta_0 \sim \mathrm{N}(0, Q_0),$$

where Q_0 is nonsingular. Equation (7.12) is still valid except that $\alpha_1 - a_1 = \eta_0$ is now replaced by $\alpha_1 - a = A\delta + R_0\eta_0$ and $P_1 = \kappa P_\infty + P_*$ where $P_* = R_0 Q_0 R_0'$. Thus for finite κ the term

$$-\frac{1}{2}\frac{\partial}{\partial \psi}(q\log\kappa + \kappa^{-1}\mathrm{tr}\{\hat{\delta}\hat{\delta}' + \mathrm{Var}(\delta|y)\})$$

must be included in (7.13). Defining

$$\frac{\partial \log L_d(y|\psi)}{\partial \psi}\bigg|_{\psi=\tilde{\psi}} = \lim_{\kappa\to\infty}\frac{\partial}{\partial \psi}\Big[\log L(y|\psi) + \frac{q}{2}\log\kappa\Big],$$

analogously to the definition of $\log L_d(y)$ in §7.2.2, and letting $\kappa \to \infty$, we have that

$$\frac{\partial \log L_d(y|\psi)}{\partial \psi}\bigg|_{\psi=\tilde{\psi}} = \frac{\partial \log L(y|\psi)}{\partial \psi}\bigg|_{\psi=\tilde{\psi}}, \qquad (7.14)$$

which is given in (7.13). In the event that α_1 consists only of diffuse elements, so the vector η_0 is null, the terms in Q_0 disappear from (7.13).

As an example consider the local level model (2.3) with η replaced by ξ, for which

$$\psi = \begin{pmatrix} \psi_\varepsilon \\ \psi_\xi \end{pmatrix} = \begin{pmatrix} \frac{1}{2} \log \sigma_\varepsilon^2 \\ \frac{1}{2} \log \sigma_\xi^2 \end{pmatrix},$$

with a diffuse initialisation for α_1. We have on substituting $y_t - \alpha_t = \varepsilon_t$ and $\alpha_{t+1} - \alpha_t = \xi_t$,

$$\log p(\alpha, y|\psi) = -\frac{2n-1}{2} \log 2\pi - \frac{n}{2} \log \sigma_\varepsilon^2 - \frac{n-1}{2} \log \sigma_\xi^2$$
$$- \frac{1}{2\sigma_\varepsilon^2} \sum_{t=1}^n \varepsilon_t^2 - \frac{1}{2\sigma_\xi^2} \sum_{t=2}^n \xi_{t-1}^2,$$

and

$$\tilde{\mathrm{E}}[\log p(\alpha, y|\psi)] = -\frac{2n-1}{2} \log 2\pi - \frac{n}{2} \log \sigma_\varepsilon^2 - \frac{n-1}{2} \log \sigma_\xi^2$$
$$- \frac{1}{2\sigma_\varepsilon^2} \sum_{t=1}^n \{\hat{\varepsilon}_t^2 + \mathrm{Var}(\varepsilon_t|y)\} - \frac{1}{2\sigma_\xi^2} \sum_{t=2}^n \{\hat{\xi}_{t-1}^2 + \mathrm{Var}(\xi_{t-1}|y)\},$$

where the conditional means and variances for ε_t and ξ_t are obtained from the Kalman filter and disturbance smoother with σ_ε^2 and σ_ξ^2 implied by $\psi = \tilde{\psi}$. To obtain the score vector, we differentiate both sides with respect to ψ where we note that, with $\psi_\varepsilon = \frac{1}{2} \log \sigma_\varepsilon^2$,

$$\frac{\partial}{\partial \sigma_\varepsilon^2} \left[\log \sigma_\varepsilon^2 + \frac{1}{\sigma_\varepsilon^2} \{\hat{\varepsilon}_t^2 + \mathrm{Var}(\varepsilon_t|y)\} \right] = \frac{1}{\sigma_\varepsilon^2} - \frac{1}{\sigma_\varepsilon^4} \{\hat{\varepsilon}_t^2 + \mathrm{Var}(\varepsilon_t|y)\},$$

$$\frac{\partial \sigma_\varepsilon^2}{\partial \psi_\varepsilon} = 2\sigma_\varepsilon^2.$$

The terms $\hat{\varepsilon}_t$ and $\mathrm{Var}(\varepsilon_t|y)$ do not vary with ψ since they have been calculated on the assumption that $\psi = \tilde{\psi}$. We obtain

$$\frac{\partial \log L_d(y|\psi)}{\partial \psi_\varepsilon} = -\frac{1}{2} \frac{\partial}{\partial \psi_\varepsilon} \sum_{t=1}^n \left[\log \sigma_\varepsilon^2 + \frac{1}{\sigma_\varepsilon^2} \{\hat{\varepsilon}_t^2 + \mathrm{Var}(\varepsilon_t|y)\} \right]$$
$$= -n + \frac{1}{\sigma_\varepsilon^2} \sum_{t=1}^n \{\hat{\varepsilon}_t^2 + \mathrm{Var}(\varepsilon_t|y)\}.$$

In a similar way we have

$$\frac{\partial \log L_d(y|\psi)}{\partial \psi_\xi} = -\frac{1}{2} \frac{\partial}{\partial \psi_\xi} \sum_{t=2}^n \left[\log \sigma_\xi^2 + \frac{1}{\sigma_\xi^2} \{\hat{\xi}_{t-1}^2 + \mathrm{Var}(\xi_{t-1}|y)\} \right]$$
$$= 1 - n + \frac{1}{\sigma_\xi^2} \sum_{t=2}^n \{\hat{\xi}_{t-1}^2 + \mathrm{Var}(\xi_{t-1}|y)\}.$$

The score vector for ψ of the local level model evaluated at $\psi = \tilde{\psi}$ is therefore

$$\left. \frac{\partial \log L_d(y|\psi)}{\partial \psi} \right|_{\psi = \tilde{\psi}} = \left[\begin{array}{c} \tilde{\sigma}_\varepsilon^2 \sum_{t=1}^n \left(u_t^2 - D_t \right) \\ \tilde{\sigma}_\xi^2 \sum_{t=2}^n \left(r_{t-1}^2 - N_{t-1} \right) \end{array} \right],$$

with $\tilde{\sigma}_\varepsilon^2$ and $\tilde{\sigma}_\xi^2$ from $\tilde{\psi}$. This result follows since from §§2.5.1 and 2.5.2 $\hat{\varepsilon}_t = \tilde{\sigma}_\varepsilon^2 u_t$, $\text{Var}(\varepsilon_t|y) = \tilde{\sigma}_\varepsilon^2 - \tilde{\sigma}_\varepsilon^4 D_t$, $\hat{\xi}_t = \tilde{\sigma}_\xi^2 r_t$ and $\text{Var}(\xi_t|y) = \tilde{\sigma}_\xi^2 - \tilde{\sigma}_\xi^4 N_t$.

It is very satisfactory that after so much algebra we obtain such a simple expression for the score vector which can be computed efficiently using the disturbance smoothing equations of §4.4. We can compute the score vector for the diffuse case efficiently because it is shown in §5.4 that no extra computing is required for disturbance smoothing when dealing with a diffuse initial state vector. Finally, score vector elements associated with variances or variance matrices in more complicated models such as multivariate structural time series models continue to have similar relatively simple expressions. Koopman and Shephard (1992) give for these models the score vector for parameters in H_t, R_t and Q_t as the expression

$$\left. \frac{\partial \log L_d(y|\psi)}{\partial \psi} \right|_{\psi = \tilde{\psi}} = \frac{1}{2} \sum_{t=1}^n \text{tr} \left\{ (u_t u_t' - D_t) \frac{\partial H_t}{\partial \psi} \right\}$$

$$+ \frac{1}{2} \sum_{t=2}^n \text{tr} \left\{ (r_{t-1} r_{t-1}' - N_{t-1}) \frac{\partial R_t Q_t R_t'}{\partial \psi} \right\} \Bigg|_{\psi = \tilde{\psi}}, \quad (7.15)$$

where u_t, D_t, r_t and N_t are evaluated by the Kalman filter and smoother as discussed in §§4.4 and 5.4.

7.3.4 THE EM ALGORITHM

The EM algorithm is a well-known tool for iterative maximum likelihood estimation which for many state space models has a particularly neat form. The earlier EM methods for the state space model were developed by Shumway and Stoffer (1982) and Watson and Engle (1983). The EM algorithm can be used either entirely instead of, or in place of the early stages of, direct numerical maximisation of the loglikelihood. It consists of an E-step (expectation) and M-step (maximisation) for which the former involves the evaluation of the conditional expectation $\tilde{E}[\log p(\alpha, y|\psi)]$ and the latter maximises this expectation with respect to the elements of ψ. The details of estimating unknown elements in H_t and Q_t are given by Koopman (1993) and they are close to those required for the evaluation of the score function. Taking first the case of a_1 and P_1 known and starting with (7.12), we evaluate $\tilde{E}[\log p(\alpha, y|\psi)]$ and as in (7.13) we obtain

$$\frac{\partial}{\partial \psi} \tilde{E}[\log p(\alpha, y|\psi)] = -\frac{1}{2} \frac{\partial}{\partial \psi} \sum_{t=1}^n \left[\log|H_t| + \log|Q_{t-1}| \right.$$

$$+ \text{tr}\left[\{\hat{\varepsilon}_t \hat{\varepsilon}_t' + \text{Var}(\varepsilon_t|y)\} H_t^{-1} \right]$$

$$+ \text{tr}\left[\{\hat{\eta}_{t-1} \hat{\eta}_{t-1}' + \text{Var}(\eta_{t-1}|y)\} Q_{t-1}^{-1} \right] |\psi], \quad (7.16)$$

where $\hat{\varepsilon}_t$, $\hat{\eta}_{t-1}$, $\text{Var}(\varepsilon_t|y)$ and $\text{Var}(\eta_{t-1}|y)$ are computed assuming $\psi = \tilde{\psi}$, while H_t and Q_{t-1} retain their original dependence on ψ. The equations obtained by setting (7.16) equal to zero are then solved for the elements of ψ to obtain a revised estimate of ψ. This is taken as the new trial value of ψ and the process is repeated either until adequate convergence is achieved or a switch is made to numerical maximisation of $\log L(y|\psi)$. The latter option is often used since although the EM algorithm usually converges fairly rapidly in the early stages, its rate of convergence near the maximum is frequently substantially slower than numerical maximisation; see Watson and Engle (1983) and Harvey and Peters (1990) for discussion of this point. As for the score vector in the previous section, when α_1 is diffuse we merely redefine η_0 and Q_0 in such a way that they are consistent with the initial state vector model $\alpha_1 = a + A\delta + R_0\eta_0$ where $\delta \sim \text{N}(0, \kappa I_q)$ and $\eta_0 \sim \text{N}(0, Q_0)$ and we ignore the part associated with δ. When α_1 consists only of diffuse elements the term in Q_0^{-1} disappears from (7.16).

To illustrate, we apply the EM algorithm to the local level model as in the previous section but now we take

$$\psi = \begin{pmatrix} \sigma_\varepsilon^2 \\ \sigma_\xi^2 \end{pmatrix}.$$

The E-step involves the Kalman filter and disturbance smoother to obtain $\hat{\varepsilon}_t$, $\hat{\xi}_{t-1}$, $\text{Var}(\varepsilon_t|y)$ and $\text{Var}(\xi_{t-1}|y)$ of (7.16) given $\psi = \tilde{\psi}$. The M-step solves for σ_ε^2 and σ_ξ^2 by equating (7.16) to zero. For example, in a similar way as in the previous section we have

$$-2\frac{\partial}{\partial\sigma_\varepsilon^2}\tilde{\text{E}}[\log p(\alpha, y|\psi)] = \frac{\partial}{\partial\sigma_\varepsilon^2}\sum_{t=1}^{n}\left[\log\sigma_\varepsilon^2 + \frac{1}{\sigma_\varepsilon^2}\{\hat{\varepsilon}_t^2 + \text{Var}(\varepsilon_t|y)\}\right]$$

$$= \frac{n}{\sigma_\varepsilon^2} - \frac{1}{\sigma_\varepsilon^4}\sum_{t=1}^{n}\{\hat{\varepsilon}_t^2 + \text{Var}(\varepsilon_t|y)\}$$

$$= 0,$$

and similarly for the term in σ_ξ^2. New trial values for σ_ε^2 and σ_ξ^2 are therefore obtained from

$$\bar{\sigma}_\varepsilon^2 = \frac{1}{n}\sum_{t=1}^{n}\{\hat{\varepsilon}_t^2 - \text{Var}(\varepsilon_t|y)\} = \tilde{\sigma}_\varepsilon^2 + \frac{1}{n}\tilde{\sigma}_\varepsilon^4\sum_{t=1}^{n}(u_t^2 - D_t),$$

$$\bar{\sigma}_\xi^2 = \frac{1}{n-1}\sum_{t=2}^{n}\{\hat{\xi}_{t-1}^2 - \text{Var}(\xi_{t-1}|y)\} = \tilde{\sigma}_\xi^2 + \frac{1}{n-1}\tilde{\sigma}_\xi^4\sum_{t=2}^{n}(r_{t-1}^2 - N_{t-1}),$$

since $\hat{\varepsilon}_t = \tilde{\sigma}_\varepsilon^2 u_t$, $\text{Var}(\varepsilon_t|y) = \tilde{\sigma}_\varepsilon^2 - \tilde{\sigma}_\varepsilon^4 D_t$, $\hat{\xi}_t = \tilde{\sigma}_\xi^2 r_t$ and $\text{Var}(\xi_t|y) = \tilde{\sigma}_\xi^2 - \tilde{\sigma}_\xi^4 N_t$. The disturbance smoothing values u_t, D_t, r_t and N_t are based on $\tilde{\sigma}_\varepsilon^2$ and $\tilde{\sigma}_\xi^2$. The new values $\bar{\sigma}_\varepsilon^2$ and $\bar{\sigma}_\xi^2$ replace $\tilde{\sigma}_\varepsilon^2$ and $\tilde{\sigma}_\xi^2$ and the procedure is repeated until either convergence has been attained or until a switch is made to numerical optimisation. Similar elegant results are obtained for more general time series models where unknown parameters occur only in the H_t and Q_t matrices.

7.3.5   PARAMETER ESTIMATION WHEN DEALING WITH DIFFUSE INITIAL CONDITIONS

It was shown in previous sections that only minor adjustments are required for parameter estimation when dealing with a diffuse initial state vector. The diffuse loglikelihood requires either the exact initial Kalman filter or the augmented Kalman filter. In both cases the diffuse loglikelihood is calculated in much the same way as for the non-diffuse case. No real new complications arise when computing the score vector or when estimating parameters via the EM algorithm. There is a compelling argument however for using the exact initial Kalman filter of §5.2 rather than the augmented Kalman filter of §5.7 for the estimation of parameters. For most practical models, the matrix $P_{\infty,t}$ and its associated matrices $F_{\infty,t}$, $M_{\infty,t}$ and $K_{\infty,t}$ do not depend on parameter vector ψ. This may be surprising but, for example, by studying the illustration given in §5.6.1 for the local linear trend model we see that the matrices $P_{\infty,t}$, $K_t^{(0)} = T_t M_{\infty,t} F_{\infty,t}^{-1}$ and $L_t^{(0)} = T_t - K_t^{(0)} Z_t$ do not depend on σ_ε^2, σ_ξ^2 or on σ_ζ^2. On the other hand, we see that all the matrices reported in §5.7.4, which deals with the augmentation approach to the same example, depend on $q_\xi = \sigma_\xi^2 / \sigma_\varepsilon^2$ and $q_\zeta = \sigma_\zeta^2 / \sigma_\varepsilon^2$. Therefore, every time that the parameter vector ψ changes during the estimation process we need to recalculate the augmented part of the augmented Kalman filter while we do not have to re-calculate the matrices related to $P_{\infty,t}$ for the exact initial Kalman filter.

First we consider the case where only the system matrices H_t, R_t and Q_t depend on the parameter vector ψ. The matrices $F_{\infty,t} = Z_t P_{\infty,t} Z_t'$ and $M_{\infty,t} = P_{\infty,t} Z_t'$ do not depend on ψ since the update equation for $P_{\infty,t}$ is given by

$$P_{\infty,t+1} = T_t P_{\infty,t} \big(T_t - K_t^{(0)} Z_t \big)',$$

where $K_t^{(0)} = T_t M_{\infty,t} F_{\infty,t}^{-1}$ and $P_{\infty,1} = AA'$, for $t = 1, \ldots, d$. Thus for all quantities related to $P_{\infty,t}$ the parameter vector ψ does not play a role. The same holds for computing a_{t+1} for $t = 1, \ldots, d$ since

$$a_{t+1} = T_t a_t + K_t^{(0)} v_t,$$

where $v_t = y_t - Z_t a_t$ and $a_1 = a$. Here again no quantity depends on ψ. The update equation

$$P_{*,t+1} = T_t P_{*,t} \big(T_t - K_t^{(0)} Z_t \big)' - K_t^{(0)} F_{\infty,t} K_t^{(1)'} + R_t Q_t R_t',$$

where $K_t^{(1)} = T_t M_{*,t} F_{\infty,t}^{-1} - K_t^{(0)} F_{*,t} F_{\infty,t}^{-1}$ depends on ψ. Thus we compute vector v_t and matrices $K_t^{(0)}$ and $F_{\infty,t}$ for $t = 1, \ldots, d$ once at the start of parameter estimation and we store them. When the Kalman filter is called again for likelihood evaluation we do not need to re-compute these quantities and we only need to update the matrix $P_{*,t}$ for $t = 1, \ldots, d$. This implies considerable computational savings during parameter estimation using the EM algorithm or maximising the diffuse loglikelihood using a variant of Newton's method.

For the case where ψ also affects the system matrices Z_t and T_t we achieve the same computational savings for all nonstationary models we have considered in this book. The matrices Z_t and T_t may depend on ψ but the parts of Z_t and T_t which affect the computation of $P_{\infty,t}$, $F_{\infty,t}$, $M_{\infty,t}$ and $K_{\infty,t}$ for $t = 1, \ldots, d$ do not depend on ψ. It should be noted that the rows and columns of $P_{\infty,t}$ associated with elements of α_1 which are not elements of δ are zero for $t = 1, \ldots, d$. Thus the columns of Z_t and the rows and columns of T_t related to stationary elements of the state vector do not influence the matrices $P_{\infty,t}$, $F_{\infty,t}$, $M_{\infty,t}$ and $K_{\infty,t}$. In the nonstationary time series models of Chapter 3 such as the ARIMA and structural time series models, all elements of ψ which affect Z_t and T_t only relate to the stationary part of the model, for $t = 1, \ldots, d$. The parts of Z_t and T_t associated with δ only have values equal to zero and unity. For example, the ARIMA(2,1,1) model of §3.3 shows that $\psi = (\phi_1\ \phi_2\ \theta_1\ \sigma^2)'$ does not influence the elements of Z_t and T_t associated with the first element of the state vector.

7.3.6   LARGE SAMPLE DISTRIBUTION OF MAXIMUM LIKELIHOOD ESTIMATES

It can be shown that under reasonable assumptions about the stability of the model over time, the distribution of $\hat{\psi}$ for large n is approximately

$$\hat{\psi} \sim \mathrm{N}(\psi, \Omega), \tag{7.17}$$

where

$$\Omega = \left[-\frac{\partial^2 \log L}{\partial \psi \partial \psi'} \right]^{-1}. \tag{7.18}$$

This distribution has the same form as the large sample distribution of maximum likelihood estimators from samples of independent and identically distributed observations. The result (7.17) is discussed by Hamilton (1994) in §5.8 for general time series models and in §13.4 for the special case of linear Gaussian state space models. In his discussion, Hamilton gives a number of references to theoretical work on the subject.

7.3.7   EFFECT OF ERRORS IN PARAMETER ESTIMATION

Up to this point we have followed standard classical statistical methodology by first deriving estimates of quantities of interest on the assumption that the parameter vector ψ is known and then replacing ψ in the resulting formulae by its maximum likelihood estimate $\hat{\psi}$. We now consider the estimation of the biases in the estimates that might arise from following this procedure. Since an analytical solution in the general case seems intractable, we employ simulation. We deal with cases where $\mathrm{Var}(\hat{\psi}) = O(n^{-1})$ so the biases are also of order n^{-1}.

   The technique that we propose is simple. Pretend that $\hat{\psi}$ is the true value of ψ. From (7.17) and (7.18) we know that the approximate large sample distribution of the maximum likelihood estimate of ψ given that the true ψ is $\hat{\psi}$ is $\mathrm{N}(\hat{\psi}, \hat{\Omega})$, where $\hat{\Omega}$ is Ω given by (7.18) evaluated at $\psi = \hat{\psi}$. Draw a simulation sample of N

independent values $\psi^{(i)}$ from $N(\hat{\psi}, \hat{\Omega})$, $i = 1, \ldots, N$. Denote by e a scalar, vector or matrix quantity that we wish to estimate from the sample y and let

$$\hat{e} = E(e|y)|_{\psi=\hat{\psi}}$$

be the estimate of e obtained by the methods of Chapter 4. For simplicity we focus on smoothed values, though an essentially identical technique holds for filtered estimates. Let

$$e^{(i)} = E(e|y)|_{\psi=\psi^{(i)}}$$

be the estimate of e obtained by taking $\psi = \psi^{(i)}$, for $i = 1, \ldots, N$. Then estimate the bias by

$$\hat{B}_e = \frac{1}{N} \sum_{i=1}^{N} e^{(i)} - \hat{e}. \tag{7.19}$$

The accuracy of $\hat{B}_e$ can be improved significantly by the use of antithetic variables, which are discussed in detail in §11.9.3 in connection with the use of importance sampling in the treatment of non-Gaussian models. For example, we can balance the sample of $\psi^{(i)}$'s for location by taking only $N/2$ draws from $N(\hat{\psi}, \hat{\Omega})$, where N is even, and defining $\psi^{(N-i+1)} = 2\hat{\psi} - \psi^{(i)}$ for $i = 1, \ldots, N/2$. Since $\psi^{(N-i+1)} - \hat{\psi} = -(\psi^{(i)} - \hat{\psi})$ and the distribution of $\psi^{(i)}$ is symmetric about $\hat{\psi}$, the distribution of $\psi^{(N-i+1)}$ is the same as that of $\psi^{(i)}$. In this way we not only reduce the numbers of draws required from the $N(\hat{\psi}, \hat{\Omega})$ distribution by a half, but we introduce negative correlation between the $\psi^{(i)}$'s which will reduce sample variation and we have arranged the simulation sample so that the sample mean $(\psi^{(1)} + \cdots + \psi^{(N)})/N$ is equal to the population mean $\hat{\psi}$. We can balance the sample for scale by a technique described in §11.9.3 using the fact that

$$\left(\psi^{(i)} - \hat{\psi}\right)'\hat{\Omega}^{-1}\left(\psi^{(i)} - \hat{\psi}\right) \sim \chi_w^2,$$

where w is the dimensionality of ψ; however, our expectation is that in most cases balancing for location only would be sufficient. The mean square error matrix due to simulation can be estimated in a manner similar to that described in §12.5.6.

Of course, we are not proposing that bias should be estimated as a standard part of routine time series analysis. We have included a description of this technique in order to assist workers in investigating the degree of bias in particular types of problems; in most practical cases we would expect the bias to be small enough to be neglected.

Simulation for correcting for bias due to errors in parameter estimates has previously been suggested by Hamilton (1994, §13.7). His methods differ from ours in two respects. First he uses simulation to estimate the entire function under study, which in his case is a mean square error matrix, rather than just the bias, as in our treatment. Secondly, he has omitted a term of the same order as the bias, namely n^{-1}, as demonstrated for the local level model that we considered in Chapter 2 by Quenneville and Singh (1997). This latter paper corrects Hamilton's method and provides interesting analytical and simulation results but it only gives details for the local level model. A different method based on parametric bootstrap samples has been proposed by Pfefferman and Tiller (2000).

7.4 Goodness of fit

Given the estimated parameter vector $\hat{\psi}$, we may want to measure the fit of the model under consideration for the given time series. Goodness of fit measures for time series models are usually associated with forecast errors. A basic measure of fit is the forecast variance F_t which could be compared with the forecast variance of a naive model. For example, when we analyse a time series with time-varying trend and seasonal, we could compare the forecast variance of this model with the forecast variance of the time series after adjusting it with fixed trend and seasonal.

When dealing with competing models, we may want to compare the loglikelihood value of a particular fitted model, as denoted by $\log L(y|\hat{\psi})$ or $\log L_d(y|\hat{\psi})$, with the corresponding loglikelihood values of competing models. Generally speaking, the larger the number of parameters that a model contains the larger its loglikelihood. In order to have a fair comparison between models with different numbers of parameters, information criteria such as the Akaike information criterion (AIC) and the Bayesian information criterion (BIC) are used. For a univariate series they are given by

$$\text{AIC} = n^{-1}[-2\log L(y|\hat{\psi}) + 2w], \qquad \text{BIC} = n^{-1}[-2\log L(y|\hat{\psi}) + w\log n],$$

and with diffuse initialisation they are given by

$$\text{AIC} = n^{-1}[-2\log L_d(y|\hat{\psi}) + 2(q + w)],$$
$$\text{BIC} = n^{-1}[-2\log L_d(y|\hat{\psi}) + (q + w)\log n],$$

where w is the dimension of ψ. Models with more parameters or more nonstationary elements obtain a larger penalty. More details can be found in Harvey (1989, §2.6.3 and §5.5.6). In general, a model with a smaller value of AIC or BIC is preferred.

7.5 Diagnostic checking

The diagnostic statistics and graphics discussed in §2.12 for the local level model (2.3) can be used in the same way for all univariate state space models. The basic diagnostics of §2.12.1 for normality, heteroscedasticity and serial correlation are applied to the one-step forecast errors defined in (4.4) after standardisation by dividing by the standard deviation $F_t^{1/2}$. In the case of multivariate models, we can consider the standardised individual elements of the vector

$$v_t \sim N(0, F_t), \qquad t = d + 1, \ldots, n,$$

but the individual elements are correlated since matrix F_t is not diagonal. The innovations can be transformed such that they are uncorrelated:

$$v_t^s = B_t v_t, \qquad F_t^{-1} = B_t' B_t.$$

It is then appropriate to apply the basic diagnostics to the individual elements of v_t^s. Another possibility is to apply multivariate generalisations of the diagnostic tests to the full vector v_t^s. A more detailed discussion on diagnostic checking can be found in Harvey (1989, §§5.4 & 8.4) and throughout the *STAMP* manual of Koopman *et al.* (2000).

Auxiliary residuals for the general state space model are constructed by

$$\hat{\varepsilon}_t^s = B_t^\varepsilon \hat{\varepsilon}_t, \qquad [\text{Var}(\hat{\varepsilon}_t)]^{-1} = B_t^{\varepsilon\prime} B_t^\varepsilon,$$
$$\hat{\eta}_t^s = B_t^\eta \hat{\eta}_t, \qquad [\text{Var}(\hat{\eta}_t)]^{-1} = B_t^{\eta\prime} B_t^\eta,$$

for $t = 1, \ldots, n$. The auxiliary residual $\hat{\varepsilon}_t^s$ can be used to identify outliers in the y_t series. Large absolute values in $\hat{\varepsilon}_t^s$ indicate that the behaviour of the observed value cannot be appropriately represented by the model under consideration. The usefulness of $\hat{\eta}_t^s$ depend on the interpretation of the state elements in α_t implied by the design of the system matrices T_t, R_t and Q_t. The way these auxiliary residuals can be exploited depends on their interpretation. For the local level model considered in §§7.3.3 and 7.3.4 it is clear that the state is the time-varying level and ξ_t is the change of the level for time $t + 1$. It follows that structural breaks in the series y_t can be identified by detecting large absolute values in the series for $\hat{\xi}_t^s$. In the same way, for the univariate local linear trend model (3.2), the second element of $\hat{\xi}_t^s$ can be exploited to detect slope changes in the series y_t. Harvey and Koopman (1992) have formalised these ideas further for the structural time series models of §3.2 and they constructed some diagnostic normality tests for the auxiliary residuals.

It is argued by de Jong and Penzer (1998) that such auxiliary residuals can be computed for any element of the state vector and that they can be considered as t-tests for the hypotheses

$$H_0 : (\alpha_{t+1} - T_t \alpha_t - R_t \eta_t)_i = 0,$$

the appropriate large-sample statistic for which is computed by

$$r_{it}^s = r_{it} / \sqrt{N_{ii,t}},$$

for $i = 1, \ldots, m$, where $(\cdot)_i$ is the ith element of the vector within brackets, r_{it} is the ith element of the vector r_t and $N_{ij,t}$ is the (i, j)th element of the matrix N_t; the recursions for evaluating r_t and N_t are given in §4.4.4. The same applies to the measurement equation for which t-test statistics for the hypotheses

$$H_0 : (y_t - Z_t \alpha_t - \varepsilon_t)_i = 0,$$

are computed by

$$e_{it}^s = e_{it} / \sqrt{D_{ii,t}},$$

for $i = 1, \ldots, p$, where the equations for computing e_t and D_t are given in §4.4.4. These diagnostics can be regarded as model specification tests. Large values in r_{it}^s and e_{it}^s, for some values of i and t, may reflect departures from the overall model and they may indicate specific adjustments to the model.

8
Bayesian analysis

8.1 Introduction

In this chapter we consider the analysis of observations generated by the linear Gaussian state space model from a Bayesian point of view. There have been many disputes in statistics about whether the classical or the Bayesian is the more valid mode of inference. Our approach to such questions is eclectic and pragmatic: we regard both approaches as valid in appropriate circumstances. If software were equally available for both standpoints, a practical worker might wish to try them both. Hence, although our starting point in both parts of the book is the classical approach, we shall include treatment of Bayesian methodology and, in separate publications, we will provide software which will enable practical Bayesian analyses to be performed. For discussions of this eclectic approach to statistical inference see Durbin (1987) and Durbin (1988).

8.2 Posterior analysis of state vector

8.2.1 POSTERIOR ANALYSIS CONDITIONAL ON PARAMETER VECTOR

For the linear Gaussian state space model (3.1), with parameter vector ψ specified, the posterior analysis of the model from a Bayesian standpoint is straightforward. The Kalman filter and smoother provide the posterior means, variances and covariances of the state vector α_t given the data. Since the model is Gaussian, posterior densities are normal, so these together with quantiles can be estimated easily from standard properties of the normal distribution.

8.2.2 POSTERIOR ANALYSIS WHEN PARAMETER VECTOR IS UNKNOWN

We now consider Bayesian analysis for the usual situation where the parameter vector ψ is not fixed and known; instead, we treat ψ as a random vector with a known prior density $p(\psi)$, which to begin with we take as a proper prior, leaving the non-informative case until later. For discussions of choice of prior see Gelman, Carlin, Stern and Rubin (1995) and Bernardo and Smith (1994). The problems we shall consider amount essentially to the estimation of the posterior mean

$$\bar{x} = E[x(\alpha)|y] \tag{8.1}$$

of a function $x(\alpha)$ of the stacked state vector α. Let

$$\bar{x}(\psi) = E[x(\alpha)|\psi, y]$$

be the conditional expectation of $x(\alpha)$ given ψ and y. We shall restrict consideration in this chapter to those functions $x(\alpha)$ for which $\bar{x}(\psi)$ can be readily calculated by the Kalman filter and smoother, leaving aside the treatment of other functions until Chapter 13. This restricted class of functions still, however, includes many important cases such as the posterior mean and variance matrix of α_t and forecasts $E(y_{n+j}|y)$ for $j = 1, 2, \ldots$. The treatment of initialisation in Chapter 5 permits elements of the initial state vector α_1 to have either proper or diffuse prior densities.

Obviously,

$$\bar{x} = \int \bar{x}(\psi)p(\psi|y)\,d\psi. \tag{8.2}$$

By Bayes theorem $p(\psi|y) = Kp(\psi)p(y|\psi)$ where K is the normalising constant defined by

$$K^{-1} = \int p(\psi)p(y|\psi)\,d\psi. \tag{8.3}$$

We therefore have

$$\bar{x} = \frac{\int \bar{x}(\psi)p(\psi)p(y|\psi)\,d\psi}{\int p(\psi)p(y|\psi)\,d\psi}. \tag{8.4}$$

Now $p(y|\psi)$ is the likelihood, which for the linear Gaussian model is easily calculated by the Kalman filter as shown in §7.2. We have already restricted $x(\alpha)$ to cases which can be calculated by Kalman filtering and smoothing operations. Thus the integrands of both integrals in (8.4) can be computed relatively easily by standard numerical methods in cases where the dimensionality of ψ is not large.

The main technique we shall employ in this book for Bayesian analysis however is simulation. As will be seen in Chapters 11 and 13, simulation produces methods that are not only effective for linear Gaussian models of any reasonable size but which also provide accurate Bayesian analyses for models that are non-Gaussian or nonlinear. The starting point for our simulation approach is importance sampling, for a discussion of which see Ripley (1987, pp. 122–123) or Geweke (1989). We shall outline here the basic idea of importance sampling applied to the linear Gaussian model but will defer completion of the numerical treatment until Chapter 13 when full details of the simulation techniques are worked out.

In principle, simulation could be applied directly to formula (8.4) by drawing a random sample $\psi^{(1)}, \ldots, \psi^{(N)}$ from the distribution with density $p(\psi)$ and then estimating the numerator and denominatior of (8.4) by the sample means of $\bar{x}(\psi)p(y|\psi)$ and $p(y|\psi)$ respectively. However, this estimator is inefficient in cases of practical interest. By using importance sampling we are able to achieve greater efficiency in a way that we now describe.

Suppose that simulation from density $p(\psi|y)$ is impractical. Let $g(\psi|y)$ be a density which is as close as possible to $p(\psi|y)$ while at the same time permitting simulation. We call this an *importance density*. From (8.2) we have

$$\bar{x} = \int \bar{x}(\psi) \frac{p(\psi|y)}{g(\psi|y)} g(\psi|y) \, d\psi$$

$$= E_g \left[\bar{x}(\psi) \frac{p(\psi|y)}{g(\psi|y)} \right]$$

$$= K E_g [\bar{x}(\psi) z^g(\psi, y)] \tag{8.5}$$

by Bayes theorem where E_g denotes expectation with respect to density $g(\psi|y)$,

$$z^g(\psi, y) = \frac{p(\psi) p(y|\psi)}{g(\psi|y)}, \tag{8.6}$$

and K is a normalising constant. By replacing $\bar{x}(\psi)$ by 1 in (8.5) we obtain

$$K^{-1} = E_g [z^g(\psi, y)],$$

so the posterior mean of $x(\alpha)$ can be expressed as

$$\bar{x} = \frac{E_g [\bar{x}(\psi) z^g(\psi, y)]}{E_g [z^g(\psi, y)]}. \tag{8.7}$$

This expression is evaluated by simulation. We choose random samples of N draws of ψ, denoted by $\psi^{(i)}$, from the importance density $g(\psi|y)$ and estimate $\bar{x}$ by

$$\hat{x} = \frac{\sum_{i=1}^N \bar{x}(\psi^{(i)}) z_i}{\sum_{i=1}^N z_i}, \tag{8.8}$$

where

$$z_i = \frac{p(\psi^{(i)}) p(y|\psi^{(i)})}{g(\psi^{(i)}|y)}. \tag{8.9}$$

As an importance density for $p(\psi|y)$ we take its large sample normal approximation

$$g(\psi|y) = N(\hat{\mu}, \hat{\Omega}),$$

where $\hat{\psi}$ is the solution to the equation

$$\frac{\partial \log p(\psi|y)}{\partial \psi} = \frac{\partial \log p(\psi)}{\partial \psi} + \frac{\partial \log p(y|\psi)}{\partial \psi} = 0, \tag{8.10}$$

and

$$\hat{\Omega}^{-1} = - \frac{\partial^2 \log p(\psi)}{\partial \psi \partial \psi'} - \frac{\partial^2 \log p(y|\psi)}{\partial \psi \partial \psi'} \bigg|_{\psi = \hat{\psi}}. \tag{8.11}$$

For a discussion of this large sample approximation to $p(\psi|y)$ see Gelman et al. (1995, Chapter 4) and Bernardo and Smith (1994, §5.3). Since $p(y|\psi)$ can easily be computed by the Kalman filter for $\psi = \psi^{(i)}$, $p(\psi)$ is given and $g(\psi|y)$ is Gaussian, the value of z_i is easy to compute. The draws for $\psi^{(i)}$ are independent and therefore $\hat{x}$ converges probabilistically to $\bar{x}$ as $N \to \infty$ under very general conditions.

The value $\hat{\psi}$ is computed iteratively by an obvious extension of the technique of maximum likelihood estimation as discussed in Chapter 7 while the second derivatives can be calculated numerically. Once $\hat{\psi}$ and $\hat{\Omega}$ are computed, it is straightforward to generate samples from $g(\psi|y)$ by use of a standard normal random number generator. Where needed, efficiency can be improved by the use of antithetic variables, which we discuss in §11.9.3. For example, for each draw $\psi^{(i)}$ we could take another value $\tilde{\psi}^{(i)} = 2\hat{\psi} - \psi^{(i)}$, which is equiprobable with $\psi^{(i)}$. The use of $\psi^{(i)}$ and $\tilde{\psi}^{(i)}$ together introduces balance in the sample.

The posterior mean of the parameter vector ψ is $\bar{\psi} = \mathrm{E}(\psi|y)$. An estimate $\tilde{\psi}$ of $\bar{\psi}$ is obtained by putting $\bar{x}(\psi^{(i)}) = \psi^{(i)}$ in (8.8) and taking $\tilde{\psi} = \hat{x}$. Similarly, an estimate $\tilde{\mathrm{V}}(\psi|y)$ of the posterior variance matrix $\mathrm{Var}(\psi|y)$ is obtained by putting $\bar{x}(\psi^{(i)}) = \psi^{(i)}\psi^{(i)\prime}$ in (8.8), taking $\tilde{S} = \hat{x}$ and then taking $\tilde{\mathrm{V}}(\psi|y) = \tilde{S} - \tilde{\psi}\tilde{\psi}'$.

To estimate the posterior distribution function of an element ψ_1 of ψ, which is not necessarily the first element of ψ, we introduce the indicator function $I_1(\psi_1^{(i)})$ which equals one if $\psi_1^{(i)} \le \psi_1$ and zero otherwise, where $\psi_1^{(i)}$ is the value of ψ_1 in the ith simulated value of ψ and ψ_1 is fixed. Then $\mathrm{F}(\psi_1|y) = \mathrm{Pr}(\psi_1^{(i)} \le \psi_1) = \mathrm{E}[I_1(\psi_1^{(i)})|y]$ is the posterior distribution function of ψ_1. Putting $\bar{x}(\psi^{(i)}) = I_1(\psi^{(i)})$ in (8.8), we estimate $\mathrm{F}(\psi_1|y)$ by $\tilde{\mathrm{F}}(\psi_1|y) = \hat{x}$. This is equivalent to taking $\tilde{\mathrm{F}}(\psi_1|y)$ as the sum of values of z_i for which $\psi_1^{(i)} \le \psi_1$ divided by the sum of all values of z_i. Similarly, if δ is the interval $(\psi_1 - \frac{1}{2}d, \psi_1 + \frac{1}{2}d)$ where d is small and positive then we can estimate the posterior density of ψ_1 by $\tilde{p}(\psi_1|y) = d^{-1}S^\delta / \sum_{i=1}^{N} z_i$ where S^δ is the sum of the values of z_i for which $\psi_1^{(i)} \in \delta$.

8.2.3   NON-INFORMATIVE PRIORS

For cases where a proper prior is not available we may wish to use a non-informative prior in which we assume that the prior density is proportional to a specified function $p(\psi)$ in a domain of ψ of interest even though the integral $\int p(\psi)d\psi$ does not exist. For a discussion of non-informative priors see, for example, Chapters 3 and 4 of Gelman et al. (1995). Where it exists, the posterior density is $p(\psi|y) = Kp(\psi)p(y|\psi)$ as in the proper prior case, so all the previous formulae apply without change. This is why we use the same symbol $p(\psi)$ for both cases even though in the non-informative case $p(\psi)$ is not a density. An important special case is the diffuse prior for which $p(\psi) = 1$ for all ψ.

Our intention in this chapter has been to describe the basic ideas underlying Bayesian analysis of linear Gaussian state space models by means of simulation based on importance sampling. Practical implementation of the analysis will be dealt with in Chapter 13 and illustrated by an example after we have developed further techniques, including antithetic variables for state space models, which we consider in §11.9.3.

8.3 Markov chain Monte Carlo methods

Another approach to Bayesian analysis based on simulation is provided by the *Markov chain Monte Carlo* (MCMC) method which has received a considerable amount of interest recently in the statistical and econometric literature on time series. Fruhwirth-Schnatter (1994) was the first to give a full Bayesian treatments of the linear Gaussian model using MCMC techniques. The proposed algorithms for simulation sample selection were later refined by de Jong and Shephard (1995). This work resulted in the simulation smoother which we discussed in §4.7. We showed there how to generate random draws from the conditional densities $p(\varepsilon|y, \psi)$, $p(\eta|y, \psi)$ and $p(\alpha|y, \psi)$ for a given parameter vector ψ. Now we briefly discuss how this technique can be incorporated into a Bayesian MCMC analysis in which we treat the parameter vector as stochastic.

The basic idea is as follows. We evaluate the posterior mean of $x(\alpha)$ or of the parameter vector ψ via simulation by choosing samples from an augmented joint density $p(\psi, \alpha|y)$. In the MCMC procedure, the sampling from this joint density is implemented as a Markov chain. After initialisation for ψ, say $\psi = \psi^{(0)}$ we repeatedly cycle through the two simulation steps:

(1) sample $\alpha^{(i)}$ from $p(\alpha|y, \psi^{(i-1)})$;
(2) sample $\psi^{(i)}$ from $p(\psi|y, \alpha^{(i)})$;

for $i = 1, 2, \ldots$. After a number of 'burning-in' iterations we are allowed to treat the samples from step (2) as being generated from the density $p(\psi|y)$. The attraction of this MCMC scheme is that sampling from conditional densities is easier than sampling from the marginal density $p(\psi|y)$. The circumstances under which subsequent samples from the marginal densities $p(\alpha|y, \psi^{(i-1)})$ and $p(\psi|y, \alpha^{(i)})$ converge to samples from the joint density $p(\psi, \alpha|y)$ are considered in books on MCMC, for example, Gamerman (1997). It is not straightforward to develop appropriate diagnostics which indicate whether convergence within the MCMC process has taken place, as is discussed, for example, in Gelman (1995).

There exist various implementations of the basic MCMC algorithm for the state space model. For example, Carlin, Polson and Stoffer (1992) propose to sample individual state vectors from $p(\alpha_t|y, \alpha^t, \psi)$ where α^t is equal to α excluding α_t. It turns out that this approach to sampling is inefficient. It is argued by Fruhwirth-Schnatter (1994) that it is more efficient to sample all the state vectors directly from the density $p(\alpha|y, \psi)$. She provides the technical details of implementation. de Jong and Shephard (1995) have developed this approach further by concentrating on the disturbance vectors ε_t and η_t instead of the state vector α_t. The details of the resulting simulation smoother were given in §4.7.

Implementing the two steps of the MCMC is not as straightforward as suggested so far. Sampling from the density $p(\alpha|y, \psi)$ for a given ψ is done by using the simulation smoother of §4.7. Sampling from $p(\psi|y, \alpha)$ depends partly on the model for ψ and is usually only possible up to proportionality. To sample under such circumstances, *accept-reject* algorithms have been developed; for example, the Metropolis algorithm is often used for this purpose. Details and an excellent general review of these matters are given by Gilks, Richardson and Spiegelhalter (1996). Applications to state space models have been developed by Carter and Kohn (1994), Shephard (1994) and Gamerman (1998).

In the case of structural time series models of §3.2 for which the parameter vector consists only of variances of disturbances associated with the components, the distribution of the parameter vector can be modelled such that sampling from $p(\psi|y, \alpha)$ in step (2) is relatively straightforward. For example, a model for a variance can be based on the inverse gamma distribution with logdensity

$$\log p(\sigma^2|c, s) = -\log \Gamma\left(\frac{c}{2}\right) - \frac{c}{2}\log\frac{s}{2} - \frac{c+2}{2}\log\sigma^2 - \frac{s}{2\sigma^2}, \qquad \text{for } \sigma^2 > 0,$$

and $p(\sigma^2|c, s) = 0$ for $\sigma^2 \leq 0$; see, for example, Poirier (1995, Table 3.3.1). We denote this density by $\sigma^2 \sim IG(c/2, s/2)$ where c determines the shape and s determines the scale of the distribution. It has the convenient property that if we take this as the prior density of σ^2 and we take a sample $u_1, \ldots, u_n$ of independent $N(0, \sigma^2)$ variables, the posterior density of σ^2 is

$$p(\sigma^2|u_1, \ldots, u_n) = IG\left[(c+n)/2, \left(s + \sum_{i=1}^{n} u_i^2\right)\Big/2\right];$$

for further details see, for example, Poirier (1995, Chapter 6). For the implementation of step (2) a sample value of σ^2 is chosen from this density. We can take u_t as an element of ε_t or η_t obtained by the simulation smoother in step (1). Further details of this approach are given by Fruhwirth-Schnatter (1994) and Carter and Kohn (1994).

Our general view is that, at least for practitioners who are not simulation experts, the methods of §8.2 are more transparent and computationally more convenient than MCMC for the practical problems we consider in this book.

9
Illustrations of the use of the linear Gaussian model

9.1 Introduction

In this chapter we will give some illustrations which show how the use of the linear Gaussian model works in practice. State space methods are usually employed for time series problems and most of our examples will be from this area but we also will treat a smoothing problem which is not normally regarded as part of time series analysis and which we solve using cubic splines.

The first example is an analysis of road accident data to estimate the reduction in car drivers killed and seriously injured in the UK due to the introduction of a law requiring the wearing of seat belts. In the second example we consider a bivariate model in which we include data on numbers of front seat passengers killed and seriously injured and on numbers of rear seat passengers killed and seriously injured and we estimate the effect that the inclusion of the second variable has on the accuracy of the estimation of the drop in the first variable. The third example shows how state space methods can be applied to Box-Jenkins ARMA models employed to model series of users logged onto the Internet. In the fourth example we consider the state space solution to the spline smoothing of motorcycle acceleration data. The final example provides an approximate analysis based on the linear Gaussian model of a stochastic volatility series of the logged exchange rate between the British pound and the American dollar. The software we have used for most of the calculations is *SsfPack* and is described in §6.6.

9.2 Structural time series models

The study by Durbin and Harvey (1985) and Harvey and Durbin (1986) on the effect of the seat belt law on road accidents in Great Britain provides an illustration of the use of structural time series models for the treatment of problems in applied time series analysis. They analysed data sets which contained numbers of casualties in various categories of road accidents to provide an independent assessment on behalf of the Department of Transport of the effects of the British seat belt law on road casualties. Most series were analysed by means of linear Gaussian state space models. We concentrate here on monthly numbers of drivers, front seat passengers

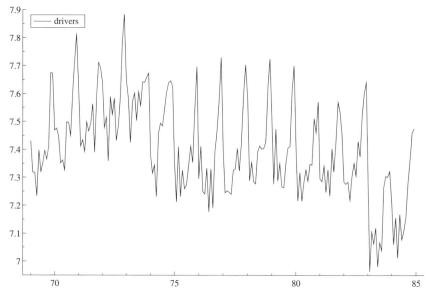

Fig. 9.1. Monthly numbers (logged) of drivers who were killed or seriously injured (KSI) in road accidents in cars in Great Britain.

and rear seat passengers who were killed or seriously injured in road accidents in cars in Great Britain from January 1969 to December 1984. Data were transformed into logarithms since logged values fitted the model better. Data on the average number of kilometres travelled per car per month and the real price of petrol are included as possible explanatory variables. We start with a univariate analysis of the drivers series. In the next section we perform a bivariate analysis using the front and rear seat passengers.

The log of monthly number of car drivers killed or seriously injured is displayed in Figure 9.1. The graph shows a seasonal pattern which may be due to weather conditions and festive celibrations. The overall trend of the series is basically constant over the years with breaks in the mid-seventies, probably due to the oil crisis, and in February 1983 after the introduction of the seat belt law. The model that we shall consider initially is the basic structural time series model which is given by

$$y_t = \mu_t + \gamma_t + \varepsilon_t,$$

where μ_t is the local level component modelled as the random walk $\mu_{t+1} = \mu_t + \xi_t$, γ_t is the trigonometric seasonal component (3.6) and (3.7), and ε_t is a disturbance term with mean zero and variance σ_ε^2. Note that for illustrative purposes we do not at this stage include an intervention component to measure the effect of the seat belt law.

The model is estimated by maximum likelihood using the techniques described in Chapter 7. The iterative method of finding the estimates for σ_ε^2, σ_ξ^2 and σ_ω^2 is implemented in *STAMP* 6.0 based on the concentrated diffuse loglikelihood as discussed in §2.10.2; the estimation output is given below where the first element of the parameter vector is $\phi_1 = 0.5 \log q_\eta$ and the second element is $\phi_2 = 0.5 \log q_\omega$ where $q_\xi = \sigma_\xi^2/\sigma_\varepsilon^2$ and $q_\omega = \sigma_\omega^2/\sigma_\varepsilon^2$. We present the results of the successive iterations in the parameter estimation process as displayed on the computer output of the *STAMP* package (version 6) of Koopman *et al.* (2000). Here 'it' indicates the iteration number and 'f' is the loglikelihood times $1/n$.

```
MaxLik iterating...
it   f          parameter vector      score vector
0    2.26711   -0.63488   -4.2642   -0.00077507   -0.00073413
1    2.26715   -0.64650   -4.3431    7.9721e-005   -0.00030219
2    2.26716   -0.64772   -4.3937    7.3021e-005   -7.7059e-005
3    2.26716   -0.64742   -4.4106    9.2115e-006   -1.1057e-005
4    2.26716   -0.64738   -4.4135    3.6016e-008   -5.0568e-007
5    2.26716   -0.64739   -4.4136   -3.2285e-008   -5.5955e-009
6    2.26716   -0.64739   -4.4136   -3.2507e-008   -5.3291e-009
```

The initial estimates are determined by an ad hoc method which is described in detail by Koopman *et al.* (2000, §10.6.3.1). It amounts to some cycles of univariate optimisations with respect to one parameter and for which the other parameters are kept fixed to their current values, starting with arbitrary values for the univariate optimisations. Usually the initial parameter values are close to their final maximum likelihood estimates as is the case in this illustration. Note that during estimation one parameter is concentrated out. The resulting parameter estimates are given below. The estimate for σ_ω^2 is very small and may be set equal to zero, implying a fixed seasonal component.

```
Estimated variances of disturbances

Component            drivers (q-ratio)
Irr               0.0034160  ( 1.0000)
Lvl               0.00093585 ( 0.2740)
Sea               5.0109e-007 ( 0.0001)
```

The estimated components are displayed in Figure 9.2. We conclude that the seasonal effect hardly changes over time and therefore can be treated as fixed. The estimated level does pick up the underlying movements in the series and the estimated irregular does not cause much concern to us.

In Figure 9.3 the estimated level is displayed; the filtered estimator is based on only the past data, that is $E(\mu_t|Y_{t-1})$, and the smoothed estimator is based on all the data, that is $E(\mu_t|Y_n)$. It can be seen that the filtered estimator lags the shocks in the series as is to be expected since this estimator does not take account of current and future observations.

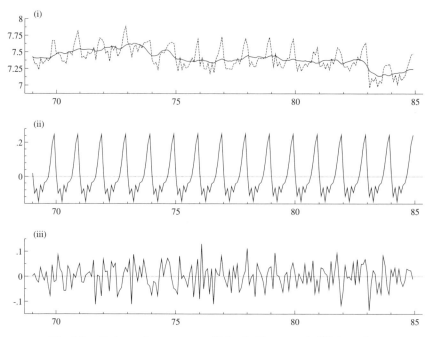

Fig. 9.2.   Estimated components: (i) level; (ii) seasonal; (iii) irregular.

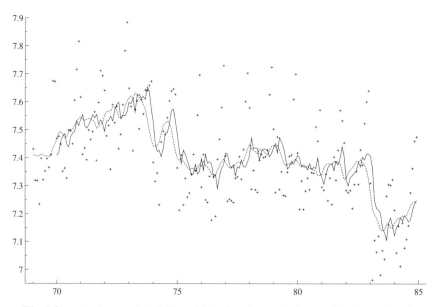

Fig. 9.3.   Data (crosses) with filtered (solid) and smoothed (dotted) estimated level.

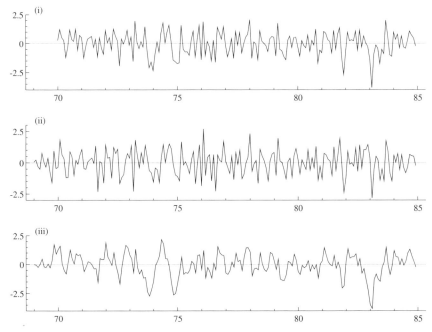

Fig. 9.4. (i) one-step ahead prediction residuals; (ii) auxiliary irregular residuals; (iii) auxiliary level residuals.

The model fit of this series and the basic diagnostics initially appear satisfactory. The standard output provided by *STAMP* is given by

```
Diagnostic summary report

Estimation sample is 69. 1 - 84.12. (T = 192, n = 180).
Log-Likelihood is 435.295 (-2 LogL = -870.59).
Prediction error variance is 0.00586998


Summary statistics
             drivers
Std.Error    0.076616
N              4.6692
H( 60)         1.0600
c( 1)        0.038623
c(12)        0.014140
Q(12,10)      11.610
```

The definitions of the diagnostics can be found in §2.12.

When we inspect the graphs of the residuals in Figure 9.4, however, in particular the auxiliary level residuals, we see a large negative value for February 1983. This suggests a need to incorporate an intervention variable to measure the level shift in February 1983. We have performed this analysis without inclusion of such a variable purely for illustrative purposes; obviously, in a real analysis the variable would be included since a drop in casualties was expected to result from the introduction of the seat belt law.

By introducing an intervention which equals one from February 1983 and is zero prior to that and the price of petrol as a further explanatory variable, we re-estimate the model and obtain the regression output

```
Estimated coefficients of explanatory variables.

Variable    Coefficient   R.m.s.e.   t-value
petrol      -0.29140      0.098318   -2.9638 [ 0.0034]
Lvl 83. 2   -0.23773      0.046317   -5.1328 [ 0.0000]
```

The estimated components, when the intervention variable and the regression effect due to the price of petrol are included, are displayed in Figure 9.5.

The estimated coefficient of a break in the level after January 1983 is -0.238, indicating a fall of 21%, that is, $1 - \exp(-0.238) = 0.21$, in car drivers killed and seriously injured after the introduction of the law. The t-value of -5.1 indicates that the break is highly significant. The coefficient of petrol price is also significant.

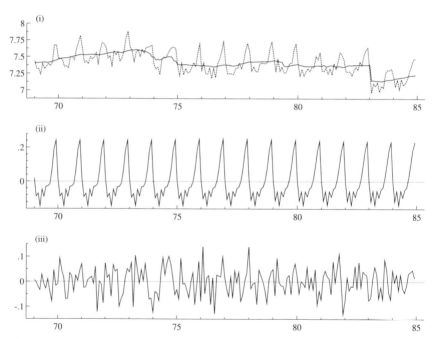

Fig. 9.5. Estimated components for model with intervention and regression effects: (i) level; (ii) seasonal; (iii) irregular.

9.3   Bivariate structural time series analysis

Multivariate structural time series models are introduced in §3.2.2. To illustrate state space methods for a multivariate model we analyse a bivariate monthly time series of front seat passengers and rear seat passengers killed and seriously injured in road accidents in Great Britain which were included in the assessment study by Harvey and Durbin (1986).

The graphs in Figure 9.6 indicate that the local level specification is appropriate for the trend component and that we need to include a seasonal component. We start by estimating a bivariate model with level, trigonometric seasonal and irregular components together with explanatory variables for the number of kilometers travelled and the real price of petrol. We estimate the model only using observations available before 1983, the year in which the seat belt law was introduced. The variance matrices of the three disturbance vectors are estimated by maximum likelihood in the way described in §7.3. The maximum likelihood estimates of the three variance matrices, associated with the level, seasonal and irregular components, respectively, are reported below where the upper off-diagonal elements are transformed to correlations:

```
Irregular disturbance    Level disturbance
0.0050306      0.67387    0.00025301      0.93059
0.0043489  0.0082792      0.00022533  0.00023173
                Seasonal disturbance
                  7.5475e-007      -1.0000
                -1.6954e-007   3.8082e-008
```

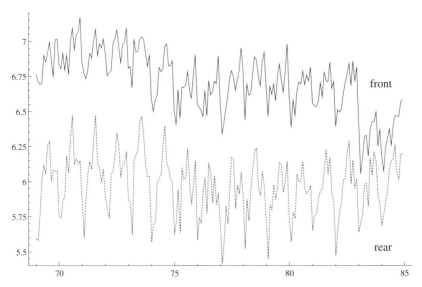

Fig. 9.6.   Front seat and rear seat passengers killed and seriously injured in road accidents in Great Britain.

The variance matrix of the seasonal component is small which lead us to model the seasonal component as fixed and to re-estimate the remaining two variances matrices:

```
Irregular disturbance   Level disturbance
0.0051672    0.66126    0.00025411    0.92359
0.0043261  0.0082832    0.00022490  0.00023334
```

The loglikelihood value of the estimated model is 761.85 with AIC equal to −4.344. The correlation between the two level disturbances is close to one. It may therefore be interesting to re-estimate the model with the restriction that the rank of the level variance matrix is one:

```
Irregular disturbance   Level disturbance
0.0051840    0.63746    0.00024909    1.0000
0.0042717  0.0086623    0.00023132  0.00021483
```

The loglikelihood value of this estimated model is 758.34 with AIC equal to −4.329 A comparison of the two AIC's shows only a marginal preference for the unrestricted model.

We now assess the effect of the introduction of the seat belt law as we have done for the drivers series using a univariate model in §9.2. We concentrate on the effect of the law on front seat passengers. We also have the rear seat series which is highly correlated with the front seat series. However, the law did not apply to rear seat passengers for which the data should therefore not be affected by the introduction of the law. Under such circumstances the rear seat series may be used as a *control group* which may result in a more precise measure of the effect of the seat belt law on front seat passengers; for the reasoning behind this idea see the discussion by Harvey (1996) to whom this approach is due.

We consider the unrestricted bivariate model but with a level intervention for February 1983 added to both series. This model is estimated using the whole data-set giving the parameter estimates:

```
Irregular disturbance   Level disturbance
0.0054015    0.65404    0.00024596    0.92292
0.0044490  0.0085664    0.00022498  0.00023216
```

```
Estimates level intervention February 1983
              Coefficient   R.m.s.e.   t-value
front seat      -0.33704    0.049243   -6.8445 [ 0.0000]
rear seat     0.00089514    0.051814   0.017276 [ 0.9862]
```

From these results and the time series plot of casualties in rear seat passengers in Figure 9.6, it is clear that they are unaffected by the introduction of the seat belt as we expect. By removing the intervention effect from the rear seat equation of the model we obtain the estimation results:

```
Irregular disturbance  Level disturbance
0.0054047     0.65395  0.00024575     0.92041
0.0044474  0.0085578   0.00021333  0.00021859


Estimates level intervention February 1983
           Coefficient  R.m.s.e.  t-value
front seat    -0.33797  0.028151  -12.005  [ 0.0000]
```

The almost two-fold decrease of the root mean squared error for the estimated intervention coefficient for the front seat series is remarkable. Enforcing the rank of Σ_η to be one, such that the levels are proportional to each other, also leads to a large increase of the t-value:

```
Irregular disturbance  Level disturbance
0.0054446     0.63057  0.00023128     1.0000
0.0043819  0.0088694   0.00021988  0.00020903


Estimates level intervention February 1983
           Coefficient  R.m.s.e.  t-value
front seat    -0.33640  0.019711  -17.066  [ 0.0000]
```

The graphs of the estimated (non-seasonal) signals and the estimated levels for the last model are presented in Figure 9.7. The substantial drop of the underlying level in front seat passenger casualties at the introduction of the seat belt law is clearly visible.

9.4 Box-Jenkins analysis

In this section we will show that fitting of ARMA models, which is an important part of the Box-Jenkins methodology, can be done using state space methods. Moreover, we will show that missing observations can be handled within the state space framework without problems whereas this is difficult within the Box-Jenkins methodology; see the discussion in §3.5. Finally, since an important objective of the Box-Jenkins methodology is forecasting, we also present forecasts of the series under investigation. In this illustration we use the series which is analysed by Makridakis, Wheelwright and Hyndman (1998): the number of users logged on to an Internet server each minute over 100 minutes. The data are differenced in order to get them closer to stationarity and these 99 observations are presented in Figure 9.8(i).

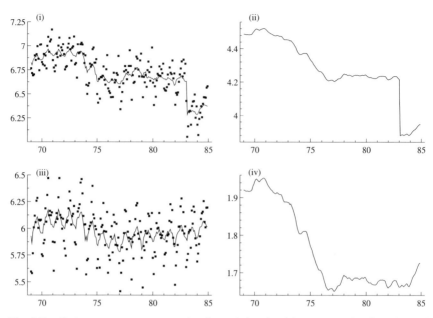

Fig. 9.7. (i) front seat passengers and estimated signal (without seasonal); (ii) estimated front seat casualties level; (iii) rear seat passengers and estimated signal (without seasonal); (iv) estimated rear seat casualties level.

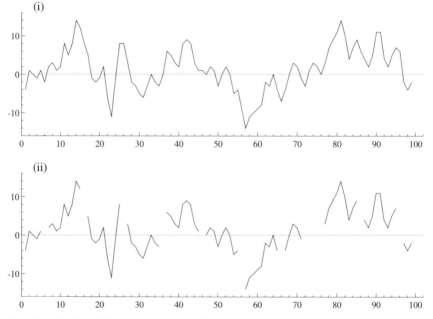

Fig. 9.8. (i) First difference of number of users logged on to Internet server each minute; (ii) The same series with 14 observations omitted.

Table 9.1. AIC for different ARMA models.

q	0	1	2	3	4	5
p						
0		2.777	2.636	2.648	2.653	2.661
1	2.673	2.608	2.628	2.629 (1)	2.642	2.658
2	2.647	2.628	2.642	2.657	2.642 (1)	2.660 (4)
3	2.606	2.626	2.645	2.662	2.660 (2)	2.681 (4)
4	2.626	2.646 (8)	2.657	2.682	2.670 (1)	2.695 (1)
5	2.645	2.665 (2)	2.654 (9)	2.673 (10)	2.662 (12)	2.727 (A)

The value between parentheses indicates the number of times the loglikelihood could not be evaluated during optimisation. The symbol A indicates that the maximisation process was automatically aborted due to numerical problems.

We have estimated a range of ARMA model (3.14) with different choices for p and q. They were estimated in state space from based on (3.17). Table 9.1 presents the Akaike information criteria (AIC), which is defined in §7.4, for these different ARMA models. We see that the ARMA models with (p, q) equal to (1, 1) and (3, 0) are optimal according to the AIC values. We prefer the ARMA(1, 1) model because it is more parsimonious. A similar table was produced for the same series by Makridakis *et al.* (1998) but the AIC statistic was computed differently. They concluded that the ARMA(3, 0) model was best.

We repeat the calculations for the same differenced series but now with 14 observations treated as missing (these are observation numbers 6, 16, 26, 36, 46, 56, 66, 72, 73, 74, 75, 76, 86, 96). The graph of the amended series is produced in Figure 9.8(ii). The reported AIC's in Table 9.2 lead to the same conclusion as for the series without missing observations: the preferred model is ARMA(1, 1) although its case is less strong now. We also learn from this illustration that estimation of higher order ARMA models with missing observations lead to more numerical problems.

Table 9.2. AIC for different ARMA models with missing observations.

q	0	1	2	3	4	5
p						
0		3.027	2.893	2.904	2.908	2.926
1	2.891	2.855	2.877	2.892	2.899	2.922
2	2.883	2.878	2.895 (6)	2.915	2.912	2.931
3	2.856 (1)	2.880	2.909	2.924	2.918 (12)	2.940 (1)
4	2.880	2.901	2.923	2.946	2.943	2.957 (2)
5	2.901	2.923	2.877 (A)	2.897 (A)	2.956 (26)	2.979

The value between parentheses indicates the number of times the loglikelihood could not be evaluated during optimisation. The symbol A indicates that the maximisation process was automatically aborted due to numerical problems.

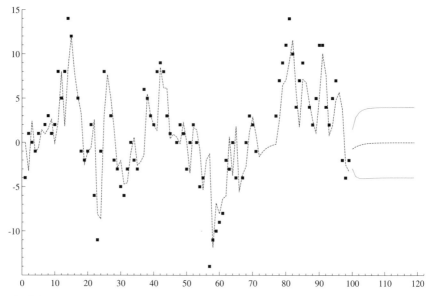

Fig. 9.9.   Internet series (solid blocks) with in-sample one-step forecasts and out-of-sample forecasts with 50% confidence interval.

Finally, we present in Figure 9.9 the in-sample one-step-ahead forecasts for the series with missing observations and the out-of-sample forecasts with 50% confidence intervals. It is one of the many advantages of state space modelling that it allows for missing observations without difficulty.

9.5   Spline smoothing

The connection between smoothing splines and the local linear trend model has been known for many years; see, for example, Wecker and Ansley (1983). In §3.11 we showed that the equivalence is with a local linear trend formulated in continuous time with the variance of the level disturbance equal to zero.

Consider a set of observations $y_1, \ldots, y_n$ which are irregularly spaced and associated with an ordered series $\tau_1, \ldots, \tau_n$. The variable τ_t can also be a measure for age, length or income, for example, as well as time. The discrete time model implied by the underlying continuous time model is the local linear trend model with

$$\text{Var}(\eta_t) = \sigma_\zeta^2 \delta_t^3 / 3, \qquad \text{Var}(\zeta_t) = \sigma_\zeta^2 \delta_t, \qquad E(\eta_t \zeta_t) = \sigma_\zeta^2 \delta_t^2 / 2, \qquad (9.1)$$

as shown in §3.10, where the distance variable $\delta_t = \tau_{t+1} - \tau_t$ is the time between observation t and observation $t + 1$. We shall show how irregularly spaced data can be analysed using state space methods. With evenly spaced observations the δ_t's are set to one.

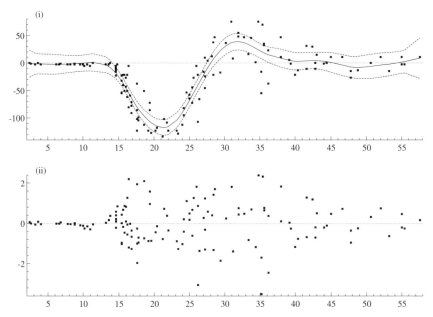

Fig. 9.10. Motorcycle acceleration data analysed by a cubic spline. (i) observations against time with spline and 95% confidence intervals, (ii) standardised irregular.

We consider 133 observations of acceleration against time (measured in milliseconds) for a simulated motorcycle accident. This data set was originally analysed by Silverman (1985) and is often used as an example of curve fitting techniques; see, for example, Hardle (1990) and Harvey and Koopman (2000). The observations are not equally spaced and at some time points there are multiple observations; see Figure 9.10. Cubic spline and kernel smoothing techniques depend on a choice of a smoothness parameter. This is usually determined by a technique called *cross-validation*. However, setting up a cubic spline as a state space model enables the smoothness parameter to be estimated by maximum likelihood and the spline to be computed by the Kalman filter and smoother. The model can easily be extended to include other unobserved components and explanatory variables, and it can be compared with alternative models using standard statistical criteria.

We follow here the analysis given by Harvey and Koopman (2000). The smoothing parameter $\lambda = \sigma_\zeta^2/\sigma_\varepsilon^2$ is estimated by maximum likelihood (assuming normally distributed disturbances) using the transformation $\lambda = \exp(\psi)$. The estimate of ψ is -3.59 with asymptotic standard error 0.22. This implies that the estimate of λ is 0.0275 with an asymmetric 95% confidence interval of 0.018 to 0.043. Silverman (1985) estimates λ by cross-validation, but does not report its value. In any case, it is not clear how one would compute a standard error for an estimate obtained by cross-validation. The Akaike information criterion (AIC) is 9.43. Figure 9.10 (i) presents the cubic spline. One of the advantages of representing the cubic spline by means of a statistical model is that, with little additional computing, we can obtain variances of our estimates and, therefore, standardised residuals defined as the residuals divided by the overall standard deviation. The 95% confidence intervals for the fitted spline are also given in Figure 9.10 (i). These are based on the root mean square errors of the smoothed estimates of μ_t, obtained from V_t as computed by (4.31), but without an allowance for the uncertainty arising from the estimation of λ as discussed in §7.3.7.

In Figure 9.10 (ii) the standardised irregular is presented for the cubic spline model and it is evident that the errors are heteroscedastic. Harvey and Koopman (2000) correct for heteroscedasticity by fitting a local level signal through the absolute values of the smoothed estimates of the irregular component. Subsequently, the measurement error variance, σ_ε^2, of the original cubic spline model is replaced by $\sigma_\varepsilon^2 h_t^{*2}$ where h_t^* is the smoothed estimate of the local level signal, scaled so that $h_1^* = 1$. The h_t^*'s are always positive because the weights of a local level model with uncorrelated disturbances are always positive. The absolute values of the smoothed irregular and the h_t^*'s are presented in Figure 9.11 (i). Estimating the heteroscedastic model, that is with measurement error variances proportional to the h_t^*'s gives an AIC of 8.74 (treating the $h_t^{*'}s$ as given). The resulting spline, shown in Figure 9.11 (ii), is not too different to the one shown in Figure 9.10 but the confidence interval is much narrower at the beginning and at the end. The smoothed irregular component in Figure 9.11 (iii) is now closer to being homoscedastic.

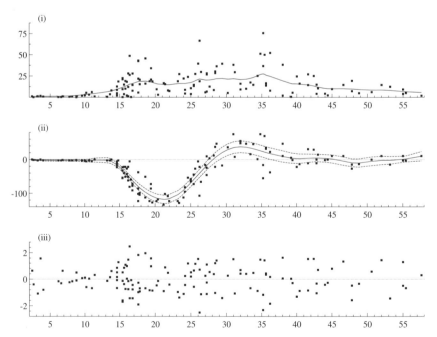

Fig. 9.11. Motorcycle acceleration data analysed by a cubic spline corrected for heteroscedasticity. (i) absolute values of smoothed irregular and h_t^*, (ii) data with signal and confidence intervals, (iii) standardised irregular.

9.6   Approximate methods for modelling volatility

Let y_t be the first difference of log prices of some portfolio of stocks, bonds or foreign currencies. Such a financial time series will normally be approximately serially uncorrelated. It may not be serially independent, however, because of serial dependence in the variance. This behaviour can be modelled by

$$y_t = \sigma_t \varepsilon_t = \sigma \varepsilon_t \exp(h_t/2), \qquad \varepsilon_t \sim N(0, 1), \qquad t = 1, \ldots, n, \quad (9.2)$$

where

$$h_{t+1} = \phi h_t + \eta_t, \qquad \eta_t \sim N(0, \sigma_\eta^2), \qquad |\phi| \leq 1, \quad (9.3)$$

and the disturbances ε_t and η_t are mutually and serially uncorrelated. In this exposition we assume the y_t has zero mean. The term σ is a scale factor, ϕ is an unknown coefficient, and η_t is a disturbance term which in the simplest model is uncorrelated with ε_t. This model is called a *stochastic volatility* (SV) model; it is dicussed further in §10.6.1. The model can be regarded as the discrete time analogue of the continuous time model used in papers on option pricing, such as Hull and White (1987). The statistical properties of y_t are easy to determine. However, the model as it stands is not linear and therefore the techniques described in Part I of this book cannot provide an exact solution. We require the methods of Part II for a full analysis of this model.

To obtain an approximate solution based on a linear model, we transform the observations y_t as follows:

$$\log y_t^2 = \kappa + h_t + \xi_t, \qquad t = 1, \ldots, n, \quad (9.4)$$

where

$$\xi_t = \log \varepsilon_t^2 - E\left(\log \varepsilon_t^2 \right)$$

and

$$\kappa = \log \sigma^2 + E\left(\log \varepsilon_t^2 \right). \quad (9.5)$$

The noise term ξ_t is not normally distributed but the model for $\log y_t^2$ is linear and therefore we can proceed approximately with the linear techniques of Part I. This approach is taken by Harvey, Ruiz and Shephard (1994) who call it quasi-maximum likelihood (QML). Parameter estimation is done via the Kalman filter; smoothed estimates of the volatility component, h_t, can be constructed and forecasts of volatility can be generated. One of the attractions of the QML approach is that it can be carried out straightforwardly using *STAMP*. This is an advantage compared with the more complex methods of Part II.

In our illustration of the QML method we use the same data as analysed by Harvey *et al.* (1994) in which y_t is the first difference of the logged exchange rate between pound sterling and US dollar. Parameter estimation is carried out by *STAMP* and some estimation results are given below. The high value for the

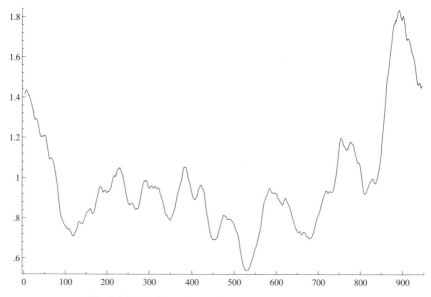

Fig. 9.12.   Estimated $\exp(h_{t|T}/2)$ for the Pound series.

normality statistic confirms that the observation error of the SV model is not Gaussian. The smoothed estimated volatility, measured as $\exp(h_t/2)$, is presented in Figure 9.12. A further discussion of the SV model is given in §10.6.1.

```
Summary statistics
                SVDLPound
Std.Error        1.7549
Normality       30.108
H(314)           1.1705
c( 1)           -0.019915
c(29)            0.044410
Q(29,27)        19.414


Estimated variances of disturbances.

Component  SVDLPound  (q-ratio)
Irr           2.9278  ( 1.0000)
Ar1           0.011650 ( 0.0040)

Estimated autoregressive coefficient.

The AR(1) rho coefficient is 0.986631.
```

Part II

Non-Gaussian and nonlinear state space models

In Part I we presented a comprehensive treatment of the construction and analysis of linear Gaussian state space models, and we discussed the software required for implementing the related methodology. Methods based on these models, possibly after transformation of the observations, are appropriate for a wide range of problems in practical time series analysis.

There are situations, however, where the linear Gaussian model fails to provide an acceptable representation of the behaviour of the data. For example, if the observations are monthly numbers of people killed in road accidents in a particular region, and if the numbers concerned are relatively small, the Poisson distribution will generally provide a more appropriate model for the data than the normal distribution. We therefore need to seek a suitable model for the development over time of a Poisson variable rather than a normal variable. Similarly, there are cases where a linear model fails to represent the behaviour of the data to an adequate extent. For example, if the trend and seasonal terms of a series combine multiplicatively but the disturbance term is additive, a linear model is inappropriate. In Part II we develop a unified methodology, based on efficient simulation techniques, for handling broad classes of non-Gaussian and nonlinear state space models.

10
Non-Gaussian and nonlinear state space models

10.1 Introduction

In this chapter we present the classes of non-Gaussian and nonlinear models that we shall consider in this book; we leave aside the analysis of observations generated by these models until later chapters. In §10.2 we specify the general forms of non-Gaussian models that we shall study and in §§10.3, 10.4 and 10.6 we consider special cases of the three main classes of models of interest, namely exponential family models, heavy-tailed models and financial models. In §10.5 we describe a class of nonlinear models that we shall investigate.

10.2 The general non-Gaussian model

The general multivariate non-Gaussian model that we shall consider has a similar state space structure to (3.1) in the sense that observational vectors y_t are determined by a relation of the form

$$p(y_t|\alpha_1, \ldots, \alpha_t, y_1, \ldots, y_{t-1}) = p(y_t|Z_t\alpha_t), \tag{10.1}$$

while the state vectors are determined independently of previous observations by the relation

$$\alpha_{t+1} = T_t\alpha_t + R_t\eta_t, \qquad \eta_t \sim p(\eta_t), \tag{10.2}$$

for $t = 1, \ldots, n$, where the η_t's are serially independent and where either $p(y_t|Z_t\alpha_t)$ or $p(\eta_t)$ or both can be non-Gaussian. The matrix Z_t has a role and a form analogous to those in the linear Gaussian models considered in Part I. We denote $Z_t\alpha_t$ by θ_t and refer to it as the *signal*. While we begin by considering a general form for $p(y_t|\theta_t)$, we shall pay particular attention to two special cases:

(1) observations which come from exponential family distributions with densities of the form

$$p(y_t|\theta_t) = \exp[y_t'\theta_t - b_t(\theta_t) + c_t(y_t)], \qquad -\infty < \theta_t < \infty, \tag{10.3}$$

where $b_t(\theta_t)$ is twice differentiable and $c_t(y_t)$ is a function of y_t only;

(2) observations generated by the relation

$$y_t = \theta_t + \varepsilon_t, \qquad \varepsilon_t \sim p(\varepsilon_t), \qquad (10.4)$$

where the ε_t's are non-Gaussian and serially independent.

The model (10.3) together with (10.2) where η_t is assumed to be Gaussian was introduced by West, Harrison and Migon (1985) under the name the *dynamic generalised linear model*. The origin of this name is that in the treatment of non-time series data the model (10.3), with $\theta_t = Z_t \alpha$ where α does not depend on t, is called a *generalised linear model*. In this context θ_t is called the *link function*; for a treatment of generalised linear models see McCullagh and Nelder (1989). Further development of the West, Harrison and Migon model is described in West and Harrison (1997). Smith (1979), Smith (1981) and Harvey and Fernandes (1989) gave an exact treatment for the special case of a Poisson observation with mean modelled as a local level model. Their approach however does not lend itself to generalisation.

10.3   Exponential family models

For model (10.3), let

$$\dot{b}_t(\theta_t) = \frac{\partial b_t(\theta_t)}{\partial \theta_t} \quad \text{and} \quad \ddot{b}_t(\theta_t) = \frac{\partial^2 b_t(\theta_t)}{\partial \theta_t \, \partial \theta_t'}. \qquad (10.5)$$

For brevity, we will write $\dot{b}_t(\theta_t)$ as $\dot{b}_t$ and $\ddot{b}_t(\theta_t)$ as $\ddot{b}_t$ in situations where it is unnecessary to emphasise the dependence on θ_t. Using the standard results

$$\mathrm{E}\left[\frac{\partial \log p(y_t|\theta_t)}{\partial \theta_t}\right] = 0,$$

$$\mathrm{E}\left[\frac{\partial^2 \log p(y_t|\theta_t)}{\partial \theta_t \, \partial \theta_t'}\right] + \mathrm{E}\left[\frac{\partial \log p(y_t|\theta_t)}{\partial \theta_t} \frac{\partial \log p(y_t|\theta_t)}{\partial \theta_t'}\right] = 0, \qquad (10.6)$$

it follows immediately from (10.3) and (10.6) that

$$\mathrm{E}(y_t) = \dot{b}_t \quad \text{and} \quad \mathrm{Var}(y_t) = \ddot{b}_t.$$

Consequently $\ddot{b}_t$ must be positive definite for non-degenerate models.

Results (10.6) are easily proved, assuming that relevant regularity conditions are satisfied, by differentiating the relation

$$\int p(y_t|\theta_t) \, dy_t = 1$$

twice with respect to θ_t and using the result

$$\frac{\partial \log p(\cdot)}{\partial \theta_t} = [p(\cdot)]^{-1} \frac{\partial p(\cdot)}{\partial \theta_t}.$$

10.3.1 POISSON DENSITY

For our first example of an exponential family distribution, suppose that the univariate observation y_t comes from a Poisson distribution with mean μ_t. For example, y_t could be the number of road accidents in a particular area during the month. Observations of this kind are called *count data*.

The logdensity of y_t is

$$\log p(y_t|\mu_t) = y_t \log \mu_t - \mu_t - \log(y_t!). \tag{10.7}$$

Comparing (10.7) with (10.3) we see that we need to take $\theta_t = \log \mu_t$ and $b_t = \exp \theta_t$ with $\theta_t = Z_t \alpha_t$, so the density of y_t given the signal θ_t is

$$p(y_t|\theta_t) = \exp[y_t\theta_t - \exp \theta_t - \log y_t!], \qquad t = 1, \dots, n. \tag{10.8}$$

It follows that the mean $\dot{b}_t = \exp \theta_t = \mu_t$ equals the variance $\ddot{b}_t = \mu_t$. Mostly we will assume that η_t in (10.2) is generated from a Gaussian distribution but all or some elements of η_t may come from other continuous distributions.

10.3.2 BINARY DENSITY

An observation y_t has a binary distribution if the probability that $y_t = 1$ has a specified probability, say π_t, and the probability that $y_t = 0$ is $1 - \pi_t$. For example, we could score 1 if Cambridge won the Boat Race in a particular year and 0 if Oxford won.

Thus the density of y_t is

$$p(y_t|\pi_t) = \pi_t^{y_t}(1 - \pi_t)^{1-y_t}, \qquad y_t = 1, 0, \tag{10.9}$$

so we have

$$\log p(y_t|\pi_t) = y_t[\log \pi_t - \log(1 - \pi_t)] + \log(1 - \pi_t). \tag{10.10}$$

To put this in form (10.3) we take $\theta_t = \log[\pi_t/(1 - \pi_t)]$ and $b_t(\theta_t) = \log(1 + e^{\theta_t})$, and the density of y_t given the signal θ_t is

$$p(y_t|\theta_t) = \exp[y_t\theta_t - \log(1 + \exp \theta_t)], \tag{10.11}$$

for which $c_t = 0$. It follows that mean and variance are given by

$$\dot{b}_t = \frac{\exp \theta_t}{1 + \exp \theta_t} = \pi_t, \qquad \ddot{b}_t = \frac{\exp \theta_t}{(1 + \exp \theta_t)^2} = \pi_t(1 - \pi_t),$$

as is well-known.

10.3.3 BINOMIAL DENSITY

Observation y_t has a binomial distribution if it is equal to the number of successes in k_t independent trials with a given probability of success, say π_t. As in the binary case we have

$$\log p(y_t|\pi_t) = y_t[\log \pi_t - \log(1 - \pi_t)] + k_t \log(1 - \pi_t) + \log \binom{k_t}{y_t}, \quad (10.12)$$

with $y_t = 0, \ldots, k_t$. We therefore take $\theta_t = \log[\pi_t/(1 - \pi_t)]$ and $b_t(\theta_t) = k_t \log(1 + \exp \theta_t)$ giving for the density of y_t in form (10.3),

$$p(y_t|\theta_t) = \exp\left[y_t\theta_t - k_t \log(1 + \exp \theta_t) + \log \binom{k_t}{y_t}\right]. \quad (10.13)$$

10.3.4 NEGATIVE BINOMIAL DENSITY

There are various ways of defining the negative binomial density; we consider the case where y_t is the number of independent trials, each with a given probability π_t of success, that are needed to reach a specified number k_t of successes. The density of y_t is

$$p(y_t|\pi_t) = \binom{k_t - 1}{y_t - 1} \pi_t^{k_t}(1 - \pi_t)^{y_t - k_t}, \qquad y_t = k_t, k_{t+1}, \ldots, \quad (10.14)$$

and the logdensity is

$$\log p(y_t|\pi_t) = y_t \log(1 - \pi_t) + k_t[\log \pi_t - \log(1 - \pi_t)] + \log \binom{k_t - 1}{y_t - 1}. \quad (10.15)$$

We take $\theta_t = \log(1 - \pi_t)$ and $b_t(\theta_t) = k_t[\theta_t - \log(1 - \exp \theta_t)]$ so the density in the form (10.3) is

$$p(y_t|\theta_t) = \exp\left[y_t\theta_t - k_t\{\theta_t - \log(1 - \exp \theta_t)\} + \log \binom{k_t - 1}{y_t - 1}\right]. \quad (10.16)$$

Since in non-trivial cases $1 - \pi_t < 1$ we must have $\theta_t < 0$ which implies that we cannot use the relation $\theta_t = Z_t\alpha_t$ since $Z_t\alpha_t$ can be negative. A way around the difficulty is to take $\theta_t = -\exp \theta_t^*$ where $\theta_t^* = Z_t\alpha_t$. The mean $E(y_t)$ is given by

$$\dot{b}_t = k_t\left[1 + \frac{\exp \theta_t}{1 - \exp \theta_t}\right] = k_t\left[1 + \frac{1 - \pi_t}{\pi_t}\right] = \frac{k_t}{\pi_t},$$

as is well-known.

10.3.5 MULTINOMIAL DENSITY

Suppose that we have $h > 2$ cells for which the probability of falling in the ith cell is π_{it} and suppose also that in k_t independent trials the number observed in the ith cell is y_{it} for $i = 1, \ldots, h$. For example, monthly opinion polls of voting preference: labour, conservative, liberal democrat, others.

Let $y_t = (y_{1t}, \ldots, y_{h-1,t})'$ and $\pi_t = (\pi_{1t}, \ldots, \pi_{h-1,t})'$ with $\sum_{j=1}^{h-1} \pi_{jt} < 1$. Then y_t is multinomial with logdensity

$$\log p(y_t|\pi_t) = \sum_{i=1}^{h-1} y_{it} \left[\log \pi_{it} - \log \left(1 - \sum_{j=1}^{h-1} \pi_{jt} \right) \right]$$

$$+ k_t \log \left(1 - \sum_{j=1}^{h-1} \pi_{jt} \right) + \log C_t, \qquad (10.17)$$

for $0 \le \sum_{i=1}^{h-1} y_{it} \le k_t$ where

$$C_t = k_t! / \left[\prod_{i=1}^{h-1} y_{it}! \left(k_t - \sum_{j=1}^{h-1} y_{jt} \right)! \right].$$

We therefore take $\theta_t = (\theta_{1t}, \ldots, \theta_{h-1,t})'$ where $\theta_{it} = \log[\pi_{it}/(1 - \sum_{j=1}^{h-1} \pi_{jt})]$, and

$$b_t(\theta_t) = k_t \log \left(1 + \sum_{i=1}^{h-1} \exp \theta_{it} \right),$$

so the density of y_t in form (10.3) is

$$p(y_t|\theta_t) = \exp \left[y_t'\theta_t - k_t \log \left(1 + \sum_{i=1}^{h-1} \exp \theta_{it} \right) \right] + \log C_t. \qquad (10.18)$$

10.4 Heavy-tailed distributions

10.4.1 t-DISTRIBUTION

A common way to introduce error terms into a model with heavier tails than those of the normal distribution is to use Student's t. We therefore consider modelling ε_t of (10.4) by the t-distribution with logdensity

$$\log p(\varepsilon_t) = \log a(\nu) + \frac{1}{2} \log \lambda - \frac{\nu + 1}{2} \log \left(1 + \lambda \varepsilon_t^2 \right), \qquad (10.19)$$

where ν is the number of degrees of freedom and

$$a(\nu) = \frac{\Gamma\left(\frac{\nu}{2} + \frac{1}{2}\right)}{\Gamma\left(\frac{\nu}{2}\right)}, \quad \lambda^{-1} = (\nu - 2)\sigma_\varepsilon^2, \quad \sigma_\varepsilon^2 = \text{Var}(\varepsilon_t), \quad \nu > 2, \quad t = 1, \ldots, n.$$

The mean of ε_t is zero and the variance is σ_ε^2 for any ν degrees of freedom which need not be an integer. The quantities ν and σ_ε^2 can be permitted to vary over time, in which case λ also varies over time.

10.4.2   MIXTURE OF NORMALS

A second common way to represent error terms with tails that are heavier than those of the normal distribution is to use a mixture of normals with density

$$p(\varepsilon_t) = \frac{\lambda^*}{\left(2\pi\sigma_\varepsilon^2\right)^{\frac{1}{2}}} \exp\left(\frac{-\varepsilon_t^2}{2\sigma_\varepsilon^2}\right) + \frac{1 - \lambda^*}{\left(2\pi\chi\sigma_\varepsilon^2\right)^{\frac{1}{2}}} \exp\left(\frac{-\varepsilon_t^2}{2\chi\sigma_\varepsilon^2}\right), \quad (10.20)$$

where λ^* is near to one, say 0.95 or 0.99, and χ is large, say from 10 to 100. This is a realistic model for situations when outliers are present, since we can think of the first normal density of (10.20) as the basic error density which applies $100\lambda^*$ per cent of the time, and the second normal density of (10.20) as representing the density of the outliers. Of course, λ^* and χ can be made to depend on t if appropriate. The investigator can assign values to λ^* and χ but they can also be estimated when the sample is large enough.

10.4.3   GENERAL ERROR DISTRIBUTION

A third heavy-tailed distribution that is sometimes used is the general error distribution with density

$$p(\varepsilon_t) = \frac{w(\ell)}{\sigma_\varepsilon} \exp\left[-c(\ell)\left|\frac{\varepsilon_t}{\sigma_\varepsilon}\right|^\ell\right], \quad 1 < \ell < 2, \quad (10.21)$$

where

$$w(\ell) = \frac{2[\Gamma(3\ell/4)]^{\frac{1}{2}}}{\ell[\Gamma(\ell/4)]^{\frac{3}{2}}}, \quad c(\ell) = \left[\frac{\Gamma(3\ell/4)}{\Gamma(\ell/4)}\right]^{\frac{\ell}{2}}.$$

Some details about this distribution are given by Box and Tiao (1973, §3.2.1), from which it follows that $\text{Var}(\varepsilon_t) = \sigma_\varepsilon^2$ for all ℓ.

10.5   Nonlinear models

In this section we introduce a class of nonlinear models which is obtained from the standard linear Gaussian model (3.1) in a natural way by permitting y_t to depend nonlinearly on α_t in the observation equation and α_{t+1} to depend nonlinearly on α_t in the state equation. Thus we obtain the model

$$y_t = Z_t(\alpha_t) + \varepsilon_t, \qquad \varepsilon_t \sim N(0, H_t), \qquad (10.22)$$

$$\alpha_{t+1} = T_t(\alpha_t) + R_t\eta_t, \qquad \eta_t \sim N(0, Q_t), \qquad (10.23)$$

for $t = 1, \ldots, n$, where $Z_t(\cdot)$ and $T_t(\cdot)$ are differentiable vector functions of α_t with dimensions p and m respectively. In principle it would be possible to extend this model by permitting ε_t and η_t to be non-Gaussian but we shall not pursue this extension in this book. Models with general forms similar to this were considered by Anderson and Moore (1979), but only for filtering, and their solutions were approximate only, whereas our treatment is exact. A simple example of the relation

(10.22) is a structural time series model in which the trend μ_t and seasonal γ_t combine multiplicatively and the observation error ε_t is additive, giving

$$y_t = \mu_t \gamma_t + \varepsilon_t;$$

a model of this kind has been considered by Shephard (1994).

10.6 Financial models

10.6.1 STOCHASTIC VOLATILITY MODELS

In the standard state space model (3.1) the variance of the observational error ε_t is frequently assumed to be constant. In the analysis of financial time series, such as daily fluctuations in stock prices and exchange rates, it is often found that the observational error variance is subject to substantial variability over time. This phenomenon is usually referred to as *volatility clustering*. An allowance for this variability in models for such series may be achieved via the *stochastic volatility* (SV) *model*. The SV model has a strong theoretical foundation in the financial theory on option pricing based on the work of the economists Black and Scholes; for a discussion see Taylor (1986). Further, the SV model has a strong connection with the state space approach as will become apparent below. In later chapters we shall give an exact treatment of this model based on simulation, in contrast to the approximate treatment of a zero-mean version of the model in §9.6 based on the linear Gaussian model. An alternative treatment of models with stochastic heterogenous error variances is provided by the *autoregressive conditional heteroscedasticity* (ARCH) *model* which will be discussed in §10.6.2.

Denote the first (daily) differences of a particular series of asset log prices by y_t. Financial time series are often constructed by first differencing log prices of some portfolio of stocks, bonds, foreign currencies, etc. A basic SV model for y_t is given by

$$y_t = a + \sigma \exp\left(\frac{1}{2}\theta_t\right)\varepsilon_t, \qquad \varepsilon_t \sim \mathrm{N}(0, 1), \tag{10.24}$$

where the mean a and the average standard deviation σ are assumed fixed and unknown. The signal θ_t is regarded as the unobserved log-volatility and it can be modelled in the usual way by $\theta_t = Z_t\alpha_t$ where α_t is generated by (10.2). In standard cases θ_t is modelled by the AR(1) with Gaussian disturbances, that is

$$\theta_{t+1} = \phi\theta_t + \eta_t, \qquad \eta_t \sim \mathrm{N}(0, \sigma_\eta^2), \qquad 0 < \phi < 1, \tag{10.25}$$

but the generality of the state equation (10.2) can be exploited. For a review of work and developments of the SV model see Shephard (1996) and Ghysels, Harvey and Renault (1996).

Various extensions of the SV model can be considered. For example, the Gaussian distributions can be replaced by distributions with heavier tails such as the t-distribution. This extension is often appropriate because many empirical studies find outliers due to unexpected jumps or downfalls in asset prices caused by 'over-heated' markets. A key example is the 'black Monday' crash in October 1987.

The basic SV model (10.24) captures only the salient features of changing volatility in financial series over time. The model becomes more precise when the mean of y_t is modelled by incorporating explanatory variables. For example, the SV model may be formulated as

$$b(L)y_t = a + c(L)'x_t + \sigma \exp\left(\frac{1}{2}\theta_t\right)\varepsilon_t,$$

where L is the lag operator defined so that $L^j z_t = z_{t-j}$ for $z_t = y_t$, x_t and where $b(L) = 1 - b_1 L - \cdots - b_{p^*} L^{p^*}$ is a scalar lag polynomial of order p^*; the column vector polynomial $c(L) = c_0 + c_1 L_1 + \cdots + c_{k^*} L^{k^*}$ contains $k^* + 1$ vectors of coeffients and x_t is a vector of exogenous explanatory variables. Note that the lagged value y_{t-j}, for $j = 1, \ldots, p^*$, can be considered as an explanatory variable to be added to exogenous explanatory variables. The signal θ_t may also depend on explanatory variables.

Another useful SV model extension is the inclusion of the volatility in the mean process which permits the measurement of risk premiums offered by the market. When risk (as measured by volatility) is higher, traders want to receive higher premiums on their transactions. Such a model is labelled as the *SV in Mean* (SVM) model and its simplest form is given by

$$y_t = a + d \exp(\theta_t) + \sigma \exp\left(\frac{1}{2}\theta_t\right)\varepsilon_t,$$

where d is the risk premium coefficient which is fixed and unknown. Other forms of the SVM model may also be considered but this one is particularly convenient; see the discussion in Koopman and Hol-Uspensky (1999).

Finally, another characteristic of financial time series is the phenomenon of *leverage*. The volatility of financial markets may adapt differently to positive and negative shocks. It is often observed that while markets might remain more or less stable when large positive earnings have been achieved but when huge losses have to be digested, markets become more unpredictable in the periods ahead. To incorporate leverage in the model we introduce $\varepsilon_{t-1}, \ldots, \varepsilon_{t-p}$, for some lag p, as explanatory variables in the generating equation for θ_t.

The extended SV model is then given by

$$b(L)y_t = a + c(L)'x_t + d \exp(\theta_t) + \sigma \exp\left(\frac{1}{2}\theta_t\right)\varepsilon_t, \qquad (10.26)$$

and

$$\phi(L)\theta_t = \beta(L)'x_t + \gamma(L)\varepsilon_t + \eta_t, \qquad (10.27)$$

with lag polynomials $\phi(L) = 1 - \phi_1 L - \cdots - \phi_p L^p$, $\beta(L) = \beta_0 + \beta_1 L + \cdots + \beta_k L^k$ and $\gamma(L) = \gamma_1 L + \cdots + \gamma_q L^q$. The disturbances ε_t and η_t are not necessarily Gaussian; they may came from densities with heavy tails.

Parameter estimation for the SV models based on maximum likelihood has been considered as a difficult problem; see the reviews in Shephard (1996) and Ghysels *et al.* (1996). Linear Gaussian techniques only offer approximate maximum likelihood estimates of the parameters as is shown in §9.6 and can only be applied to the basic SV model (10.24). The techniques we develop in the remaining chapters of this book, however, provide analyses of SV models based on importance sampling methods which are as accurate as is required.

10.6.2 GENERAL AUTOREGRESSIVE CONDITIONAL HETEROSCEDASTICITY

The general autoregressive conditional heteroscedasticity (GARCH) model, a special case of which was introduced by Engle (1982) and is known as the ARCH model, is a widely discussed model in the financial and econometrics literature. A simplified version of the GARCH(1, 1) model is given by

$$
\begin{aligned}
y_t &= \sigma_t \varepsilon_t, \qquad \varepsilon_t \sim N(0, 1), \\
\sigma_{t+1}^2 &= \alpha^* y_t^2 + \beta^* \sigma_t^2,
\end{aligned}
\qquad (10.28)
$$

where the parameters to be estimated are α^* and β^*. For a review of the GARCH model and its extensions see Bollerslev, Engle and Nelson (1994).

It is shown by Barndorff-Nielsen and Shephard (2001) that recursion (10.28) is equivalent to the steady state Kalman filter for a particular representation of the SV model. Consider, the model

$$
\begin{aligned}
y_t &= \sigma_t \varepsilon_t, \qquad &\varepsilon_t \sim N(0, 1), \\
\sigma_{t+1}^2 &= \phi \sigma_t^2 + \eta_t, \qquad &\eta_t > 0,
\end{aligned}
\qquad (10.29)
$$

for $t = 1, \ldots, n$ and where disturbances ε_t and η_t are serially and mutually independently distributed. Possible distributions for η_t are the gamma, inverse gamma or inverse Gaussian distributions. We can write the model in its squared form as follows

$$
y_t^2 = \sigma_t^2 + u_t, \qquad u_t = \sigma_t^2 (\varepsilon_t^2 - 1),
$$

which is in a linear state space form with $E(u_t) = 0$. The Kalman filter provides the minimum mean squared error estimator a_t of σ_t^2. When in steady state, the Kalman update equation for a_{t+1} can be represented as the GARCH(1, 1) recursion

$$
a_{t+1} = \alpha^* y_t^2 + \beta^* a_t,
$$

with

$$
\alpha^* = \phi \frac{\bar{P}}{\bar{P} + 1}, \qquad \beta^* = \phi \frac{1}{\bar{P} + 1},
$$

where $\bar{P}$ is the steady state value for P_t of the Kalman filter which we have defined for the local level model in §2.11 and for the general linear model in §4.2.3. We note that $\alpha^* + \beta^* = \phi$.

10.6.3   DURATIONS: EXPONENTIAL DISTRIBUTION

Consider a series of transactions in a stock market in which the tth transaction x_t is time-stamped by the time τ_t at which it took place. When studying the behaviour of traders in the market, attention may be focused on the duration between successive transactions, that is $y_t = \Delta \tau_t = \tau_t - \tau_{t-1}$. The duration y_t with mean μ_t can be modelled by a simple exponential density given by

$$p(y_t | \mu_t) = \frac{1}{\mu_t} \exp(-y_t / \mu_t), \qquad y_t, \mu_t > 0. \tag{10.30}$$

This density is a special case of the exponential family of densities and to put it in the form (10.3) we define

$$\theta_t = -\frac{1}{\mu_t} \quad \text{and} \quad b_t(\theta_t) = \log \mu_t = -\log(-\theta_t),$$

so we obtain

$$\log p(y_t | \theta_t) = y_t \theta_t + \log(-\theta_t). \tag{10.31}$$

Since $\dot{b}_t = -\theta_t^{-1} = \mu_t$ we confirm that μ_t is the mean of y_t, as is obvious from (10.30). The mean is restricted to be positive and so we model $\theta_t^* = \log(\mu_t)$ rather than μ_t directly. The durations in financial markets are typically short at the opening and closing of the daily market hours due to heavy trading in these periods. The time stamp τ_t is therefore often used as an explanatory variable in the mean function of durations and in order to smooth out the huge variations of this effect, a cubic spline is used. A simple durations model which allows for the daily seasonal is then given by

$$\theta_t = \gamma(\tau_{t-1}) + \psi_t,$$
$$\psi_t = \rho \psi_{t-1} + \chi_t, \qquad \chi_t \sim N(0, \sigma_\chi^2),$$

where $\gamma(\cdot)$ is the cubic spline function and χ_t is serially uncorrelated. Such models can be regarded as state space counterparts of the influential *autoregressive conditional duration* (ACD) model of Engle and Russell (1998).

10.6.4   TRADE FREQUENCIES: POISSON DISTRIBUTION

Another way of analysing market activity is to divide the daily market trading period up into intervals of one or five minutes and record the number of transactions in each interval. The counts in each interval can be modelled by a Poisson density for which the details are given in §10.3.1. Such a model would be a basic discrete version of what Rydberg and Shephard (1999) have labelled as BIN models.

11
Importance sampling

11.1 Introduction

In this chapter we begin the development of a comprehensive methodology for the analysis of observations from the non-Gaussian and nonlinear models that we specified in the previous chapter. Since no purely analytical techniques are available the methodology will be based on simulation. The simulation techniques we shall describe were considered for maximum likelihood estimation of the parameters in these models briefly by Shephard and Pitt (1997) and in more detail by Durbin and Koopman (1997). These techniques were extended to provide comprehensive classical and Bayesian analyses by Durbin and Koopman (2000); the resulting methods are easy to apply using publicly available software and are computationally efficient.

We shall consider inferential aspects of the analyses from both the classical and the Bayesian standpoints. From the classical point of view we shall discuss the estimation of functions of the state vector and the estimation of the error variance matrices of the resulting estimates. We shall also provide estimates of conditional densities, distribution functions and quantiles of interest, given the observations. Methods of estimating unknown parameters by maximum likelihood will be developed. From the Bayesian standpoint we shall describe methods for estimating posterior means, variances and densities.

The methods are based on standard ideas in simulation methodology, namely importance sampling and antithetic variables, as described, for example, in Ripley (1987, pp. 122–123). In this chapter we will develop the basic ideas of importance sampling that we employ in our methodology. Details of applications to particular models will be given in later chapters.

Denote the stacked vectors $(\alpha_1', \ldots, \alpha_{n+1}')'$ and $(y_1', \ldots, y_n')'$ by α and y. From the classical standpoint, most of the problems we shall consider in this book are essentially the estimation of the conditional mean

$$\bar{x} = \mathrm{E}[x(\alpha)|y] = \int x(\alpha) p(\alpha|y) \, d\alpha, \qquad (11.1)$$

of an arbitrary function $x(\alpha)$ of α given the observation vector y. This formulation includes estimates of quantities of interest such as the mean $\mathrm{E}(\alpha_t|y)$ of the state vector α_t given y and its conditional variance matrix $\mathrm{Var}(\alpha_t|y)$; it also includes estimates of the conditional density and distribution function of $x(\alpha)$ given y when $x(\alpha)$ is scalar. The conditional density $p(\alpha|y)$ depends on an unknown parameter vector ψ, but in order to keep the notation simple we shall not indicate this dependence explicitly in this chapter. In applications based on classical inference, ψ is replaced by its maximum likelihood estimator $\hat{\psi}$, while in Bayesian analyses ψ is treated as a random vector.

In theory, we could draw a random sample of values from the distribution with density $p(\alpha|y)$ and estimate $\bar{x}$ by the sample mean of the corresponding values of $x(\alpha)$. In practice, however, since explicit expressions are not available for $p(\alpha|y)$ for the models of Chapter 10, this idea is not feasible. Instead, we seek a density as close to $p(\alpha|y)$ as possible for which random draws are available, and we sample from this, making an appropriate adjustment to the integral in (11.1). This technique is called *importance sampling* and the density is referred to as the *importance density*. The techniques we shall describe will be based on Gaussian importance densities since these are available for the problems we shall consider and they work well in practice. We shall use the generic notation $g(\cdot)$, $g(\cdot, \cdot)$ and $g(\cdot|\cdot)$ for Gaussian marginal, joint and conditional densities.

Techniques for handling t-distributions and Gaussian mixture distributions without using importance sampling are considered in §§11.9.5 and 11.9.6.

11.2   Basic ideas of importance sampling

In order to keep the exposition simple we shall assume in this section and the next section that the initial density $p(\alpha_1)$ is non-degenerate and known. The case where some of the elements of α_1 are diffuse will be considered in §11.9.4, where we shall show that the treatment of partially diffuse initialisation is a good deal simpler for the non-Gaussian case than for the linear Gaussian model.

For given ψ, let $g(\alpha|y)$ be a Gaussian importance density which is chosen to resemble $p(\alpha|y)$ as closely as is reasonably possible; we have from (11.1),

$$\bar{x} = \int x(\alpha)\frac{p(\alpha|y)}{g(\alpha|y)}g(\alpha|y)\,d\alpha = \mathrm{E}_g\left[x(\alpha)\frac{p(\alpha|y)}{g(\alpha|y)}\right], \tag{11.2}$$

where E_g denotes expectation with respect to the importance density $g(\alpha|y)$. For the models of Chapter 10, $p(\alpha|y)$ and $g(\alpha|y)$ are complicated algebraically, whereas the corresponding joint densities $p(\alpha, y)$ and $g(\alpha, y)$ are straightforward. We therefore put $p(\alpha|y) = p(\alpha, y)/p(y)$ and $g(\alpha|y) = g(\alpha, y)/g(y)$ in (11.2), giving

$$\bar{x} = \frac{g(y)}{p(y)}\mathrm{E}_g\left[x(\alpha)\frac{p(\alpha, y)}{g(\alpha, y)}\right]. \tag{11.3}$$

Putting $x(\alpha) = 1$ in (11.3) we have

$$1 = \frac{g(y)}{p(y)}\mathrm{E}_g\left[\frac{p(\alpha, y)}{g(\alpha, y)}\right]. \tag{11.4}$$

Taking the ratio of (11.3) and (11.4) gives

$$\bar{x} = \frac{E_g[x(\alpha)w(\alpha, y)]}{E_g[w(\alpha, y)]}, \quad \text{where} \quad w(\alpha, y) = \frac{p(\alpha, y)}{g(\alpha, y)}. \tag{11.5}$$

In this formula,

$$p(\alpha, y) = p(\alpha_1) \prod_{t=1}^{n} p(\eta_t)p(y_t|\alpha_t), \tag{11.6}$$

with the substitution $\eta_t = R_t'(\alpha_{t+1} - T_t\alpha_t)$ for $t = 1, \ldots, n$. The formula for $\log g(\alpha, y)$ will be given in equation (11.9). Expression (11.5) provides the basis for the simulation methods in this book. We could in principle obtain a Monte Carlo estimate $\hat{x}$ of $\bar{x}$ in the following way. Choose a series of independent draws $\alpha^{(1)}, \ldots, \alpha^{(N)}$ from the distribution with density $g(\alpha|y)$ and take

$$\hat{x} = \frac{\sum_{i=1}^{N} x_i w_i}{\sum_{i=1}^{N} w_i}, \quad \text{where} \quad x_i = x(\alpha^{(i)}) \quad \text{and} \quad w_i = w(\alpha^{(i)}, y). \tag{11.7}$$

Since the draws are independent, and under assumptions which are usually satisfied in cases of practical interest, $\hat{x}$ converges to $\bar{x}$ probabilistically as $N \to \infty$. This simple estimate however is numerically inefficient and we shall improve it considerably in Chapter 12.

An important special case is where the observations are non-Gaussian but the state vector is generated by the state equation in the linear Gaussian model (3.1). We then have $p(\alpha) = g(\alpha)$ so

$$\frac{p(\alpha, y)}{g(\alpha, y)} = \frac{p(\alpha)p(y|\alpha)}{g(\alpha)g(y|\alpha)} = \frac{p(y|\alpha)}{g(y|\alpha)} = \frac{p(y|\theta)}{g(y|\theta)},$$

where as before θ is the stacked vector of the signals $\theta_t = Z_t\alpha_t$ for $t = 1, \ldots, n$. Thus (11.5) becomes the simpler formula

$$\bar{x} = \frac{E_g[x(\alpha)w^*(\theta, y)]}{E_g[w^*(\theta, y)]}, \quad \text{where} \quad w^*(\theta, y) = \frac{p(y|\theta)}{g(y|\theta)}; \tag{11.8}$$

its estimate $\hat{x}$ is given by an obvious analogue of (11.7) which we shall improve in Chapter 12. The advantage of (11.8) relative to (11.5) is that the dimensionality of θ_t is often much smaller than that of α_t. In the important case in which y_t is univariate, θ_t is a scalar.

11.3 Linear Gaussian approximating models

In this section we obtain the Gaussian importance densities that we need for simulation. We shall take these as the densities derived from the linear Gaussian models which have the same conditional modes of α given y as the non-Gaussian models.

Since we are considering the case of $p(\alpha_1)$ known, it is reasonable to suggest taking the density of α_1 in the approximating model as the Gaussian density $g(\alpha_1)$

which has the same mean vector and variance matrix as $p(\alpha_1)$. In fact, this does not have much practical significance since the assumption that $p(\alpha_1)$ is known was made for simplicity of exposition in the early stages of the development of the theory. In practice α_1 normally consists of diffuse or stationary elements and the treatment we shall give of this case in §11.9.4 does not normally require the construction of $g(\alpha_1)$.

Let $g(\alpha|y)$ and $g(\alpha, y)$ be the conditional and joint densities generated by linear Gaussian model (3.1) and let $p(\alpha|y)$ and $p(\alpha, y)$ be the corresponding densities generated by the general model (10.1) and (10.2). We will determine the approximating model by choosing H_t and Q_t in model (3.1) so that densities $g(\alpha|y)$ and $p(\alpha|y)$ have the same mode $\hat{\alpha}$. The possibility that $p(\alpha, y)$ might be multimodal will be considered in §11.8. Taking the Gaussian model first, the mode $\hat{\alpha}$ is the solution of the vector equation $\partial \log g(\alpha|y)/\partial\alpha = 0$. Now $\log g(\alpha|y) = \log g(\alpha, y) - \log g(y)$. Thus the mode is also the solution of the vector equation $\partial \log g(\alpha, y)/\partial\alpha = 0$. This version of the equation is easier to manage since $g(\alpha, y)$ has a simple form whereas $g(\alpha|y)$ does not. Since R_t consists of columns of I_m, $\eta_t = R_t'(\alpha_{t+1} - T_t\alpha_t)$. Assuming that $g(\alpha_1) = N(a_1, P_1)$, we therefore have

$$\log g(\alpha, y) = \text{constant} - \frac{1}{2}(\alpha_1 - a_1)' P_1^{-1}(\alpha_1 - a_1)$$
$$- \frac{1}{2}\sum_{t=1}^{n}(\alpha_{t+1} - T_t\alpha_t)' R_t Q_t^{-1} R_t'(\alpha_{t+1} - T_t\alpha_t)$$
$$- \frac{1}{2}\sum_{t=1}^{n}(y_t - Z_t\alpha_t)' H_t^{-1}(y_t - Z_t\alpha_t). \tag{11.9}$$

Differentiating with respect to α_t and equating to zero gives the equations

$$(d_t - 1)P_1^{-1}(\alpha_1 - a_1) - d_t R_{t-1} Q_{t-1}^{-1} R_{t-1}'(\alpha_t - T_{t-1}\alpha_{t-1})$$
$$+ T_t' R_t Q_t^{-1} R_t'(\alpha_{t+1} - T_t\alpha_t) + Z_t' H_t^{-1}(y_t - Z_t\alpha_t) = 0, \tag{11.10}$$

for $t = 1, \ldots, n$, where $d_1 = 0$ and $d_t = 1$ for $t = 2, \ldots, n$, together with the equation

$$R_n Q_n R_n'(\alpha_{n+1} - T_n\alpha_n) = 0.$$

The solution to these equations is the conditional mode $\hat{\alpha}$. Since $g(\alpha|y)$ is Gaussian the mode is equal to the mean so $\hat{\alpha}$ can be routinely calculated by the Kalman filter and smoother as described in Chapters 4 and 5. It follows that linear equations of the form (11.10) can be solved by the Kalman filter and smoother which is efficient computationally.

Assuming that the non-Gaussian model (10.1) and (10.2) is sufficiently well behaved, the mode $\hat{\alpha}$ of $p(\alpha|y)$ is the solution of the vector equation

$$\frac{\partial \log p(\alpha|y)}{\partial\alpha} = 0$$

and hence, as in the Gaussian case, of the equation

$$\frac{\partial \log p(\alpha, y)}{\partial \alpha} = 0.$$

Let $q_t(\eta_t) = -\log p(\eta_t)$ and let $h_t(y_t|\theta_t) = -\log p(y_t|\theta_t)$ where $\theta_t = Z_t\alpha_t$. Then,

$$\log p(\alpha, y) = \text{constant} + \log p(\alpha_1) - \sum_{t=1}^{n} [q_t(\eta_t) + h_t(y_t|\theta_t)], \qquad (11.11)$$

with $\eta_t = R'_t(\alpha_{t+1} - T_t\alpha_t)$, so $\hat{\alpha}$ is a solution of the vector equations

$$\frac{\partial \log p(\alpha, y)}{\partial \alpha_t} = (1 - d_t)\frac{\partial \log p(\alpha_1)}{\partial \alpha_1} - d_t R_{t-1}\frac{\partial q_{t-1}(\eta_{t-1})}{\partial \eta_{t-1}}$$

$$+ T'_t R_t \frac{\partial q_t(\eta_t)}{\partial \eta_t} - Z'_t \frac{\partial h_t(y_t|\theta_t)}{\partial \theta_t} = 0, \qquad (11.12)$$

for $t = 1, \ldots, n$, where, as before, $d_1 = 0$ and $d_t = 1$ for $t = 2, \ldots, n$, together with the equation

$$R_n \frac{\partial q_n(\eta_n)}{\partial \eta_n} = 0.$$

We solve these equations by iteration, where at each step we linearise, put the result in the form (11.10) and solve by the Kalman filter and smoother. Convergence is fast and normally only around ten iterations or less are needed. The final linearised model in the iteration is then the linear Gaussian model with the same conditional mode of α given y as the non-Gaussian model. We use the conditional density of α given y for this model as the importance density. A different method of solving these equations was given by Fahrmeir and Kaufmann (1991) but it is more cumbersome than our method; moreover, while it finds the mode, it does not find the linear approximating model directly.

11.4 Linearisation based on first two derivatives

We shall consider two methods of linearising the observation component of the mode equation (11.12). The first method is based on the first two derivatives of the observational logdensity and it enables exponential family observations, such as Poisson distributed observations, to be handled; the second method is given in §11.5 and deals with observations having the form (10.4) when $p(\varepsilon_t)$ is a function of ε_t^2; this is suitable for distributions with heavy tails such as the t-distribution.

To simplify the exposition we shall ignore the term $(d_t - 1)P_1^{-1}(\alpha_1 - a_1)$ in (11.10) and $(1 - d_t)\partial \log p(\alpha_1)/\partial \alpha_1$ in (11.12) in the derivation of the approximating model that follows. While there would be no difficulty in including them if this were thought to be necessary, we shall claim in §11.9.4 that the contributions of their analogues in the case of diffuse initialisation, which is the important case in practice, are generally negligible.

Suppose that $\tilde{\alpha} = [\tilde{\alpha}_1', \ldots, \tilde{\alpha}_{n+1}']'$ is a trial value of α, let $\tilde{\theta}_t = Z_t\tilde{\alpha}_t$ and define

$$\dot{h}_t = \frac{\partial h_t(y_t|\theta_t)}{\partial \theta_t}\Big|_{\theta_t = \tilde{\theta}_t}, \qquad \ddot{h}_t = \frac{\partial^2 h_t(y_t|\theta_t)}{\partial \theta_t \partial \theta_t'}\Big|_{\theta_t = \tilde{\theta}_t}. \tag{11.13}$$

Expanding about $\tilde{\theta}_t$ gives approximately

$$\frac{\partial h_t(y_t|\theta_t)}{\partial \theta_t} = \dot{h}_t + \ddot{h}_t(\theta_t - \tilde{\theta}_t). \tag{11.14}$$

Substituting in the final term of (11.12) gives the linearised form

$$-Z_t'(\dot{h}_t + \ddot{h}_t\theta_t - \ddot{h}_t\tilde{\theta}_t). \tag{11.15}$$

To put this in the same format as the final term of (11.10) put

$$\tilde{H}_t = \ddot{h}_t^{-1}, \qquad \tilde{y}_t = \tilde{\theta}_t - \ddot{h}_t^{-1}\dot{h}_t. \tag{11.16}$$

Then the final term becomes $Z_t'\tilde{H}_t^{-1}(\tilde{y}_t - \theta_t)$ as required.

Consider, for example, the important special case in which the state equation retains the original linear Gaussian form $\alpha_{t+1} = T_t\alpha_t + R_t\eta_t$ in which $\eta_t \sim N(0, Q_t)$. Equations (11.12) then have the linearised form

$$-d_t R_{t-1} Q_{t-1}^{-1} R_{t-1}'(\alpha_t - T_{t-1}\alpha_{t-1}) + T_t' R_t Q_t^{-1} R_t'(\alpha_{t+1} - T_t\alpha_t)$$
$$+ Z_t'\tilde{H}_t^{-1}(\tilde{y}_t - Z_t\alpha_t) = 0, \tag{11.17}$$

which has the same structure as (11.10) and therefore can be solved for α by the Kalman filter and smoother to give a new trial value; the process is repeated until convergence. The values of α and θ after convergence to the mode are denoted by $\hat{\alpha}$ and $\hat{\theta}$, respectively.

It is evident from (11.16) that this method only works when $\ddot{h}_t$ is positive definite. When $\ddot{h}_t$ is negative definite or semi-definite, the method based on one derivative, as presented in the next section, should normally be used. Finally, it is important to note that in the special case when the state equation is linear and Gaussian, the second derivative of the log of the approximating density with respect to α_t is

$$-R_{t-1}Q_{t-1}^{-1}R_{t-1}' - T_t'R_t Q_t^{-1}R_t'T_t - Z_t'\ddot{h}_t Z_t,$$

for $t = 2, \ldots, n$, and

$$-R_t Q_t^{-1} R_t',$$

for $t = n + 1$, which is the same as that of the logdensity of the non-Gaussian density. This means that not only does the approximating linear Gaussian model have the same conditional mode as the non-Gaussian model when the state equation is linear and Gaussian, it also has the same curvature at the mode. In consequence, when the state equation is given by (10.2) and $p(\eta_t)$ is not too far from normality, it is reasonable to expect that the curvatures at the mode of the non-Gaussian model and the approximating model will be approximately equal. This is a useful bonus for the technique of equalising modes.

11.4.1 EXPONENTIONAL FAMILY MODELS

The most important application of these results is to observations from exponential family distributions. For density (10.3) we have

$$h_t(y_t|\theta_t) = -\log p(y_t|\theta_t) = -[y_t'\theta_t - b_t(\theta_t) + c_t(y_t)]. \qquad (11.18)$$

Define

$$\dot{b}_t = \frac{\partial b_t(\theta_t)}{\partial \theta_t}\bigg|_{\theta_t = \tilde{\theta}_t},$$

$$\ddot{b}_t = \frac{\partial^2 b_t(\theta_t)}{\partial \theta_t \partial \theta_t'}\bigg|_{\theta_t = \tilde{\theta}_t}.$$

Then $\dot{h}_t = \dot{b}_t - y_t$ and $\ddot{h}_t = \ddot{b}_t$ so using (11.16) we take $\tilde{H}_t = \ddot{b}_t^{-1}$ and $\tilde{y}_t = \tilde{\theta}_t - \ddot{b}_t^{-1}(\dot{b}_t - y_t)$. These values can be substituted in (11.17) to obtain a solution for the case where the state equation is linear and Gaussian. Since, as shown in §10.3, $\ddot{b}_t = \text{Var}(y_t|\theta_t)$, it is positive definite in non-degenerate cases, so for the exponential family, linearisation based on first two derivatives can always be used.

As an example, for the Poisson distribution with density (10.8) we have $\dot{b}_t = \ddot{b}_t = \exp(\tilde{\theta}_t)$ so we take $\tilde{H}_t = \exp(-\tilde{\theta}_t)$ and $\tilde{y}_t = \tilde{\theta}_t - 1 + \exp(-\tilde{\theta}_t)y_t$. Other examples for the exponential family models are given in Table 11.1.

11.4.2 STOCHASTIC VOLATILITY MODEL

For the basic SV model (10.24) we have

$$h_t(y_t|\theta_t) = -\log p(y_t|\theta_t) = \frac{1}{2}\big[x_t^2 \exp(-\theta_t) + \theta_t + \log 2\pi\sigma^2\big],$$

where $x_t = (y_t - a)/\sigma$. It follows that

$$\dot{h}_t = -\frac{1}{2}\big[x_t^2 \exp(-\theta_t) - 1\big], \qquad \ddot{h}_t = \frac{1}{2}x_t^2 \exp(-\theta_t).$$

Using the definitions in (11.16), we have

$$\tilde{H}_t = 2\exp(\tilde{\theta}_t)/x_t^2, \qquad \tilde{y}_t = \tilde{\theta}_t + 1 - \exp(\tilde{\theta}_t)/x_t^2,$$

where we note that $\tilde{H}_t$ is always positive as required.

The approximating model for the SV model (10.24) is obtained by starting with putting $\tilde{\theta}_t = 0$ and computing $\tilde{H}_t$ and $\tilde{y}_t$ as given above. New $\tilde{\theta}_t$'s are obtained by applying the Kalman filter and smoother to the model implied by (11.16). The new $\tilde{\theta}_t$ values can be used to compute new $\tilde{H}_t$'s and new $\tilde{y}_t$'s. This recursive process is repeated until convergence.

11.5 Linearisation based on the first derivative

We now consider the case where the observations are generated by model (10.4) and where linearisation is based on the first derivative of the observational logdensity.

Table 11.1. Approximating model details for exponential family models.

Distribution		
Poisson	b_t	$\exp\theta_t$
	$\dot{b}_t$	$\exp\theta_t$
	$\ddot{b}_t$	$\exp\theta_t$
	$\ddot{b}_t^{-1}\dot{b}_t$	1
binary	b_t	$\log(1+\exp\theta_t)$
	$\dot{b}_t$	$\exp\theta_t(1+\exp\theta_t)^{-1}$
	$\ddot{b}_t$	$\exp\theta_t(1+\exp\theta_t)^{-2}$
	$\ddot{b}_t^{-1}\dot{b}_t$	$1+\exp\theta_t$
binomial	b_t	$k_t\log(1+\exp\theta_t)$
	$\dot{b}_t$	$k_t\exp\theta_t(1+\exp\theta_t)^{-1}$
	$\ddot{b}_t$	$k_t\exp\theta_t(1+\exp\theta_t)^{-2}$
	$\ddot{b}_t^{-1}\dot{b}_t$	$1+\exp\theta_t$
negative binomial	b_t	$k_t\{\theta_t-\log(1-\exp\theta_t)\}$
	$\dot{b}_t$	$k_t(1-\exp\theta_t)^{-1}$
	$\ddot{b}_t$	$k_t\exp\theta_t(1-\exp\theta_t)^{-2}$
	$\ddot{b}_t^{-1}\dot{b}_t$	$\exp(-\theta_t)-1$
exponential	b_t	$-\log\theta_t$
	$\dot{b}_t$	$-\theta_t^{-1}$
	$\ddot{b}_t$	θ_t^{-2}
	$\ddot{b}_t^{-1}\dot{b}_t$	$-\theta_t$

Note that $\tilde{H}_t = \ddot{b}_t^{-1}$ and $\tilde{y}_t = \tilde{\theta}_t + \tilde{H}_t y_t - \ddot{b}_t^{-1}\dot{b}_t$.

We shall assume that y_t is univariate and that $p(\varepsilon_t)$ is a function of ε_t^2; this case is important for heavy-tailed densities such as the t-distribution, and for Gaussian mixtures with zero means.

Let $\log p(\varepsilon_t) = -\frac{1}{2}h_t^*(\varepsilon_t^2)$. Then the contribution of the observation component to the equation $\partial \log p(\alpha, y)/\partial\alpha_t = 0$ is

$$-\frac{1}{2}\frac{dh_t^*\left(\varepsilon_t^2\right)}{d\left(\varepsilon_t^2\right)}\frac{d\varepsilon_t^2}{d\alpha_t} = Z_t'\frac{dh_t^*\left(\varepsilon_t^2\right)}{d\left(\varepsilon_t^2\right)}(y_t - \theta_t). \qquad (11.19)$$

Let

$$h_t^* = \left.\frac{dh_t^*\left(\varepsilon_t^2\right)}{d\left(\varepsilon_t^2\right)}\right|_{\varepsilon_t = y_t - \tilde{\theta}_t}. \qquad (11.20)$$

Then take $Z_t' h_t^*(y_t - \theta_t)$ as the linearised form of (11.19). By taking $\tilde{H}_t^{-1} = h_t^*$ we have the observation component in the correct form (11.10) so we can use the Kalman filter and smoother at each step of the solution of the equation $\partial \log p(\alpha, y)/\partial\alpha_t = 0$. We emphasise that now only the mode of the implied

Gaussian density of ε_t is equal to that of $p(\varepsilon_t)$, compared to linearisation using two derivatives with the linear Gaussian state which equalised both modes and curvatures at the mode.

It is of course necessary for this method to work that $\dot{h}_t^*$ is positive with probability one; however, this condition is satisfied for the applications we consider below. Strictly speaking it is not essential that $p(\varepsilon_t)$ is a function of ε_t^2. In other cases we could define

$$\dot{h}_t^* = -\frac{1}{\varepsilon_t}\frac{d \log p(\varepsilon_t)}{d\varepsilon_t}\bigg|\varepsilon_t = y_t - \tilde{\theta}_t, \qquad (11.21)$$

and proceed in the same way. Again, the method only works when $\dot{h}_t^*$ is positive with probability one.

11.5.1 t-DISTRIBUTION

As a first example, we take the t-distribution (10.19) for which

$$\dot{h}_t^* = (v+1)\left[\sigma_\varepsilon^2(v-2) + (y_t - \tilde{\theta}_t)^2\right]^{-1}.$$

11.5.2   MIXTURE OF NORMALS

From (10.20),

$$h_t^* = -2\log p(\varepsilon_t)$$

$$= -2\log\left[\frac{\lambda^*}{\sqrt{2\pi\sigma_\varepsilon^2}}\exp\left(-\frac{\varepsilon_t^2}{2\sigma_\varepsilon^2}\right) + \frac{1-\lambda^*}{\sqrt{2\pi\chi\sigma_\varepsilon^2}}\exp\left(-\frac{\varepsilon_t^2}{2\chi\sigma_\varepsilon^2}\right)\right].$$

So from (11.20) we have

$$\dot{h}_t^* = \frac{1}{p(\varepsilon_t)}\left\{\frac{\lambda^*}{\sigma_\varepsilon^2\sqrt{2\pi\sigma_\varepsilon^2}}\exp\left(-\tilde{\varepsilon}_t^{*2}\right) + \frac{1-\lambda^*}{\sigma_\varepsilon^2\chi\sqrt{2\pi\chi\sigma_\varepsilon^2}}\exp\left(-\tilde{\varepsilon}_t^{*2}/\chi\right)\right\},$$

where $\tilde{\varepsilon}_t^{*2} = (y_t - \tilde{\theta}_t)^2/(2\sigma_\varepsilon^2)$.

11.5.3   GENERAL ERROR DISTRIBUTION

From (10.21),

$$\log p(\varepsilon_t) = \text{constant} - c(\ell)\left|\frac{\varepsilon_t}{\sigma_\varepsilon}\right|^\ell, \qquad \ell < 2.$$

Since this is not a function of ε_t^2 we employ (11.21) which gives

$$\dot{h}_t^* = \frac{c(\ell)\ell}{\tilde{\varepsilon}_t\sigma_\varepsilon}\left|\frac{\tilde{\varepsilon}_t}{\sigma_\varepsilon}\right|^{\ell-1}, \qquad \text{for} \quad \varepsilon_t > 0,$$

$$= -\frac{c(\ell)\ell}{\tilde{\varepsilon}_t\sigma_\varepsilon}\left|\frac{\tilde{\varepsilon}_t}{\sigma_\varepsilon}\right|^{\ell-1}, \qquad \text{for} \quad \varepsilon_t < 0,$$

where $\tilde{\varepsilon}_t = y_t - \tilde{\theta}_t$. Thus $\dot{h}_t^*$ is positive with probability one so we can take $\tilde{H}_t^{-1} = \dot{h}_t^*$ and proceed with the construction of the linear Gaussian approximation as before.

11.6   Linearisation for non-Gaussian state components

We now consider the linearisation of the state component in equations (11.12) when the state disturbances η_t are non-Gaussian. Suppose that $\tilde{\eta} = [\tilde{\eta}_1', \ldots, \tilde{\eta}_n']'$ is a trial value of $\eta = (\eta_1', \ldots, \eta_n')'$ where $\tilde{\eta}_t = R_t'\,(\tilde{\alpha}_{t+1} - T_t\tilde{\alpha}_t)$. We shall confine ourselves to the situation where the elements η_{it} of η_t are mutually independent and where the density $p(\eta_{it})$ of η_{it} is a function of η_{it}^2. These assumptions are not very restrictive since they enable us to deal relatively easily with two cases of particular interest in practice, namely heavy-tailed errors and models with structural shifts for univariate series.

Let $q_{it}^*(\eta_{it}^2) = -2\log p(\eta_{it})$ and denote the ith column of R_t by R_{it}. Then the state contribution to the conditional mode equations (11.12) is

$$-\frac{1}{2}\sum_{i=1}^{r}\left[d_t R_{i,t-1}\frac{dq_{i,t-1}^*\left(\eta_{i,t-1}^2\right)}{d\eta_{i,t-1}} - T_t' R_{it}\frac{dq_{it}^*\left(\eta_{it}^2\right)}{d\eta_{it}}\right], \qquad t = 1, \ldots, n.$$

$$(11.22)$$

The linearised form of (11.22) is

$$-\sum_{i=1}^{r}[d_t R_{i,t-1}\dot{q}_{i,t-1}^*\eta_{i,t-1} - T_t' R_{it}\dot{q}_{it}^*\eta_{it}], \qquad (11.23)$$

where

$$\dot{q}_{it}^* = \left.\frac{dq_{it}^*\left(\eta_{it}^2\right)}{d\eta_{it}^2}\right|_{\eta_t = \tilde{\eta}_t}. \qquad (11.24)$$

Putting $\tilde{Q}_t^{-1} = \operatorname{diag}(\dot{q}_{1t}^*, \ldots, \dot{q}_{rt}^*)$, $\eta_t = R_t'\,(\alpha_{t+1} - T_t\alpha_t)$, and similarly for $\tilde{Q}_{t+1}$ and η_{t+1}, we see that (11.23) has the same form as the state component of (11.10). Consequently, in the iterative estimation of $\hat{\alpha}$ the Kalman filter and smoother can be used to update the trial value $\tilde{\alpha}$.

The general form for the joint density $g(\alpha, y)$ for all three types of linearisation is

$$g(\alpha, y) = \text{constant} \times g(\alpha_1)\exp\left[-\frac{1}{2}\sum_{t=1}^{n}\left(\eta_t'Q_t^{-1}\eta_t + \varepsilon_t'H_t^{-1}\varepsilon_t\right)\right], \qquad (11.25)$$

where for the approximation in §11.4 we substitute $\varepsilon_t = \hat{y}_t - \hat{\theta}_t$ and $H_t = \ddot{h}_t^{-1}$ in (11.25), where $\hat{y}_t$ and $\hat{\theta}_t$ are the values of $\tilde{y}_t$ and $\tilde{\theta}_t$ at the mode and $\ddot{h}_t$ is the value of $\ddot{h}_t$ given by (11.13) at the mode; for the approximation in §11.5 we substitute $\varepsilon_t = y_t - \hat{\theta}_t$ and $H_t = \dot{h}_t^{*-1}$, where $\hat{\theta}_t$ is as before and $\dot{h}_t^*$ is the value given by (11.20) at the mode; and finally, for the approximation in this section we substitute

$\eta_t = R_t'(\alpha_{t+1} - T_t\alpha_t)$ and $Q_t = \text{diag}(\dot{q}_{1t}^*, \ldots, \dot{q}_{rt}^*)$, where $\dot{q}_{it}^*$ is the value given by (11.24) at the mode.

This approach can be extended to functions of η_{it} other than η_{it}^2 and to non-diagonal variance matrices Q_t.

11.6.1 t-DISTRIBUTION FOR STATE ERRORS

As an illustration we consider the local level model (2.3) but with a t-distribution for the state error term η_t so that

$$y_t = \alpha_t + \varepsilon_t, \qquad \varepsilon_t \sim \text{N}(0, \sigma_\varepsilon^2),$$
$$\alpha_{t+1} = \alpha_t + \eta_t, \qquad \eta_t \sim t_\nu,$$

and we assume that $\alpha_1 \sim \text{N}(0, \kappa)$ with $\kappa \to \infty$. It follows that

$$\dot{q}_t^* = (\nu + 1)\left[\sigma_\eta^2(\nu - 2) + \tilde{\eta}_t^2\right]^{-1}.$$

Starting with initial values for $\tilde{\eta}_t$, we compute $\dot{q}_t^*$ and apply the Kalman filter and disturbance smoother to the approximating Gaussian local level model with

$$\alpha_{t+1} = \alpha_t + \eta_t, \qquad \eta_t \sim \text{N}(0, \dot{q}_t^*).$$

New values for the smoothed estimates $\tilde{\eta}_t$ are used to compute new values for $\dot{q}_t^*$ until convergence to $\hat{\eta}_t$.

11.7 Linearisation for nonlinear models

We now develop an analogous treatment of the nonlinear model (10.22) and (10.23). As with the non-Gaussian models, our objective is to find an approximating linear Gaussian model with the same conditional mode of α given y as the nonlinear model. We do this by a technique which is slightly different, though simpler, than that used for the non-Gaussian models. The basic idea is to linearise the observation and state equations (10.22) and (10.23) directly, which immediately delivers an approximating linear Gaussian model. We then iterate to ensure that this approximating model has the same conditional mode as the original nonlinear model.

Taking first the observation equation (10.22), let $\tilde{\alpha}_t$ be a trial value of α_t. Expanding about $\tilde{\alpha}_t$ gives approximately

$$Z_t(\alpha_t) = Z_t(\tilde{\alpha}_t) + \dot{Z}_t(\tilde{\alpha}_t)(\alpha_t - \tilde{\alpha}_t),$$

where $\dot{Z}(\alpha_t) = \partial Z_t(\alpha_t)/\partial \alpha_t'$, which is a $p \times m$ matrix, and $\dot{Z}_t(\tilde{\alpha}_t)$ is the value of this matrix at $\alpha_t = \tilde{\alpha}_t$. Putting $\tilde{y}_t = y_t - Z_t(\tilde{\alpha}_t) + \dot{Z}_t(\tilde{\alpha}_t)\tilde{\alpha}_t$, we approximate (10.22) by the relation

$$\tilde{y}_t = \dot{Z}_t(\tilde{\alpha}_t)\alpha_t + \varepsilon_t, \tag{11.26}$$

which is in standard form. Similarly, if we expand (10.23) about $\tilde{\alpha}_t$ we obtain approximately

$$T_t(\alpha_t) = T_t(\tilde{\alpha}_t) + \dot{T}_t(\tilde{\alpha}_t)(\alpha_t - \tilde{\alpha}_t),$$

where $\dot{T}(\alpha_t) = \partial T_t(\alpha_t)/\partial\alpha_t'$, which is a $m \times m$ matrix. Thus we obtain the linearised relation

$$\alpha_{t+1} = \tilde{T}_t + \dot{T}_t(\tilde{\alpha}_t)\alpha_t + R_t\eta_t, \tag{11.27}$$

where

$$\tilde{T}_t = T_t(\tilde{\alpha}_t) - \dot{T}_t(\tilde{\alpha}_t)\tilde{\alpha}_t.$$

This is not quite in standard form because of the presence of the non-random term $\tilde{T}_t$. Instead of approximating the nonlinear model (10.22) and (10.23) at its conditional mode by a linear Gaussian model of the standard form (3.1), we must therefore consider approximating it by a linear Gaussian model of the modified form

$$\begin{aligned} \tilde{y}_t &= \dot{Z}_t(\tilde{\alpha}_t)\alpha_t + \varepsilon_t, & \varepsilon_t &\sim N(0, H_t), \\ \alpha_{t+1} &= \tilde{T}_t + \dot{T}_t(\tilde{\alpha}_t)\alpha_t + R_t\eta_t, & \eta_t &\sim N(0, Q_t), \end{aligned} \tag{11.28}$$

where $\tilde{T}_t$ is a known vector. This does not raise any difficulties for the methods of Chapter 4. It can easily be shown that the standard Kalman filter (4.13) applies to model (11.28) except that the state update equation (4.8),

$$a_{t+1} = T_t a_t + K_t v_t,$$

is replaced by the equation

$$a_{t+1} = \tilde{T}_t + T_t a_t + K_t v_t,$$

with the appropriate change of notation from model (4.1) to model (11.28).

We use the output of the Kalman filter to define a new $\tilde{\alpha}_t$ which gives a new approximating model (11.28), and we continue to iterate as in §11.4 until convergence is achieved. Denote the resulting value of α by $\hat{\alpha}$.

To show that $\hat{\alpha}$ is the conditional mode for the nonlinear model (10.22) and (10.23), we write

$$p(\alpha, y) = \text{constant} - \frac{1}{2}\left[\sum_{t=1}^{n}\{\alpha_{t+1} - T_t(\alpha_t)\}'R_t Q_t^{-1} R_t'\{\alpha_{t+1} - T_t(\alpha_t)\} \right.$$
$$\left. + \sum_{t=1}^{n}\{y_t - Z_t(\alpha_t)\}'H_t^{-1}\{y_t - Z_t(\alpha_t)\}\right].$$

Differentiating with respect to α_t gives for the mode equation

$$\begin{aligned} &- d_t R_{t-1} Q_{t-1} R_{t-1}'[\alpha_t - T_{t-1}(\alpha_{t-1})] \\ &+ \dot{T}_t(\alpha_t)' R_t Q_t^{-1} R_t'[\alpha_{t+1} - T_t(\alpha_t)] \\ &- \dot{Z}_t(\alpha_t)' H_t^{-1}[y_t - Z_t(\alpha_t)] = 0, \end{aligned} \tag{11.29}$$

for $t = 1, \ldots, n$ with $d_1 = 0$ and $d_t = 1$ for $t > 1$. The mode equation for the approximating model (11.28) is

$$
\begin{aligned}
& - d_t R_{t-1} Q_{t-1} R'_{t-1} [\alpha_t - \tilde{T}_{t-1} - \dot{T}_{t-1}(\tilde{\alpha}_{t-1})\alpha_{t-1}] \\
& + \dot{T}_t(\alpha_t)' R_t Q_t^{-1} R'_t [\alpha_{t+1} - \tilde{T}_t - \dot{T}_t(\tilde{\alpha}_t)\alpha_t] \\
& - \dot{Z}_t(\alpha_t)' H_t^{-1} [\tilde{y}_t - \dot{Z}_t(\tilde{\alpha}_t)\alpha_t],
\end{aligned}
\tag{11.30}
$$

for $t = 1, \ldots, n$.

Now when the iteration converges we have $\alpha_t = \tilde{\alpha}_t = \hat{\alpha}_t$, so

$$
\begin{aligned}
\alpha_{t+1} - \tilde{T}_t - \dot{T}_t(\tilde{\alpha}_t)\alpha_t &= \hat{\alpha}_{t+1} - T_t(\hat{\alpha}_t) + \dot{T}_t(\hat{\alpha}_t)\hat{\alpha}_t - \dot{T}_t(\hat{\alpha}_t)\hat{\alpha}_t \\
&= \hat{\alpha}_{t+1} - T_t(\hat{\alpha}_t),
\end{aligned}
$$

and

$$
\begin{aligned}
\tilde{y}_t - \dot{Z}_t(\tilde{\alpha}_t)\alpha_t &= y_t - Z_t(\hat{\alpha}_t) + \dot{Z}_t(\hat{\alpha}_t)\hat{\alpha}_t - \dot{Z}_t(\hat{\alpha}_t)\hat{\alpha}_t \\
&= y_t - Z_t(\hat{\alpha}_t).
\end{aligned}
$$

It follows that $\alpha_t = \hat{\alpha}_t$ satisfies (11.29) so $\hat{\alpha}$ is the conditional mode for the nonlinear model. As in §11.3, the linear approximating model (11.28) at $\tilde{\alpha} = \hat{\alpha}$ is used to provide the density.

Suppose that, as in the non-Gaussian case, we wish to estimate the mean $\bar{x}$ of a function $x(\alpha)$ of the state vector α. The same formula (11.5) applies, that is,

$$
\bar{x} = \frac{E_g[x(\alpha)w(\alpha, y)]}{E_g[w(\alpha, y)]}, \quad \text{with} \quad w(\alpha, y) = \frac{p(\alpha, y)}{g(\alpha, y)},
$$

where, by neglecting $p(\alpha_1)$ as before,

$$
p(\alpha, y) = \text{constant} \times \exp\left[-\frac{1}{2} \sum_{t=1}^{n} \left(\eta'_t Q_t^{-1} \eta_t + \varepsilon'_t H_t^{-1} \varepsilon_t \right) \right], \tag{11.31}
$$

with the substitutions $\eta_t = R'_t[\alpha_{t+1} - T_t(\alpha_t)]$ and $\varepsilon_t = y_t - Z_t(\alpha_t)$. For $g(\alpha, y)$ the same formula (11.31) applies, with substitutions

$$
\eta_t = R'_t[\alpha_{t+1} - \hat{T}_t - \dot{T}_t(\hat{\alpha}_t)\alpha_t], \quad \text{where} \quad \hat{T}_t = T_t(\hat{\alpha}_t) - \dot{T}_t(\hat{\alpha}_t)\hat{\alpha}_t,
$$

and

$$
\varepsilon_t = \hat{y}_t - \dot{Z}_t(\hat{\alpha}_t)\alpha_t, \quad \text{where} \quad \hat{y}_t = y_t - Z_t(\hat{\alpha}_t) + \dot{Z}_t(\hat{\alpha}_t)\hat{\alpha}_t.
$$

11.7.1 MULTIPLICATIVE MODELS

Shephard (1994) considered the multiplicative trend and seasonal model with additive Gaussian observation noise. We consider here a simple version of this model with the trend modelled as a local level and the seasonal given by a single trigonometric term such as given in (3.6) with $s = 3$. We have

$$
y_t = \mu_t \gamma_t + \varepsilon_t, \qquad \varepsilon_t \sim N(0, \sigma_\varepsilon^2),
$$

with

$$\alpha_{t+1} = \begin{pmatrix} \mu_{t+1} \\ \gamma_{t+1} \\ \gamma_{t+1}^* \end{pmatrix} = \begin{bmatrix} 1 & 0 & 0 \\ 0 & \cos\lambda & \sin\lambda \\ 0 & -\sin\lambda & \cos\lambda \end{bmatrix} \alpha_t + \begin{pmatrix} \eta_t \\ \omega_t \\ \omega_t^* \end{pmatrix},$$

and $\lambda = 2\pi/3$. It follows that $Z_t(\alpha_t) = \mu_t\gamma_t$ and $\dot{Z}_t(\alpha_t) = (\gamma_t, \mu_t, 0)$ which lead us to the approximating model

$$\tilde{y}_t = (\tilde{\gamma}_t, \tilde{\mu}_t, 0)\alpha_t + \varepsilon_t,$$

where $\tilde{y}_t = y_t + \tilde{\mu}_t\tilde{\gamma}_t$.

Another example of a multiplicative model we consider is

$$y_t = \mu_t\varepsilon_t, \qquad \mu_{t+1} = \mu_t\xi_t,$$

where ε_t and ξ_t are mutually and serially uncorrelated Gaussian disturbance terms. For the general model (10.22) and (10.23) we have $\alpha_t = (\mu_t, \varepsilon_t, \xi_t)'$, $\eta_t = (\varepsilon_{t+1}, \xi_{t+1})'$ $Z_t(\alpha_t) = \mu_t\varepsilon_t$, $H_t = 0$, $T_t(\alpha_t) = (\mu_t\xi_t, 0, 0)'$, $R_t = [0, I_2]'$ and Q_t is a 2×2 diagonal matrix. It follows that

$$\dot{Z}_t(\alpha_t) = (\varepsilon_t, \mu_t, 0), \qquad \dot{T}_t(\alpha_t) = \begin{bmatrix} \xi_t & 0 & \mu_t \\ 0 & 0 & 0 \\ 0 & 0 & 0 \end{bmatrix}.$$

The approximating model (11.26) and (11.27) reduces to

$$\tilde{y}_t = \tilde{\varepsilon}_t\mu_t + \tilde{\mu}_t\varepsilon_t, \qquad \mu_{t+1} = -\tilde{\mu}_t\tilde{\xi}_t + \tilde{\xi}_t\mu_t + \tilde{\mu}_t\xi_t,$$

with $\tilde{y}_t = y_t + \tilde{\mu}_t\tilde{\varepsilon}_t$. Thus the Kalman filter and smoother can be applied to an approximating time-varying local level model with state vector $\alpha_t = \mu_t$.

11.8   Estimating the conditional mode

So far we have emphasised the use of the mode $\hat{\alpha}$ of $p(\alpha|y)$ to obtain a linear approximating model which we use to obtain Gaussian importance densities for simulation. If, however, the sole object of the investigation was to estimate α and if economy in computation was desired, then $\hat{\alpha}$ could be used for the purpose without recurse to simulation; indeed, this was the estimator used by Durbin and Koopman (1992) and an approximation to it was used by Fahrmeir (1992).

The property that the conditional mode is the most probable value of the state vector given the observations can be regarded as an optimality property; we now consider a further optimality property possessed by the conditional mode. To find it we examine the analogous situation in maximum likelihood estimation. The maximum likelihood estimate of a parameter ψ is the most probable value of it given the observations and is well known to be asymptotically efficient. To develop a finite-sample property analogous to asymptotic efficiency, Godambe (1960) and Durbin (1960) introduced the idea of unbiased estimating equations and Godambe showed that the maximum likelihood estimate of scalar ψ is the solution to an unbiased estimating equation which has a minimum variance property. This can be regarded as a finite-sample analogue of asymptotic efficiency. The extension to multidimensional ψ was indicated by Durbin (1960). Since that time there have been extensive developments of this basic idea, as can be seen from the collection of papers edited by Basawa, Godambe and Taylor (1997). Following Durbin (1997), we now develop a minimum-variance unbiased estimating equation property for the conditional mode estimate $\hat{\alpha}$ of the random vector α.

If α^* is the unique solution for α of the $mn \times 1$ vector equation $H(\alpha, y) = 0$ and if $E[H(\alpha, y)] = 0$, where expectation is taken with respect to the joint density $p(\alpha, y)$, we say that $H(\alpha, y) = 0$ is an unbiased estimating equation. It is obvious that the equation can be multiplied through by an arbitrary nonsingular matrix and still give the same solution α^*. We therefore standardise $H(\alpha, y)$ in the way that is usual in estimating equation theory and multiply it by $[E\{\dot{H}(\alpha, y)\}]^{-1}$, where $\dot{H}(\alpha, y) = \partial H(\alpha, y)/\partial\alpha'$, and then seek a minimum variance property for the resulting function $h(\alpha, y) = [E\{\dot{H}(\alpha, y)\}]^{-1} H(\alpha, y)$.

Let

$$\text{Var}[h(\alpha, y)] = E[h(\alpha, y)\, h(\alpha, y)'],$$

$$\mathcal{J} = E\left[\frac{\partial \log p(\alpha, y)}{\partial \alpha} \frac{\partial \log p(\alpha, y)}{\partial \alpha'}\right].$$

Under mild conditions that are likely to be satisfied in many practical cases, Durbin (1997) showed that $\text{Var}\,[h\,(\alpha, y)] - \mathcal{J}^{-1}$ is non-negative definite. If this is a zero matrix we say that the corresponding equation $H(\alpha, y) = 0$ is an *optimal esti-mating equation*. Now take $H(\alpha, y) = \partial \log p(\alpha, y)/\partial\alpha$. Then $E[\dot{H}(\alpha, y)] = -\mathcal{J}$, so $h(\alpha, y) = -\mathcal{J}^{-1}\partial \log p(\alpha, y)/\partial\alpha$. Thus $\text{Var}[h(\alpha, y)] = \mathcal{J}^{-1}$ and consequently the equation $\partial \log p(\alpha, y)/\partial\alpha = 0$ is optimal. Since $\hat{\alpha}$ is the solution of this, it is the solution of an optimal estimating equation. In this sense the conditional mode has an optimality property analogous to that of maximum likelihood estimates of fixed parameters in finite samples.

We have assumed above that there is a single mode and the question arises whether multimodality will create complications. If multimodality is suspected it can be investigated by using different starting points and checking whether iterations from them converge to the same mode. In none of the cases we have examined has multimodality of $p(\alpha|y)$ caused any difficulties. For this reason we regard it as unlikely that this will give rise to problems in routine time series analysis. If, however, multimodality were to occur in a particular case, we would suggest fitting a linear Gaussian model to the data at the outset and using this to define the first importance density $g_1(\eta|y)$ and conditional joint density $g_1(\eta, y)$. Simulation is employed to obtain a first estimate $\tilde{\eta}^{(1)}$ of $E(\eta|y)$ and from this a first estimate $\tilde{\theta}_t^{(1)}$ of θ_t is calculated for $t = 1, \ldots, n$. Now linearise the true densities at $\tilde{\eta}^{(1)}$ or $\tilde{\theta}_t^{(1)}$ to obtain a new approximating linear Gaussian model which defines a new $g(\eta|y)$, $g_2(\eta|y)$, and a new $g(\eta, y)$, $g_2(\eta, y)$. Simulation using these gives a new estimate $\tilde{\eta}^{(2)}$ of $E(\eta|y)$. This iterative process is continued until adequate convergence is achieved. We emphasise, however, that it is not necessary for the final value of α at which the model is linearised to be a precisely accurate estimate of either the mode or the mean of $p(\alpha|y)$. The only way that the choice of the value of α used as the basis for the simulation affects the final estimate $\hat{x}$ is in the variances due to simulation as we shall show later. Where necessary, the simulation sample size can be increased to reduce these error variances to any required extent. It will be noted that we are basing these iterations on the mean, not the mode. Since the mean, when it exists, is unique, no question of 'multimeanality' can arise.

11.9   Computational aspects of importance sampling

11.9.1   INTRODUCTION

In this section we describe the practical methods we use to implement importance sampling for the models of Chapter 10. The first step is to express the relevant formulae in terms of variables which are as simple as possible; we do this in §11.9.2. In §11.9.3 we describe antithetic variables, which increase the efficiency of the simulation by introducing a balanced structure into the simulation sample. Questions of initialisation of the approximating linear Gaussian model are considered in §11.9.4. In §11.9.5 and §11.9.6 we develop simulation-based methods which do not require importance sampling for special cases of linear state space models for which the errors have t-distributions or distributions which are mixtures of Gaussian distributions.

11.9.2   PRACTICAL IMPLEMENTATION OF IMPORTANCE SAMPLING

Up to this point we have based our exposition of the ideas underlying the use of importance sampling on α and y since these are the basic vectors of interest in the state space model. However, for practical computations it is important to express formulae in terms of variables that are as simple as possible. In particular, in place of the α_t's it is usually more convenient to work with the state disturbance terms $\eta_t = R_t'(\alpha_{t+1} - T_t\alpha_t)$. We therefore consider how to reformulate the previous results in terms of η rather than α.

By repeated substitution from the relation $\alpha_{t+1} = T_t\alpha_t + R_t\eta_t$, for $t = 1, \ldots, n$, we express $x(\alpha)$ as a function of α_1 and η; for notational convenience and because we intend to deal with the initialisation in §11.9.4, we suppress the dependence on α_1 and write $x(\alpha)$ as a function of η in the form $x^*(\eta)$. We next note that we could have written (11.1) in the form

$$\bar{x} = E[x^*(\eta)|y] = \int x^*(\eta)p(\eta|y)\,d\eta. \tag{11.32}$$

Analogously to (11.5) we have

$$\bar{x} = \frac{E_g[x^*(\eta)w^*(\eta, y)]}{E_g[w^*(\eta, y)]}, \qquad w^*(\eta, y) = \frac{p(\eta, y)}{g(\eta, y)}. \tag{11.33}$$

In this formula, E_g denotes expectation with respect to importance density $g(\eta|y)$, which is the conditional density of η given y in the approximating model, and

$$p(\eta, y) = \prod_{t=1}^{n} p(\eta_t)p(y_t|\theta_t),$$

where $\theta_t = Z_t \alpha_t$. In the special case where $y_t = \theta_t + \varepsilon_t$, $p(y_t|\theta_t) = p(\varepsilon_t)$. In a similar way, for the same special case,

$$g(\eta, y) = \prod_{t=1}^{n} g(\eta_t)g(\varepsilon_t).$$

For cases where the state equation is not linear and Gaussian, formula (11.33) provides the basis for the simulation estimates. When the state is linear and Gaussian, $p(\eta_t) = g(\eta_t)$ so in place of $w^*(\eta, y)$ in (11.33) we take

$$w^*(\theta, y) = \prod_{t=1}^{n} \frac{p(y_t|\theta_t)}{g(\varepsilon_t)}. \tag{11.34}$$

For the case $p(\eta_t) = g(\eta_t)$ and $y_t = \theta_t + \varepsilon_t$, we replace $w^*(\eta, y)$ by

$$w^*(\varepsilon) = \prod_{t=1}^{n} \frac{p(\varepsilon_t)}{g(\varepsilon_t)}. \tag{11.35}$$

11.9.3 ANTITHETIC VARIABLES

The simulations are based on random draws of η from the importance density $g(\eta|y)$ using the simulation smoother as described in §4.7; this computes efficiently a draw of η as a linear function of rn independent standard normal deviates where r is the dimension of vector η_t and n is the number of observations. Efficiency is increased by the use of antithetic variables. An *antithetic variable* in this context is a function of a random draw of η which is equiprobable with η and which, when included together with η in the estimate of $\bar{x}$ increases the efficiency of the estimation. We shall employ two types of antithetic variables. The first is the standard one given by $\check{\eta} = 2\hat{\eta} - \eta$ where $\hat{\eta} = E_g(\eta|y)$ is obtained from the disturbance smoother as described in §4.4. Since $\check{\eta} - \hat{\eta} = -(\eta - \hat{\eta})$ and η is normal, the two vectors η and $\check{\eta}$ are equi-probable. Thus we obtain two simulation samples from each draw of the simulation smoother; moreover, values of conditional means calculated from the two samples are negatively correlated, giving further efficiency gains. When this antithetic is used we say that the simulation sample is *balanced for location*.

The second antithetic variable was developed by Durbin and Koopman (1997). Let u be the vector of rn $N(0, 1)$ variables that is used in the simulation smoother to generate η and let $c = u'u$; then $c \sim \chi_{rn}^2$. For a given value of c let $q = \Pr(\chi_{rn}^2 < c) = F(c)$ and let $\acute{c} = F^{-1}(1 - q)$. Then as c varies, c and $\acute{c}$ have the same distribution. Now take, $\acute{\eta} = \hat{\eta} + \sqrt{\acute{c}/c}(\eta - \hat{\eta})$. Then $\acute{\eta}$ has the same distribution as η. This follows because c and $(\eta - \hat{\eta})/\sqrt{c}$ are independently distributed. Finally, take $\check{\eta} = \hat{\eta} + \sqrt{\check{c}/c}(\check{\eta} - \hat{\eta})$. When this antithetic is used we say that the simulation sample is *balanced for scale*. By using both antithetics we obtain a set of four equi-probable values of η for each run of the simulation smoother giving a simulation sample which is balanced for location and scale.

The number of antithetics can be increased without difficulty. For example, take c and q as above. Then q is uniformly distributed on $(0, 1)$ and we write $q \sim U(0, 1)$. Let $q_1 = q + 0.5$ modulo 1; then $q_1 \sim U(0, 1)$ and we have a balanced set of four $U(0, 1)$ variables, q, q_1, $1 - q$ and $1 - q_1$. Take $\acute{c} = F^{-1}(1 - q)$ as before and similarly $c_1 = F^{-1}(q_1)$ and $\acute{c}_1 = F^{-1}(1 - q_1)$. Then each of c_1 and $\acute{c}_1$ can be combined with η and $\check{\eta}$ as was $\acute{c}$ previously and we emerge with a balanced set of eight equi-probable values of η for each simulation. In principle this process could be extended indefinitely by taking $q_1 = q$ and $q_{j+1} = q_j + 2^{-k}$ modulo 1, for $j = 1, \ldots, 2^{k-1}$ and $k = 2, 3, \ldots$; however, two or four values of q are probably enough in practice. By using the standard normal distribution function applied to elements of u, the same idea could be used to obtain a new balanced value η_1 from η so by taking $\check{\eta}_1 = 2\hat{\eta} - \eta_1$ we would have four values of η to combine with the four values of c. In the following we will assume that we have generated N draws of η using the simulation smoother and the antithetic variables; this means that N is a multiple of the number of different values of η obtained from a single draw of the simulation smoother. For example, when 250 simulation samples are drawn by the smoother and the two basic antithetics are employed, one for location and the other for scale, $N = 1000$. In practice, we have found that satisfactory results are obtained by only using the two basic antithetics.

In theory, importance sampling could give an inaccurate result on a particular occasion if in the basic formulae (11.33) very high values of $w^*(\eta, y)$ are associated with very small values of the importance density $g(\eta|y)$ in such a way that together they make a significant contribution to $\bar{x}$, and if also, on this particular occasion, these values happen to be over- or under-represented; for further discussion of this point see Gelman *et al.* (1995, p. 307). In practice, we have not experienced difficulties from this source in any of the examples we have considered.

11.9.4 DIFFUSE INITIALISATION

We now consider the situation where the model is non-Gaussian and some elements of the initial state vector are diffuse, the remaining elements having a known joint density; for example, they could come from stationary series. Assume that α_1 is given by (5.2) with $\eta_0 \sim p_0(\eta_0)$ where $p_0(\cdot)$ is a known density. It is legitimate to assume that δ is normally distributed as in (5.3) since we intend to let $\kappa \to \infty$. The joint density of α and y is

$$p(\alpha, y) = p(\eta_0)g(\delta)\prod_{t=1}^{n} p(\eta_t)p(y_t|\theta_t), \qquad (11.36)$$

with $\eta_0 = R'_0(\alpha_1 - a)$, $\delta = A'(\alpha_1 - a)$ and $\eta_t = R'_t(\alpha_{t+1} - T_t\alpha_t)$ for $t = 1, \ldots, n$, since $p(y_t|\alpha_t) = p(y_t|\theta_t)$.

As in §11.3, we find the mode of $p(\alpha|y)$ by differentiating $\log p(\alpha, y)$ with respect to $\alpha_1, \ldots, \alpha_{n+1}$. For given κ the contribution from $\partial \log g(\delta)/\partial\alpha_1$ is $-A\delta/\kappa$ which $\to 0$ as $\kappa \to \infty$. Thus in the limit the mode equation is the same as (11.12) except that $\partial \log p(\alpha_1)/\partial\alpha_1$ is replaced by $\partial \log p(\eta_0)/\partial\alpha_1$. In the case that α_1 is entirely diffuse, the term $p(\eta_0)$ does not enter into (11.36), so the procedure given in §11.4 for finding the mode applies without change.

When $p(\eta_0)$ exists but is non-Gaussian, it is preferable to incorporate a normal approximation to it, $g(\eta_0)$ say, in the approximating Gaussian density $g(\alpha, y)$, rather than include a linearised form of its derivative $\partial \log p(\eta_0)/\partial\eta_0$ within the linearisation of $\partial \log p(\alpha, y)/\partial\alpha$. The reason is that we are then able to initialise the Kalman filter for the linear Gaussian approximating model by means of the standard initialisation routines developed in Chapter 5. For $g(\eta_0)$ we could take either the normal distribution with mean vector and variance matrix equal to those of $p(\eta_0)$ or with mean vector equal to the mode of $p(\eta_0)$ and variance matrix equal to $[-\partial^2 p(\eta_0)/\partial\eta_0\partial\eta'_0]^{-1}$. For substitution in the basic formula (11.5) we take

$$w(\alpha, y) = \frac{p(\eta_0)p(\alpha_2, \ldots, \alpha_{n+1}, y|\eta_0)}{g(\eta_0)g(\alpha_2, \ldots, \alpha_{n+1}, y|\eta_0)}, \qquad (11.37)$$

since the denisties $p(\delta)$ and $g(\delta)$ are the same and therefore cancel out; thus $w(\alpha, y)$ remains unchanged as $\kappa \to \infty$. The corresponding equation for (11.33) becomes simply

$$w^*(\eta, y) = \frac{p(\eta_0)p(\eta_1, \ldots, \eta_n, y)}{g(\eta_0)g(\eta_1, \ldots, \eta_n, y)}. \qquad (11.38)$$

While the expressions (11.37) and (11.38) are technically manageable, the practical worker may well believe in a particular situation that knowledge of $p(\eta_0)$ contributes such a small amount of information to the investigation that it can be simply ignored. In that event the factor $p(\eta_0)/g(\eta_0)$ disappears from (11.37), which amounts to treating the whole vector α_1 as diffuse, and this simplifies the analysis significantly. Expression (11.37) then reduces to

$$w(\alpha, y) = \prod_{t=1}^{n} \frac{p(\alpha_t)p(y_t|\alpha_t)}{g(\alpha_t)g(y_t|\alpha_t)}.$$

Expression (11.38) reduces to

$$w^*(\eta, y) = \prod_{t=1}^{n} \frac{p(\eta_t)p(y_t|\theta_t)}{g(\eta_t)g(y_t|\theta_t)},$$

with $\eta_t = R'_t(\alpha_{t+1} - T_t\alpha_t)$ for $t = 1, \ldots, n$.

For nonlinear models, the initialisation of the Kalman filter is similar and the details are handled in the same way.

11.9.5   TREATMENT OF t-DISTRIBUTION WITHOUT IMPORTANCE SAMPLING

In some cases it is possible to construct simulations by using antithetic variables without importance sampling. For example, it is well-known that if a random variable u_t has the standard t-distribution with ν degrees of freedom then u_t has the representation

$$u_t = \frac{\nu^{1/2}\varepsilon_t^*}{c_t^{1/2}}, \qquad \varepsilon_t^* \sim N(0, 1), \qquad c_t \sim \chi^2(\nu), \qquad \nu > 2, \qquad (11.39)$$

where ε_t^* and c_t are independent. In the case where ν is not an integer we take $\frac{1}{2}c_t$ as a gamma variable with parameter $\frac{1}{2}\nu$. It follows that if we consider the case where ε_t is univariate and we take ε_t in model (10.4) to have logdensity (10.19) then ε_t has the representation

$$\varepsilon_t = \frac{(\nu - 2)^{1/2}\sigma_\varepsilon\varepsilon_t^*}{c_t^{1/2}}, \qquad (11.40)$$

where ε_t^* and c_t are as in (11.39). Now take $\varepsilon_1^*, \ldots, \varepsilon_n^*$ and $c_1, \ldots, c_n$ to be mutually independent. Then conditional on $c_1, \ldots, c_n$ fixed, model (10.4) and (10.2), with $\eta_t \sim N(0, Q_t)$, is a linear Gaussian model with $H_t = \mathrm{Var}(\varepsilon_t) = (\nu - 2)\sigma_\varepsilon^2 c_t^{-1}$. Put $c = (c_1, \ldots, c_n)'$. We now show how to estimate the conditional means of functions of the state using simulation samples from the distribution of c.

Suppose first that α_t is generated by the linear Gaussian model $\alpha_{t+1} = T_t\alpha_t + R_t\eta_t$, $\eta_t \sim N(0, Q_t)$, and that as in (11.32) we wish to estimate

$$\bar{x} = \mathrm{E}[x^*(\eta)|y]$$
$$= \int x^*(\eta)p(c, \eta|y)\,dc\,d\eta$$
$$= \int x^*(\eta)p(\eta|c, y)p(c|y)\,dc\,d\eta$$
$$= \int x^*(\eta)p(\eta|c, y)p(c, y)p(y)^{-1}\,dc\,d\eta$$
$$= p(y)^{-1}\int x^*(\eta)p(\eta|c, y)p(y|c)p(c)\,dc\,d\eta. \qquad (11.41)$$

For given c, the model is linear and Gaussian. Let

$$\bar{x}(c) = \int x^*(\eta)p(\eta|c, y)\,d\eta.$$

For many cases of interest, $\bar{x}(c)$ is easily calculated by the Kalman filter and smoother, as in Chapters 4 and 5; to begin with, let us restrict attention to these cases. We have

$$p(y) = \int p(y, c)\,dc = \int p(y|c)p(c)\,dc,$$

where $p(y|c)$ is the likelihood given c which is easily calculated by the Kalman filter as in 7.2. Denote expectation with respect to density $p(c)$ by E_c Then from (11.41),

$$\bar{x} = \frac{E_c[\bar{x}(c)p(y|c)]}{E_c[p(y|c)]}.$$

(11.42)

We estimate this by simulation. Independent simulation samples $c^{(1)}, c^{(2)}, \ldots$ of c are easily obtained since c is a vector of independent χ_ν^2 variables. We suggest that antithetic values of χ_ν^2 are employed for each element of c, either in balanced pairs or balanced sets of four as described in §11.9.3. Suppose that values $c^{(1)}, \ldots, c^{(N)}$ have been selected. Then estimate $\bar{x}$ by

$$\hat{x} = \frac{\sum_{i=1}^N \bar{x}(c^{(i)})p(y|c^{(i)})}{\sum_{i=1}^N p(y|c^{(i)})}.$$

(11.43)

When $\bar{x}(c)$ cannot be computed by the Kalman filter and smoother we first draw a value of $c^{(i)}$ as above and then for the associated linear Gaussian model that is obtained when this value $c^{(i)}$ is fixed we draw a simulated value $\eta^{(i)}$ of η using the simulated smoother of §4.7, employing antithetics independently for both $c^{(i)}$ and $\eta^{(i)}$. The value $x^*(\eta^{(i)})$ is then calculated for each $\eta^{(i)}$. If there are N pairs of values $c^{(i)}, \eta^{(i)}$ we estimate $\bar{x}$ by

$$\hat{x}^* = \frac{\sum_{i=1}^N x^*(\eta^{(i)})p(y|c^{(i)})}{\sum_{i=1}^N p(y|c^{(i)})}.$$

(11.44)

Since we now have sampling variation arising from the drawing of values of η as well as from drawing values of c, the variance of $\hat{x}^*$ will be larger than of $\hat{x}$ for a given value of N. We present formulae (11.43) and (11.44) at this point for expository convenience in advance of the general treatment of analogous formulae in §12.2.

Now consider the case where the error term ε_t in the observation equation is $N(0, \sigma_\varepsilon^2)$ and where the elements of the error vector η_t in the state equation are independently distributed as Student's t. For simplicity assume that the number of degrees of freedom in these t-distributions are all equal to ν, although there is no difficulty in extending the treatment to the case where some of the degrees of freedom differ or where some elements are normally distributed. Analogously to (11.40) we have the representation

$$\eta_{it} = \frac{(\nu - 2)^{1/2}\sigma_{\eta i}\eta_{it}^*}{c_{it}^{1/2}}, \qquad \eta_{it}^* \sim N(0, 1), \qquad c_{it} \sim \chi_\nu^2, \qquad \nu > 2, \quad (11.45)$$

for $i = 1, \ldots, r$ and $t = 1, \ldots, n$, where $\sigma_{\eta i}^2 = \text{Var}(\eta_{it})$. Conditional on $c_{11}, \ldots,$ c_{rn} held fixed, the model is linear and Gaussian with $H_t = \sigma_\varepsilon^2$ and $\eta_t \sim N(0, Q_t)$ where $Q_t = \text{diag}[(\nu - 2)\sigma_{\eta 1}^2 c_{1t}^{-1}, \ldots, (\nu - 2)\sigma_{\eta r}^2 c_{rt}^{-1}]$. Formulae (11.43) and (11.44) remain valid except that $c^{(i)}$ is now a vector with r elements. The extension to the case where both ε_t and elements of η_t have t-distributions is straightforward.

The idea of using representation (11.39) for dealing with disturbances with t-distributions in the local level model by means of simulation was proposed by Shephard (1994) in the context of MCMC simulation.

11.9.6 TREATMENT OF GAUSSIAN MIXTURE DISTRIBUTIONS WITHOUT IMPORTANCE SAMPLING

An alternative to the t-distribution for representing error distributions with heavy tails is to employ the Gaussian mixture density (10.20), which for univariate ε_t we write in the form

$$p(\varepsilon_t) = \lambda^* N\big(0, \sigma_\varepsilon^2\big) + (1 - \lambda^*) N\big(0, \chi\sigma_\varepsilon^2\big), \qquad 0 < \lambda^* < 1. \quad (11.46)$$

It is obvious that values of ε_t with this density can be realised by means of a two-stage process in which we first select the value of a binomial variable b_t such that $\Pr(b_t = 1) = \lambda^*$ and $\Pr(b_t = 0) = 1 - \lambda^*$, and then take $\varepsilon_t \sim N(0, \sigma_\varepsilon^2)$ if $b_t = 1$ and $\varepsilon_t \sim N(0, \chi\sigma_\varepsilon^2)$ if $b_t = 0$. Assume that the state vector α_t is generated by the linear Gaussian model $\alpha_{t+1} = T_t\alpha_t + R_t\eta_t$, $\eta_t \sim N(0, Q_t)$. Putting $b = (b_1, \ldots, b_n)'$, it follows that for b given, the state space model is linear and Gaussian. We can therefore employ the same approach for the mixture distribution that we used for the t-distribution in the previous subsection, giving as in (11.41),

$$\bar{x} = p(y)^{-1} M^{-1} \sum_{j=1}^{M} \int x^*(\eta) p\big(\eta | b_{(j)}, y\big) p\big(y | b_{(j)}\big) p\big(b_{(j)}\big) d\eta, \quad (11.47)$$

where $b_{(1)}, \ldots, b_{(M)}$ are the $M = 2^n$ possible values of b. Let

$$\bar{x}(b) = \int x^*(\eta) p(\eta | b, y) d\eta,$$

and consider cases where this can be calculated by the Kalman filter and smoother. Denote expectation over the distribution of b by E_b. Then

$$p(y) = M^{-1} \sum_{j=1}^{M} p\big(y | b_{(j)}\big) p\big(b_{(j)}\big) = E_b[p(y|b)],$$

and analogously to (11.42) we have,

$$\bar{x} = \frac{E_b[\bar{x}(b) p(y|b)]}{E_b[p(y|b)]}. \quad (11.48)$$

We estimate this by simulation. A simple way to proceed is to choose a sequence $b^{(1)}, \ldots, b^{(N)}$ of random values of b and then estimate $\bar{x}$ by

$$\hat{x} = \frac{\sum_{i=1}^{N} \bar{x}\big(b^{(i)}\big) p\big(y | b^{(i)}\big)}{\sum_{i=1}^{N} p\big(y | b^{(i)}\big)}. \quad (11.49)$$

Variability in this formula arises only from the random selection of b. To construct antithetic variables for the problem we consider how this variability can be restricted while preserving correct overall probabilities. We suggest the following approach. Consider the situation where the probability $1 - \lambda^*$ in (11.46) of taking $N(0, \chi \sigma_\varepsilon^2)$ is small. Take $1 - \lambda^* = 1/B$ where B is an integer, say $B = 10$ or 20. Divide the simulation sample of values of b into K blocks of B, with $N = KB$. Within each block, and for each $t = 1, \ldots, n$, choose integer j randomly from 1 to B, put the jth value in the block as $b_t = 0$ and the remaining $B - 1$ values in the block as $b_t = 1$. Then take $b^{(i)} = (b_1, \ldots, b_n)'$ with $b_1, \ldots, b_n$ defined in this way for $i = 1, \ldots, N$ and use formula (11.49) to estimate $\bar{x}$. With this procedure we have ensured that for each i, $\Pr(b_t = 1) = \lambda^*$ as desired, with b_s and b_t independent for $s \neq t$, while enforcing balance in the sample by requiring that within each block b_t has exactly $B - 1$ values of 1 and one value of 0. Of course, choosing integers at random from 1 to B is a much simpler way to select a simulation sample than using the simulation smoother.

The restriction of B to integer values is not a serious drawback since the results are insensitive to relatively small variations in the value of λ^*, and in any case the value of λ^* is normally determined on a trial-and-error basis. It should be noted that for purposes of estimating mean square errors due to simulation, the numerator and denominator of (11.49) should be treated as composed of M independent values.

The idea of using the binomial representation of (11.46) in MCMC simulation for the local level model was proposed by Shephard (1994).

12
Analysis from a classical standpoint

12.1 Introduction

In this chapter we will discuss classical inference methods based on importance sampling for analysing data from non-Gaussian and nonlinear models. In §12.2 we show how to estimate means and variances of functions of the state using simulation and antithetic variables. We also derive estimates of the additional variances of estimates due to simulation. We use these results in §12.3 to obtain estimates of conditional densities and distribution functions of scalar functions of the state. In §12.4 we investigate the complications due to missing observations and we also consider the related question of forecasting. Finally, in §12.5 we show how to estimate unknown parameters by maximum likelihood and we examine the effect of parameter estimation errors on estimates of variance.

12.2 Estimating conditional means and variances

Following up the treatment in §11.9.2, we now consider details of the estimation of conditional means $\bar{x}$ of functions $x^*(\eta)$ of the stacked state error vector and the estimation of error variances of our estimates. Let

$$w^*(\eta) = \frac{p(\eta, y)}{g(\eta, y)},$$

taking the dependence of $w^*(\eta)$ on y as implicit since y is constant from now on. Then (11.33) gives

$$\bar{x} = \frac{\mathrm{E}_g\left[x^*(\eta)w^*(\eta)\right]}{\mathrm{E}_g\left[w^*(\eta)\right]}, \tag{12.1}$$

which is estimated by

$$\hat{x} = \frac{\sum_{i=1}^{N} x_i w_i}{\sum_{i=1}^{N} w_i}, \tag{12.2}$$

where

$$x_i = x^*\left(\eta^{(i)}\right), \qquad w_i = w^*\left(\eta^{(i)}\right) = \frac{p\left(\eta^{(i)}, y\right)}{g\left(\eta^{(i)}, y\right)},$$

and $\eta^{(i)}$ is the ith draw of η from the importance density $g(\eta|y)$ for $i = 1, \ldots, N$. Note that by 'the ith draw of η' here we include antithetic variables as described in §11.9.3. For the case where $x^*(\eta)$ is a vector we could at this point present formulae for estimating the matrix $\mathrm{Var}[x^*(\eta)|y]$ and also the variance matrix due to simulation of $\hat{x} - \bar{x}$. However, from a practical point of view the covariance terms are of little interest so it seems sensible to focus on variance terms by taking $x^*(\eta)$ as a scalar for estimation of variances; extension to include covariance terms is straightforward. We estimate $\mathrm{Var}[x^*(\eta)|y]$ by

$$\widehat{\mathrm{Var}}[x^*(\eta)|y] = \frac{\sum_{i=1}^{N} x_i^2 w_i}{\sum_{i=1}^{N} w_i} - \hat{x}^2. \qquad (12.3)$$

The estimation error due to the simulation is

$$\hat{x} - \bar{x} = \frac{\sum_{i=1}^{N} w_i(x_i - \bar{x})}{\sum_{i=1}^{N} w_i}.$$

To estimate the variance of this, consider the introduction of the antithetic variables as described in §11.9.3 and for simplicity will restrict the exposition to the case of the two basic antithetics for location and scale; the extension to a larger number of antithetics is straightforward. Denote the sum of the four values of $w_i(x_i - \bar{x})$ that come from the jth run of the simulation smoother by v_j and the sum of the corresponding values of $w_i(x_i - \hat{x})$ by $\hat{v}_j$. For N large enough, since the draws from the simulation smoother are independent, the variance due to simulation is, to a good approximation,

$$\mathrm{Var}_s(\hat{x}) = \frac{1}{4N} \frac{\mathrm{Var}(v_j)}{[E_g\{w^*(\eta)\}^2]}, \qquad (12.4)$$

which we estimate by

$$\widehat{\mathrm{Var}}_s(\hat{x}) = \frac{\sum_{j=1}^{N/4} \hat{v}_j^2}{\left(\sum_{i=1}^{N} w_i\right)^2}. \qquad (12.5)$$

The ability to estimate simulation variances so easily is an attractive feature of our methods.

12.3 Estimating conditional densities and distribution functions

When $x^*(\eta)$ is a scalar function the above technique can be used to estimate the conditional distribution function and the conditional density function of x given y. Let $G[x|y] = \Pr[x^*(\eta) \le x|y]$ and let $I_x(\eta)$ be an indicator which is unity if $x^*(\eta) \le x$ and is zero if $x^*(\eta) > x$. Then $G(x|y) = E_g(I_x(\eta)|y)$. Since $I_x(\eta)$ is a function of η we can treat it in the same way as $x^*(\eta)$. Let S_x be the sum of the values of w_i for which $x_i \le x$, for $i = 1, \ldots, N$. Then estimate $G(x|y)$ by

$$\hat{G}(x|y) = \frac{S_x}{\sum_{i=1}^{N} w_i}. \tag{12.6}$$

This can be used to estimate quantiles. We order the values of x_i and we order the corresponding values of w_i accordingly. The ordered sequences for x_i and w_i are denoted by $x_{[i]}$ and $w_{[i]}$, respectively. The $100k\%$ quantile is given by $x_{[m]}$ which is chosen such that

$$\frac{\sum_{i=1}^{m} w_{[i]}}{\sum_{i=1}^{N} w_{[i]}} \approx k.$$

We may interpolate between the two closest values for m in this approximation to estimate the $100k\%$ quantile. The approximation error becomes smaller as N increases.

Similarly, if δ is the interval $(x - \frac{1}{2}d, x + \frac{1}{2}d)$ where d is suitably small and positive, let S^δ be the sum of the values of w_i for which $x^*(\eta) \in \delta$. Then the estimate of the conditional density $p(x|y)$ of x given y is

$$\hat{p}(x|y) = d^{-1} \frac{S^\delta}{\sum_{i=1}^{N} w_i}. \tag{12.7}$$

This estimate can be used to construct a histogram.

We now show how to generate a sample of M independent values from the estimated conditional distribution of $x^*(\eta)$ using importance resampling; for further details of the method see Gelfand and Smith (1999) and Gelman *et al.* (1995). Take $x^{[k]} = x_j$ with probability $w_j / \sum_{i=1}^{N} w_i$ for $j = 1, \ldots, N$. Then

$$\Pr\left(x^{[k]} \le x\right) = \frac{\sum_{x_j \le x} w_j}{\sum_{i=1}^{N} w_i} = \hat{G}(x|y).$$

Thus $x^{[k]}$ is a random draw from the distribution function given by (12.6). Doing this M times with replacement gives a sample of $M \le N$ independent draws. The sampling can also be done without replacement but the values are not then independent.

12.4   Forecasting and estimating with missing observations

The treatment of missing observations and forecasting by the methods of this chapter is straightforward. For missing observations, our objective is to estimate $\bar{x} = \int x^*(\eta) p(\eta|y) d\eta$ where the stacked vector y contains only those observational elements actually observed. We achieve this by omitting from the linear Gaussian approximating model the observational components that correspond to the missing elements in the original model. Only the Kalman filter and smoother algorithms are needed in the determination of the approximating model and we described in §4.8 how the filter is modified when observational vectors or elements are missing. For the simulation, the simulation smoother of §4.7 must be similarly modified to allow for the missing elements.

For forecasting, our objective is to estimate $\bar{y}_{n+j} = E(y_{n+j}|y)$, $j = 1, \ldots, J$, where we assume that $y_{n+1}, \ldots y_{n+J}$ and $\alpha_{n+2}, \ldots, \alpha_{n+J}$ have been generated by model (10.1) and (10.2), noting that α_{n+1} has already been generated by (10.2) with $t = n$. It follows from (10.1) that

$$\bar{y}_{n+j} = E[E(y_{n+j}|\theta_{n+j})|y], \tag{12.8}$$

for $j = 1, \ldots, J$, where $\theta_{n+j} = Z_{n+j}\alpha_{n+j}$, with $Z_{n+1}, \ldots, Z_{n+J}$ assumed known. We estimate this as in §12.2 with $x^*(\eta) = E(y_{n+j}|\theta_{n+j})$, extending the simulation smoother for $t = n + 1, \ldots, n + J$.

For exponential families,

$$E(y_{n+j}|\theta_{n+j}) = \dot{b}_{n+j}(\theta_{n+j}),$$

as in §10.3 for $t \le n$, so we take $x^*(\eta) = \dot{b}_{n+j}(\theta_{n+j})$, for $j = 1, \ldots, J$. For the model $y_t = \theta_t + \varepsilon_t$ in (10.4) we take $x^*(\eta) = \theta_t$.

12.5 Parameter estimation

12.5.1 INTRODUCTION

In this section we consider the estimation of the parameter vector ψ by maximum likelihood. Since analytical methods are not feasible we employ techniques based on simulation using importance sampling. We shall find that the techniques we develop are closely related to those we employed earlier in this chapter for estimation of the mean of $x(\alpha)$ given y, with $x(\alpha) = 1$. Estimation of ψ by maximum likelihood using importance sampling was considered briefly by Shephard and Pitt (1997) and in more detail by Durbin and Koopman (1997) for the special case where $p(y_t|\theta_t)$ is non-Gaussian but α_t is generated by a linear Gaussian model. In this section we will begin by considering first the general case where both $p(y_t|\theta_t)$ and the state error density $p(\eta_t)$ in (10.2) are non-Gaussian and will specialise later to the simpler case where $p(\eta_t)$ is Gaussian. We will also consider the case where the state space models are nonlinear. Our approach will be to estimate the loglikelihood by simulation and then to estimate ψ by maximising the resulting value numerically.

12.5.2 ESTIMATION OF LIKELIHOOD

The likelihood $L(\psi)$ is defined by $L(\psi) = p(y|\psi)$, where for convenience we suppress the dependence of $L(\psi)$ on y, so we have

$$L(\psi) = \int p(\alpha, y) \, d\alpha.$$

Dividing and multiplying by the importance density $g(\alpha|y)$ as in §11.2 gives

$$
\begin{aligned}
L(\psi) &= \int \frac{p(\alpha, y)}{g(\alpha|y)} g(\alpha|y) \, d\alpha \\
&= g(y) \int \frac{p(\alpha, y)}{g(\alpha, y)} g(\alpha|y) \, d\alpha \\
&= L_g(\psi) \, E_g \left[w(\alpha, y) \right],
\end{aligned}
\tag{12.9}
$$

where $L_g(\psi) = g(y)$ is the likelihood of the approximating linear Gaussian model that we employ to obtain the importance density $g(\alpha|y)$, E_g denotes expectation with respect to density $g(\alpha|y)$, and $w(\alpha, y) = p(\alpha, y)/g(\alpha, y)$ as in (11.5). Indeed we observe that (12.9) is essentially equivalent to (11.4). We note the elegant feature of (12.9) that the non-Gaussian likelihood $L(\psi)$ has been obtained as an adjustment to the linear Gaussian likelihood $L_g(\psi)$, which is easily calculated by the Kalman filter; moreover, the adjustment factor $E_g[w(\alpha, y)]$ is readily estimable by simulation. Obviously, the closer the importance joint density $g(\alpha, y)$ is to the non-Gaussian density $p(\alpha, y)$, the smaller will be the simulation sample required.

For practical computations we follow the practice discussed in §11.9.2 and §12.2 of working with the signal $\theta_t = Z_t \alpha_t$ in the observation equation and the state disturbance η_t in the state equation, rather than with α_t directly, since these lead to simpler computational procedures. In place of (12.9) we therefore use the form

$$
L(\psi) = L_g(\psi) \, E_g[w^*(\eta, y)],
\tag{12.10}
$$

where $L(\psi)$ and $L_g(\psi)$ are the same as in (12.9) but E_g and $w^*(\eta, y)$ have the interpretations discussed in §11.9.2. We then suppress the dependence on y and write $w^*(\eta)$ in place of $w^*(\eta, y)$ as in §12.2. We employ antithetic variables as in §11.9.3, and analogously to (12.2) our estimate of $L(\psi)$ is

$$
\hat{L}(\psi) = L_g(\psi) \bar{w},
\tag{12.11}
$$

where $\bar{w} = (1/N) \sum_{i=1}^{N} w_i$, with $w_i = w^*(\eta^{(i)})$ where $\eta^{(1)}, \ldots \eta^{(N)}$ is the simulation sample generated by the importance density $g(\eta|y)$.

12.5.3   MAXIMISATION OF LOGLIKELIHOOD

We estimate ψ by the value $\hat{\psi}$ of ψ that maximises $\hat{L}(\psi)$. In practice, it is numerically more stable to maximise

$$
\log \hat{L}(\psi) = \log L_g(\psi) + \log \bar{w},
\tag{12.12}
$$

rather than to maximise $\hat{L}(\psi)$ directly because the likelihood value can become very large. Moreover, the value of ψ that maximises $\log \hat{L}(\psi)$ is the same as the value that maximises $\hat{L}(\psi)$.

To calculate $\hat{\psi}$, $\log \hat{L}(\psi)$ is maximised by any convenient iterative numerical optimisation technique, as discussed, for example in §7.3.2. To ensure stability of the iterative process, it is important to use the same random numbers from the simulation smoother for each value of ψ. To start the iteration, an initial value of ψ can be obtained by maximising the approximate loglikelihood

$$\log L(\psi) \approx \log L_g(\psi) + \log w(\hat{\eta}),$$

where $\hat{\eta}$ is the mode of $g(\eta|y)$ that is determined during the process of approximating $p(\eta|y)$ by $g(\eta|y)$; alternatively, the more accurate non-simulated approximation given in expression (21) of Durbin and Koopman (1997) may be used.

12.5.4 VARIANCE MATRIX OF MAXIMUM LIKELIHOOD ESTIMATE

Assuming that appropriate regularity conditions are satisfied, the estimate of the large-sample variance matrix of $\hat{\psi}$ is given by the standard formula

$$\hat{\Omega} = \left[-\frac{\partial^2 \log L(\psi)}{\partial \psi \partial \psi'} \right]^{-1} \Bigg|_{\psi = \hat{\psi}}, \tag{12.13}$$

where the derivatives of $\log L(\psi)$ are calculated numerically from values of ψ in the neighbourhood of $\hat{\psi}$.

12.5.5 EFFECT OF ERRORS IN PARAMETER ESTIMATION

In the above treatment we have performed classical analyses in the traditional way by first assuming that the parameter vector is known and then substituting the maximum likelihood estimate $\hat{\psi}$ for ψ. The errors $\hat{\psi} - \psi$ give rise to biases in the estimates of functions of the state and disturbance vectors, but since the biases are of order n^{-1} they are usually small enough to be neglected. It may, however, be important to investigate the amount of bias in particular cases. In §7.3.7 we described techniques for estimating the bias for the case where the state space model is linear and Gaussian. Exactly the same procedure can be used for estimating the bias due to errors $\hat{\psi} - \psi$ for the non-Gaussian and nonlinear models considered in this chapter.

12.5.6 MEAN SQUARE ERROR MATRIX DUE TO SIMULATION

We have denoted the estimate of ψ that is obtained from the simulation by $\hat{\psi}$; let us denote by $\tilde{\psi}$ the 'true' maximum likelihood estimate of ψ that would be obtained by maximising the exact $\log L(\psi)$ without simulation, if this could be done. The error due to simulation is $\hat{\psi} - \tilde{\psi}$, so the mean square error matrix is

$$\text{MSE}(\hat{\psi}) = \text{E}_g[(\hat{\psi} - \tilde{\psi})(\hat{\psi} - \tilde{\psi})'].$$

Now $\hat{\psi}$ is the solution of the equation

$$\frac{\partial \log \hat{L}(\psi)}{\partial \psi} = 0,$$

which on expansion about $\tilde{\psi}$ gives approximately,

$$\frac{\partial \log \hat{L}(\tilde{\psi})}{\partial \psi} + \frac{\partial^2 \log \hat{L}(\tilde{\psi})}{\partial \psi \, \partial \psi'}(\hat{\psi} - \tilde{\psi}) = 0,$$

where

$$\frac{\partial \log \hat{L}(\tilde{\psi})}{\partial \psi} = \frac{\partial \log \hat{L}(\psi)}{\partial \psi}\bigg|_{\psi = \tilde{\psi}}, \qquad \frac{\partial^2 \log \hat{L}(\tilde{\psi})}{\partial \psi \, \partial \psi'} = \frac{\partial^2 \log \hat{L}(\psi)}{\partial \psi \, \partial \psi'}\bigg|_{\psi = \tilde{\psi}},$$

giving

$$\hat{\psi} - \tilde{\psi} = \left[-\frac{\partial^2 \log \hat{L}(\tilde{\psi})}{\partial \psi \, \partial \psi'} \right]^{-1} \frac{\partial \log \hat{L}(\tilde{\psi})}{\partial \psi}.$$

Thus to a first approximation we have

$$\mathrm{MSE}(\hat{\psi}) = \hat{\Omega} \, \mathrm{E}_g \left[\frac{\partial \log \hat{L}(\tilde{\psi})}{\partial \psi} \frac{\partial \log \hat{L}(\tilde{\psi})}{\partial \psi'} \right] \hat{\Omega}, \qquad (12.14)$$

where $\hat{\Omega}$ is given by (12.13).

From (12.12) we have

$$\log \hat{L}(\tilde{\psi}) = \log L_g(\tilde{\psi}) + \log \bar{w},$$

so

$$\frac{\partial \log \hat{L}(\tilde{\psi})}{\partial \psi} = \frac{\partial \log L_g(\tilde{\psi})}{\partial \psi} + \frac{1}{\bar{w}} \frac{\partial \bar{w}}{\partial \psi}.$$

Similarly, for the true loglikelihood $\log L(\tilde{\psi})$ we have

$$\frac{\partial \log L(\tilde{\psi})}{\partial \psi} = \frac{\partial \log L_g(\tilde{\psi})}{\partial \psi} + \frac{\partial \log \mu_w}{\partial \psi},$$

where $\mu_w = \mathrm{E}_g(\bar{w})$. Since $\tilde{\psi}$ is the 'true' maximum likelihood estimator of ψ,

$$\frac{\partial \log L(\tilde{\psi})}{\partial \psi} = 0.$$

Thus

$$\frac{\partial \log L_g(\tilde{\psi})}{\partial \psi} = -\frac{\partial \log \mu_w}{\partial \psi} = -\frac{1}{\mu_w} \frac{\partial \mu_w}{\partial \psi},$$

so we have

$$\frac{\partial \log \hat{L}(\tilde{\psi})}{\partial \psi} = \frac{1}{\bar{w}} \frac{\partial \bar{w}}{\partial \psi} - \frac{1}{\mu_w} \frac{\partial \mu_w}{\partial \psi}.$$

It follows that, to a first approximation,

$$\frac{\partial \log \hat{L}(\tilde{\psi})}{\partial \psi} = \frac{1}{\bar{w}} \frac{\partial}{\partial \psi} (\bar{w} - \mu_w),$$

and hence

$$E_g \left[\frac{\partial \log \hat{L}(\tilde{\psi})}{\partial \psi} \frac{\partial \log \hat{L}(\tilde{\psi})}{\partial \psi'} \right] = \frac{1}{\bar{w}^2} \text{Var} \left(\frac{\partial \bar{w}}{\partial \psi} \right).$$

Taking the case of two antithetics, denote the sum of the four values of w obtained from each draw of the simulation smoother by w_j^* for $j = 1, \ldots, N/4$. Then $\bar{w} = N^{-1} \sum_{j=1}^{N/4} w_j^*$, so

$$\text{Var} \left(\frac{\partial \bar{w}}{\partial \psi} \right) = \frac{4}{N} \text{Var} \left(\frac{\partial w_j^*}{\partial \psi} \right).$$

Let $q^{(j)} = \partial w_j^* / \partial \psi$, which we calculate numerically at $\psi = \hat{\psi}$, and let $\bar{q} = (4/N) \sum_{j=1}^{N/4} q^{(j)}$. Then estimate (12.14) by

$$\widehat{\text{MSE}}(\hat{\psi}) = \hat{\Omega} \left[\left(\frac{4}{N\bar{w}} \right)^2 \sum_{j=1}^{N/4} (q^{(j)} - \bar{q})(q^{(j)} - \bar{q})' \right] \hat{\Omega}. \tag{12.15}$$

The square roots of the diagonal elements of (12.15) may be compared with the square roots of the diagonal elements of $\hat{\Omega}$ in (12.13) to obtain relative standard errors due to simulation.

12.5.7 ESTIMATION WHEN THE STATE DISTURBANCES ARE GAUSSIAN

When the state disturbance vector is distributed as $\eta_t \sim N(0, Q_t)$, the calculations are simplified substantially. Denote the normal density of η by $g(\eta)$ and arrange the approximating linear Gaussian model to have the same state disturbance density. Then $p(\eta, y) = g(y)p(y|\theta)$ and $g(\eta, y) = g(\eta)g(y|\theta)$ so for $w^*(\eta, y)$ in (12.10) we have $w^*(\eta, y) = p(\eta, y)/g(\eta, y) = p(y|\theta)/g(y|\theta)$ giving

$$L(\psi) = L_g(\psi) E_g \left[\frac{p(y|\theta)}{g(y|\theta)} \right], \tag{12.16}$$

which is the same as (8) of Durbin and Koopman (1997). Since in many important applications y_t is univariate, or at least has dimensionality significantly smaller than that of η_t, (12.16) is normally substantially easier to handle than (12.10). Other aspects of the analysis proceed as in the general case considered in earlier sections.

12.5.8 CONTROL VARIABLES

Another traditional device for improving the efficiency of simulations is to use *control variables*. A control variable is a variable whose mean is exactly known and for which an estimate can be calculated from the simulation sample. The idea is then to use the difference between the sample estimate and the true mean to construct an adjustment to the initial estimate of a quantity of interest which enhances efficiency. Although we do not make use of control variables in the illustrations considered in Chapter 14, we believe it is worth including a brief treatment of them here in order to stimulate readers to consider their potential for other problems. We shall base our discussion on the use we made of control variables for the estimation of the likelihood function in Durbin and Koopman (1997) for the case where observations have non-Gaussian density $p(y_t|\theta_t)$, where $\theta_t = Z_t \alpha_t$ and where the state equation has the linear Gaussian form $\alpha_{t+1} = T_t \alpha_t + R_t \eta_t$, $\eta_t \sim N(0, Q_t)$ for $t = 1, \ldots, n$. For simplicity, we assume that y_t is univariate and that the distribution of α_1 is known.

In (12.16) let

$$w(\theta) = \frac{p(y|\theta)}{g(y|\theta)}, \qquad l(\theta) = \log w(\theta), \tag{12.17}$$

and put

$$w = w(\theta), \qquad w'_t = \frac{\partial w(\theta)}{\partial \theta_t}, \qquad w''_t = \frac{\partial^2 w(\theta)}{\partial \theta_t^2}, \ldots,$$

$$l = l(\theta), \qquad l'_t = \frac{\partial l(\theta)}{\partial \theta_t}, \qquad l''_t = \frac{\partial^2 l(\theta)}{\partial \theta_t^2}, \ldots.$$

We start with the idea that a way to obtain a control variable for $w(\theta)$ is to expand it as a Taylor series and then take the difference between exact and simulation means of the first few terms. We therefore expand $w(\theta)$ about $\hat{\theta} = E_g(\theta|y)$. We have

$$w'_t = w l'_t,$$
$$w''_t = w[l''_t + (l'_t)^2],$$
$$w'''_t = w[l'''_t + 3l'_t l''_t + (l'_t)^3],$$
$$w''''_t = w[l''''_t + 4l'_t l'''_t + 3(l''_t)^2 + 6l''_t(l'_t)^2 + (l'_t)^4].$$

Since the state equation is linear and Gaussian it follows from §11.4 that $l'_t = 0$ and $l''_t = 0$ at $\theta = \hat{\theta}$. Denote the values of w, l'''_t and l''''_t at $\theta = \hat{\theta}$ by $\hat{w}, \hat{l}'''_t$ and $\hat{l}''''_t$. The required Taylor series as far as the term in $(\theta_t - \hat{\theta}_t)^4$ is therefore

$$w(\theta) = \hat{w}\left[1 + \frac{1}{6}\sum_{t=1}^{n}\hat{l}'''_t(\theta_t - \hat{\theta}_t)^3 + \frac{1}{24}\sum_{t=1}^{n}\hat{l}''''_t(\theta_t - \hat{\theta}_t)^4 + \cdots\right].$$

Now draw a sample of N values of $\varepsilon_t^{(i)}$ given y, and hence of $\theta_t^{(i)} = y_t - \varepsilon_t^{(i)}$, for the approximating linear Gaussian model using the simulation smoother. Take as the control variable $\hat{w}\bar{f}$ where $\bar{f} = N^{-1}\sum_{i=1}^{N} f_i$ with

$$f_i = \frac{1}{6}\sum_{t=1}^{n}\hat{l}'''_t\left(\theta_t^{(i)} - \hat{\theta}_t\right)^3 + \frac{1}{24}\sum_{t=1}^{n}\hat{l}''''_t\left(\theta_t^{(i)} - \hat{\theta}_t\right)^4,$$

and assume that the location balanced antithetic of §11.9.3 is used. The effect of this is that $\sum_{i=1}^{N}(\theta_t^{(i)} - \hat{\theta}_t)^3 = 0$ so we can neglect the first term in the expression for f_i. The antithetic therefore reduces to $\hat{w}\bar{f}$ where $\bar{f} = N^{-1}\sum_{i=1}^{N} f_i$ with

$$f_i = \frac{1}{24}\sum_{t=1}^{n}\hat{l}_t''''(\theta_t^{(i)} - \hat{\theta}_t)^4.$$

Let $g_t = \text{Var}(\varepsilon_t|y)$, which is obtained from the disturbance smoother. Then

$$E_g(f_i) = \frac{1}{8}\sum_{t=1}^{n}\hat{l}_t''''g_t^2.$$

Without the use of the control variable, the estimate of (12.16) that we would have used is

$$\hat{L}(\psi) = L_g(\psi)\bar{w}, \qquad (12.18)$$

where $\bar{w} = N^{-1}\sum_{i=1}^{N} w_i$ with $w_i = w(\theta^{(i)})$ given by (12.17). Using the control variable, we estimate $L(\psi)$ by

$$\hat{L}^\dagger(\psi) = L_g(\psi)\bar{w}^\dagger, \qquad (12.19)$$

where $\bar{w}^\dagger = N^{-1}\sum_{i=1}^{N} w_i^\dagger$ with $w_i^\dagger = w_i - \hat{w}(f_i - \frac{1}{8}\sum_{t=1}^{n}\hat{l}_t''''g_t^2)$. What we have set out to achieve by the use of the control variable is to take out a substantial amount of the variation in $\bar{w}$ from expression (12.18).

We applied the technique in Durbin and Koopman (1997) to two illustrations. Our overall conclusion was stated in the following terms: 'The location and scale balancing variables (by this we meant the antithetics) together are so efficient that the extra variance reduction provided by the new control variable is small. Nevertheless, the control variable is so cheap computationally relative to the cost of extra simulation samples that it is worthwhile using them in practical applications.' Since we wrote these words our views have shifted somewhat due to the rapid reduction in computing costs. We therefore decided not to highlight control variables in our presentation of simulation methods in Chapter 11. Nevertheless, as we stated above, we believe it is worthwhile including this brief presentation here in the belief that control variables might prove to be of value in specific time series applications.

13
Analysis from a Bayesian standpoint

13.1 Introduction

In this chapter we discuss the analysis of non-Gaussian and nonlinear state space models from the standpoint of Bayesian inference. As we made clear in the introduction to Chapter 8, which deals with the Bayesian approach to the analysis of the linear Gaussian model, we regard both the classical and Bayesian approaches as providing valid modes of inferences in appropriate circumstances.

In the next section we develop Bayesian techniques for estimating posterior means and posterior variance matrices of functions of the state vector. We also show how to estimate posterior distribution and density functions of scalar functions of the state vector. Remarkably, it turns out that the basic ideas of importance sampling and antithetic variables developed for classical analysis in Chapter 11 can be applied with little essential change to the Bayesian case. Different considerations apply to questions regarding the posterior distribution of the parameter vector and we deal with these in §13.4. The treatment is based on the methods developed by Durbin and Koopman (2000). We have found these methods to be transparent and computationally efficient.

Previous work on Bayesian analysis of non-Gaussian state space models has been based almost entirely on Markov chain Monte Carlo (MCMC) methods: we note in particular here the contributions by Carlin *et al.* (1992), Shephard (1994), Carter and Kohn (1994), Carter and Kohn (1996), Carter and Kohn (1997), Shephard and Pitt (1997), Cargnoni, Muller and West (1997) and Gamerman (1998). General accounts of Bayesian methodology and computation are given by Gelman *et al.* (1995) and Bernardo and Smith (1994). In §13.5 we give a brief overview of MCMC methods as applied to state space models. References to software will also be given so that workers who wish to compare the methods based on importance sampling with the MCMC approach will be able to do so.

13.2 Posterior analysis of functions of the state vector

In the Bayesian approach the parameter vector ψ is treated as random with a prior density $p(\psi)$, which to begin with we shall take to be a proper prior. We first obtain some basic formulae analogous to those derived in §11.2 for the classical case. Suppose that we wish to calculate the posterior mean

$$\bar{x} = E[x(\alpha)|y],$$

of a function $x(\alpha)$ of the stacked state vector α given the stacked observation vector y. As we shall show, this is a general formulation which enables us not only to estimate posterior means of quantities of interest such as the trend or seasonal, but also posterior variance matrices and posterior distribution functions and densities of scalar functions of the state. We shall estimate $\bar{x}$ by simulation techniques based on importance sampling and antithetic variables analogous to those developed in Chapter 11 for the classical case.

We have

$$\bar{x} = \int x(\alpha) p(\psi, \alpha|y) \, d\psi \, d\alpha$$

$$= \int x(\alpha) p(\psi|y) p(\alpha|\psi, y) \, d\psi \, d\alpha. \tag{13.1}$$

As an importance density for $p(\psi|y)$ we take its large sample normal approximation

$$g(\psi|y) = N(\hat{\psi}, \hat{V}),$$

where $\hat{\psi}$ is the solution of the equation

$$\frac{\partial \log p(\psi|y)}{\partial \psi} = \frac{\partial \log p(\psi)}{\partial \psi} + \frac{\partial \log p(y|\psi)}{\partial \psi} = 0, \tag{13.2}$$

and

$$\hat{V}^{-1} = -\left. \frac{\partial^2 \log p(\psi)}{\partial \psi \partial \psi'} - \frac{\partial^2 \log p(y|\psi)}{\partial \psi \partial \psi'} \right|_{\psi=\hat{\psi}}. \tag{13.3}$$

For a discussion of this large sample approximation to $p(\psi|y)$ see Gelman *et al.* (1995, Chapter 4) and Bernardo and Smith (1994, §5.3).

Let $g(\alpha|\psi, y)$ be a Gaussian importance density for α given ψ and y which is obtained from an approximating linear Gaussian model in the way described in Chapter 11. From (13.1),

$$\bar{x} = \int x(\alpha) \frac{p(\psi|y) p(\alpha|\psi, y)}{g(\psi|y) g(\alpha|\psi, y)} g(\psi|y) g(\alpha|\psi, y) \, d\psi \, d\alpha$$

$$= \int x(\alpha) \frac{p(\psi|y) g(y|\psi) p(\alpha, y|\psi)}{g(\psi|y) p(y|\psi) g(\alpha, y|\psi)} g(\psi, \alpha|y) \, d\psi \, d\alpha.$$

By Bayes theorem,

$$p(\psi|y) = K p(\psi) p(y|\psi),$$

in which K is a normalising constant, so we have

$$\bar{x} = K \int x(\alpha) \frac{p(\psi)g(y|\psi)}{g(\psi|y)} \frac{p(\alpha, y|\psi)}{g(\alpha, y|\psi)} g(\psi, \alpha|y) \, d\psi \, d\alpha$$
$$= K \mathrm{E}_g \left[x(\alpha) z(\psi, \alpha, y) \right], \tag{13.4}$$

where E_g denotes expectation with respect to the importance joint density

$$g(\psi, \alpha|y) = g(\psi|y)g(\alpha|\psi, y),$$

and where

$$z(\psi, \alpha, y) = \frac{p(\psi)g(y|\psi)}{g(\psi|y)} \frac{p(\alpha, y|\psi)}{g(\alpha, y|\psi)}. \tag{13.5}$$

In this formula, $g(y|\psi)$ is the likelihood for the approximating Gaussian model, which is easily calculated by the Kalman filter.

Taking $x(\alpha) = 1$ in (13.4) gives

$$K^{-1} = \mathrm{E}_g[z(\psi, \alpha, y)],$$

so we have finally

$$\bar{x} = \frac{\mathrm{E}_g[x(\alpha)z(\psi, \alpha, y)]}{\mathrm{E}_g[z(\psi, \alpha, y)]}. \tag{13.6}$$

We note that (13.6) differs from the corresponding formula (11.5) in the classical inference case only in the replacement of $w(\alpha, y)$ by $z(\psi, \alpha, y)$ and the inclusion of ψ in the importance density $g(\psi, \alpha|y)$.

In the important special case in which the state equation error η_t is $\mathrm{N}(0, Q_t)$, then α is Gaussian so we can write its density as $g(\alpha)$ and use this as the state density for the approximating model. This gives $p(\alpha, y|\psi) = g(\alpha)p(y|\theta, \psi)$ and $g(\alpha, y|\psi) = g(\alpha)g(y|\theta, \psi)$, where θ is the stacked vector of signals $\theta_t = Z_t\alpha_t$, so (13.5) simplifies to

$$z(\psi, \alpha, y) = \frac{p(\psi)g(y|\psi)}{g(\psi|y)} \frac{p(y|\theta, \psi)}{g(y|\theta, \psi)}. \tag{13.7}$$

For cases where a proper prior is not available, we may wish to use a non-informative prior in which we assume that the prior density is proportional to a specified function $p(\psi)$ in a domain of ψ of interest even though the integral $\int p(\psi)d\psi$ does not exist. The posterior density, where it exists, is

$$p(\psi|y) = Kp(\psi)p(y|\psi),$$

which is the same as in the proper prior case, so all the previous formulae apply without change. This is why we can use the same symbol $p(\psi)$ in both cases even when $p(\psi)$ is not a proper density. An important special case is the diffuse prior for which $p(\psi) = 1$ for all ψ. For a general discussion of non-informative priors, see, for example, Gelman *et al.* (1995, Chapters 2 and 3).

13.3 Computational aspects of Bayesian analysis

For practical computations based on these ideas we express the formulae in terms of variables that are as simple as possible as in §§11.9, 12.3 and 12.4 for the classical analysis. This means that to the maximum feasible extent we employ formulae based on the disturbance terms $\eta_t = R'_t(\alpha_{t+1} - T_t\alpha_t)$ and $\varepsilon_t = y_t - \theta_t$ for $t = 1, \ldots, n$. By repeated substitution for α_t we first obtain $x(\alpha)$ as a function $x^*(\eta)$ of η. We then note that in place of (13.1) we obtain the posterior mean of $x^*(\eta)$,

$$\bar{x} = \int x^*(\eta)p(\psi|y)p(\eta|\psi, y)\,d\psi\,d\eta. \tag{13.8}$$

By reductions analogous to those above we obtain in place (13.6)

$$\bar{x} = \frac{E_g[x^*(\eta)z^*(\psi, \eta, y)]}{E_g[z^*(\psi, \eta, y)]}, \tag{13.9}$$

where

$$z^*(\psi, \eta, y) = \frac{p(\psi)g(y|\psi)}{g(\psi|y)} \frac{p(\eta, y|\psi)}{g(\eta, y|\psi)}, \tag{13.10}$$

and E_g denotes expectation with respect to the importance density $g(\psi, \eta|y)$.

Let $\psi^{(i)}$ be a random draw from the importance density for ψ, $g(\psi|y) = N(\hat{\psi}, \hat{V})$, where $\hat{\psi}$ satisfies (13.2) and $\hat{V}$ is given by (13.3), and let $\eta^{(i)}$ be a random draw from density $g(\eta|\psi^{(i)}, y)$ for $i = 1, \ldots, N$. To obtain this we need an approximation to the mode $\hat{\eta}^{(i)}$ of density $g(\eta|\psi^{(i)}, y)$ but this is rapidly obtained in a few iterations from the mode of $g(\eta|\hat{\psi}, y)$. Let

$$x_i = x^*\big(\eta^{(i)}\big), \qquad z_i = z^*\big(\psi^{(i)}, \eta^{(i)}, y\big), \tag{13.11}$$

and consider as an estimate of $\bar{x}$ the ratio

$$\hat{x} = \frac{\sum_{i=1}^{N} x_i z_i}{\sum_{i=1}^{N} z_i}. \tag{13.12}$$

The efficiency of this estimate can obviously be improved by the use of antithetic variables. For $\eta^{(i)}$ we can use the location and scale antithetics described in §11.9.3. Antithethics may not be needed for $\psi^{(i)}$ since $\hat{V} = O(n^{-1})$ but it is straightforward to allow for them if their use is worthwhile; for example, it would be an easy matter to employ the location antithetic $\tilde{\psi}^{(i)} = 2\hat{\psi} - \psi^{(i)}$.

There is flexibility in the way the pairs $\psi^{(i)}, \eta^{(i)}$ are chosen, depending on the number of antithetics employed and the way the values of ψ and η are combined. For example, one could begin by making a random selection ψ^s of ψ from $N(\hat{\psi}, \hat{V})$. Next we compute the antithetic value $\tilde{\psi}^s = 2\hat{\psi} - \psi^s$. For each of the values ψ^s and

$\tilde{\psi}^s$ one could draw separate values of η from $g(\eta|\psi, y)$, and then employ the two antithetics for each η that are described in §11.9.3. Thus in the sample there are four values of η combined with each value of ψ so N is a multiple of four and the number of draws of η from the simulation smoother is $N/4$. For estimation of variances due to simulation we need however to note that, since ψ^s and $\tilde{\psi}^s$ are related, there are only $N/8$ independent draws from the joint importance density $g(\psi, \eta|y)$.

For the purpose of estimating posterior variances of scalar quantities, assume that $x^*(\eta)$ is a scalar. Then, as in (12.3), the estimate of its posterior variance is

$$\widehat{\text{Var}}[x^*(\eta)|y] = \frac{\sum_{i=1}^N x_i^2 z_i}{\sum_{i=1}^N z_i} - \hat{x}^2. \tag{13.13}$$

Let us now consider the estimation of variance of the estimate $\hat{x}$ of the posterior mean of scalar $x^*(\eta)$ due to simulation. As indicated above the details depend on the way values of ψ and η are combined. For the example we considered, with a single antithetic for ψ and two antithetics for η, combined in the way described, let $\hat{v}_j^\dagger$ be the sum of the eight associated values of $z_i(x_i - \hat{x})$. Then as in (12.5), the estimate of the variance of $\hat{x}$ due to errors of simulation is

$$\widehat{\text{Var}}_s(\hat{x}) = \frac{\sum_{j=1}^{N/8} \hat{v}_j^{\dagger 2}}{\left(\sum_{i=1}^N z_i\right)^2}. \tag{13.14}$$

For the estimation of posterior distribution functions and densities of scalar $x^*(\eta)$, let $I_x(\eta)$ be an indicator which is unity if $x^*(\eta) \le x$ and is zero if $x^*(\eta) > x$. Then the posterior distribution function is estimated by (12.6) provided that w_i is replaced by z_i. With the same proviso, the posterior density of $x^*(\eta)$ is estimated by (12.7). Samples of independent values from the estimated posterior distribution can be obtained by a method analogous to that described by a method at the end of §12.3.

13.4   Posterior analysis of parameter vector

In this section we consider the estimation of posterior means, variances, distribution functions and densities of functions of the parameter vector ψ. Denote by $v(\psi)$ the function of ψ whose posterior properties we wish to investigate. Using Bayes theorem, the posterior mean of $v(\psi)$ is

$$\begin{aligned} \bar{v} &= \text{E}[v(\psi)|y] \\ &= \int v(\psi) p(\psi|y) \, d\psi \\ &= K \int v(\psi) p(\psi) p(y|\psi) \, d\psi \\ &= K \int v(\psi) p(\psi) p(\eta, y|\psi) \, d\psi \, d\eta, \end{aligned} \tag{13.15}$$

where K is a normalising constant. Introducing importance densities $g(\psi|y)$ and $g(\eta|\psi, y)$ as in §13.3, we have

$$\bar{v} = K \int v(\psi) \frac{p(\psi)g(y|\psi)}{g(\psi|y)} \frac{p(\eta, y|\psi)}{g(\eta, y|\psi)} g(\psi, \eta|y) \, d\psi \, d\eta$$
$$= K E_g[v(\psi)z^*(\psi, \eta, y)], \tag{13.16}$$

where E_g denotes expectation with respect to the joint importance density $g(\psi, \eta|y)$ and

$$z^*(\psi, \eta, y) = \frac{p(\psi)g(y|\psi)}{g(\psi|y)} \frac{p(\eta, y|\psi)}{g(\eta, y|\psi)}.$$

Putting $v(\psi) = 1$ in (13.16) we obtain as in (13.9),

$$\bar{v} = \frac{E_g[v(\psi)z^*(\psi, \eta, y)]}{E_g[z^*(\psi, \eta, y)]}. \tag{13.17}$$

In the simulation, take $\psi^{(i)}$ and $\eta^{(i)}$ as in §13.3 and let $v_i = v(\psi^{(i)})$. Then the estimates $\hat{v}$ of $\bar{v}$ and $\widehat{\mathrm{Var}}[v(\psi)|y]$ of $\mathrm{Var}[v(\psi)|y]$ are given by (13.12) and (13.13) by replacing x_i by v_i. Similarly, the variance of $\hat{v}$ due to simulation can, for the antithetics considered in §12.3, be calculated by defining $v_j^\dagger$ as the sum of the eight associated values of $z_i(v_i - \bar{v})$ and using (13.14) to obtain the estimate $\widehat{\mathrm{Var}}_s(\hat{v})$. Estimates of the posterior distribution and density functions are obtained by the indicator function techniques described at the end of §13.3. While $\hat{v}$ can be a vector, for the remaining estimates $v(\psi)$ has to be a scalar quantity.

The estimate of the posterior density $p[v(\psi)|y]$ obtained in this way is essentially a histogram estimate, which is accurate at values of $v(\psi)$ near the midpoint of the intervals containing them. An alternative estimate of the posterior density of a particular element of ψ, which is accurate at any value of the element, was proposed by Durbin and Koopman (2000). Without loss of generality take this element to be the first element of ψ and denote it by ψ_1. Denote the remaining elements by ψ_2. Let $g(\psi_2|\psi_1, y)$ be the approximate conditional density of ψ_2 given ψ_1 and y, which is easily obtained by applying standard regression theory to $g(\psi|y)$, where $g(\psi|y) = N(\hat{\mu}, \hat{V})$. We take $g(\psi_2|\psi_1, y)$ as an importance density in place of $g(\psi|y)$. Then

$$p(\psi_1|y) = \int p(\psi|y) \, d\psi_2$$
$$= K \int p(\psi)p(y|\psi) \, d\psi_2$$
$$= K \int p(\psi)p(\eta, y|\psi) \, d\psi_2 d\eta$$
$$= K E_g[\bar{z}(\psi, \eta, y)], \tag{13.18}$$

where E_g denotes expectation with respect to importance density $g(\psi_2|\psi_1, y)$

and

$$\tilde{z}(\psi, \eta, y) = \frac{p(\psi)g(y|\psi)}{g(\psi_2|\psi_1, y)} \frac{p(\eta, y|\psi)}{g(\eta, y|\psi)}. \tag{13.19}$$

Let $\tilde{\psi}_2^{(i)}$ be a draw from $g(\psi_2|\psi_1, y)$, let $\tilde{\psi}^{(i)} = (\psi_1, \tilde{\psi}_2^{(i)\prime})\prime$ and let $\tilde{\eta}^{(i)}$ be a draw from $g(\eta|\tilde{\psi}^{(i)}, y)$. Then take

$$\tilde{z}_i = \frac{p(\tilde{\psi}^{(i)})g(y|\tilde{\psi}^{(i)})}{g(\tilde{\psi}_2^{(i)}|\psi_1, y)} \frac{p(\tilde{\eta}^{(i)}, y|\tilde{\psi}^{(i)})}{g(\tilde{\eta}^{(i)}, y|\tilde{\psi}^{(i)})}. \tag{13.20}$$

Now as in (13.17),

$$K^{-1} = E_g[z^*(\psi, \eta, y)],$$

where E_g denotes expectation with respect to importance denisty $g(\psi|y)g(\eta|\psi, y)$ and

$$z^*(\psi, \eta, y) = \frac{p(\psi)g(y|\psi)}{g(\psi|y)} \frac{p(\eta, y|\psi)}{g(\eta, y|\psi)}.$$

Let ψ_i^* be a draw from $g(\psi|y)$ and let η_i^* be a draw from $g(\eta|\psi_i^*, y)$. Then take

$$z_i^* = \frac{p(\psi_i^*)g(y|\psi_i^*)}{g(\psi_i^*|y)} \frac{p(\eta_i^*, y|\psi_i^*)}{g(\eta_i^*, y|\psi_i^*)}, \tag{13.21}$$

and estimate $p(\psi_i|y)$ by the simple form

$$\hat{p}(\psi_i|y) = \sum_{i=1}^N \tilde{z}_i / \sum_{i=1}^N z_i^*. \tag{13.22}$$

The simulations for the numerator and denominator of (13.22) are different since for the numerator only ψ_2 is drawn, whereas for the denominator the whole vector ψ is drawn. The variability of the ratio can be reduced however by employing the same set of N(0, 1) deviates employed for choosing η from $p(\eta|\tilde{\psi}^{(i)}, y)$ in the simulation smoother as for choosing η from $p(\eta|\psi_i^*, y)$. The variability can be reduced further by first selecting ψ_{1i}^* from $g(\psi_1|y)$ and then using the same set of N(0, 1) deviates to select ψ_{2i}^* from $g(\psi_2|\psi_{1i}^*, y)$ as were used to select $\tilde{\psi}_2^{(i)}$ from $g(\psi_2|\psi_1, y)$ when computing $\tilde{z}_i$; in this case $g(\psi^*|y)$ in (13.21) is replaced by $g(\psi_1^*)g(\psi_2^*|\psi_{1i}^*, y)$.

To improve efficiency, antithetics may be used for draws of ψ and η in the way suggested in §13.3.

13.5 Markov chain Monte Carlo methods

A substantial number of publications have appeared on Markov chain Monte Carlo (MCMC) methods for non-Gaussian and nonlinear state space models in the statistical, econometric and engineering literatures. It is beyond the scope of this book to give a treatment of MCMC methods for these models. An introductory and accessible book on MCMC methods has been written by Gamerman (1997). Other contributions in the development of MCMC methods for non-Gaussian and nonlinear state space models are referred to in §13.1. In addition, MCMC methods have been developed for the stochastic volatility model as introduced in §10.6.1; we mention in this connection the contributions of Shephard (1993) and Jacquier, Polson and Rossi (1994) which more recently have been improved by Kim, Shephard and Chib (1998) and Shephard and Pitt (1997).

In the next chapter we will present numerical illustrations of the methods of Part II for which we use the package *SsfNong.ox* developed by Koopman, Shephard and Doornik (1998). This is a collecion of *Ox* functions which use *SsfPack*, as discussed in §6.6, and it provides the tools for the computational implementation of the methods of Part II. However, *SsfNong.ox* can also be used to do MCMC computations and it provides a basis for comparing the MCMC methods with the simulation techniques developed in Chapters 11 to 13 on specific examples. We reiterate our view that, at least for practitioners who are not simulation experts, the methods presented in the earlier sections of this chapter are more transparent and computationally more convenient than MCMC for the type of applications that we are concerned with in this book.

14
Non-Gaussian and nonlinear illustrations

14.1 Introduction

In this chapter we illustrate the methodology of Part II by applying it to four real data sets. In the first example we examine the effects of seat belt legislation on deaths of van drivers due to road accidents in Great Britain modelled by a Poisson distribution. In the second we consider the usefulness of the t-distribution for modelling observation errors in a gas consumption series containing outliers. The third example fits a stochastic volatility model to a series of pound/dollar exchange rates. In the fourth example we fit a binary model to the results of the Oxford-Cambridge boat race over a long period with many missing observations and we forecast the probability that Cambridge will win in 2001. We discuss the software needed to perform these and similar calculations for non-Gaussian and nonlinear models in §14.6.

14.2 Poisson density: van drivers killed in Great Britain

The assessment for the Department of Transport of the effects of seat belt legislation on road traffic accidents in Great Britain, described by Harvey and Durbin (1986) and also discussed in §9.2, was based on linear Gaussian methods as described in Part I. One series that was excluded from this study was the monthly numbers of light goods vehicle (van) drivers killed in road accidents from 1969 to 1984. The numbers of deaths of van drivers were too small to justify the use of the linear Gaussian model. A better model for the data is based on the Poisson distribution with mean $\exp(\theta_t)$ and density

$$p(y_t|\theta_t) = \exp\{\theta_t' y_t - \exp(\theta_t) - \log y_t!\}, \qquad t = 1, \ldots, n, \qquad (14.1)$$

as discussed in §10.3.1. We model θ_t by the relation

$$\theta_t = \mu_t + \gamma_t + \lambda x_t,$$

where the trend μ_t is the random walk

$$\mu_{t+1} = \mu_t + \eta_t, \qquad \eta_t \sim N(0, \sigma_\eta^2), \qquad (14.2)$$

λ is the intervention parameter which measures the effects of the seat belt law, x_t is an indicator variable for the post legislation period and the monthly seasonal γ_t is generated by

$$\sum_{j=0}^{11} \gamma_{t+1-j} = \omega_t, \qquad \omega_t \sim N(0, \sigma_\omega^2). \qquad (14.3)$$

The disturbances η_t and ω_t are mutually independent Gaussian white noise terms with variances $\sigma_\eta^2 = \exp(\psi_\eta)$ and $\sigma_\omega^2 = \exp(\psi_\omega)$, respectively. The parameter estimates are reported by Durbin and Koopman (1997) as $\hat{\sigma}_\eta = \exp(\hat{\psi}_\eta) = \exp(-3.708) = 0.0245$ and $\hat{\sigma}_\omega = 0$. The fact that $\hat{\sigma}_\omega = 0$ implies that the seasonal is constant over time.

For the Poisson model we have $b_t(\theta_t) = \exp(\theta_t)$. As in §11.4.1 we have $\dot{b}_t = \ddot{b}_t = \exp(\tilde{\theta}_t)$, so we take

$$\tilde{H}_t = \exp(-\tilde{\theta}_t), \qquad \tilde{y}_t = \tilde{\theta}_t + \tilde{H}_t y_t - 1,$$

where $\tilde{\theta}_t$ is some trial value for θ_t with $t = 1, \ldots, n$. The iterative process for determining the approximating model as described in §11.4 converges quickly; usually, between three and five iterations are needed for the Poisson model. For a classical analysis, the conditional value $\mu_t + \lambda x_t$ for ψ_η fixed at $\hat{\psi}_\eta$ is computed and exponentiated values of this mean are plotted together with the raw data in Figure 14.1. The posterior mean from a Bayesian perspective with ψ_η diffuse was also calculated and its exponentiated values are also plotted in Figure 14.1. The difference between the graphs is virtually imperceptible. Conditional and posterior standard deviations of $\mu_t + \lambda x_t$ are plotted in Figure 14.2. The posterior standard deviations are about 12% larger than the conditional standard deviations; this is due to the fact that in the Bayesian analysis ψ_η is random. The ratios of simulation standard deviations to actual standard deviations never exceeded the 9% level before the break and never exceed the 7% level after the break. The ratios for a Bayesian analysis are slightly greater at 10% and 8%, respectively.

The main objective of the analysis is the estimation of the effect of the seat belt law on the number of deaths. Here, this is measured by λ which in the Bayesian analysis has a posterior mean of -0.280; this corresponds to a reduction in the number of deaths of 24.4%. The posterior standard deviation is 0.126 and the standard error due to simulation is 0.0040. The corresponding values for the classical analysis are -0.278, 0.114 and 0.0036, which are not very different. It is clear that the value of λ is significant as is obvious visually from Figure 14.1. The posterior distribution of λ is presented in Figure 14.3 in the form of a histogram. This is based on the estimate of the posterior distribution function calculated as indicated in §13.3. The distribution is slightly skewed to the right. All the above calculations were based on a sample of 500 generated draws from the simulation smoother with four antithetics per draw. The reported results show that this relatively small number of samples is adequate for this particular example.

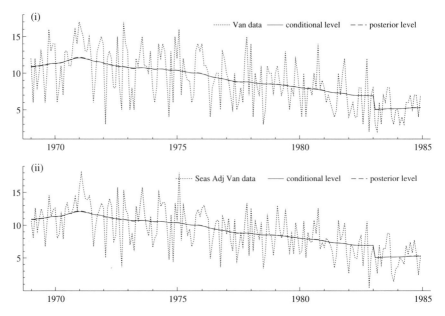

Fig. 14.1. (i) Numbers of van drivers killed and estimated level including intervention; (ii) Seasonally adjusted numbers of van drivers killed and estimated level including intervention.

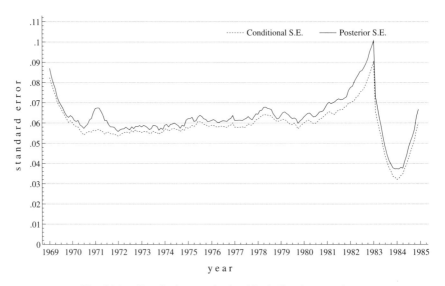

Fig. 14.2. Standard errors for level including intervention.

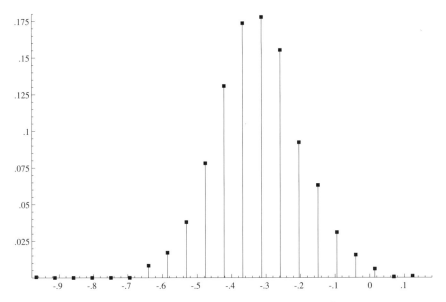

Fig. 14.3. Posterior distribution of intervention effect.

What we learn from this exercise so far as the underlying real investigation is concerned is that up to the point where the law was introduced there was a slow regular decline in the number of deaths coupled with a constant multiplicative seasonal pattern, while at that point there was an abrupt drop in the trend of around 25%; afterwards, the trend appeared to flatten out, with the seasonal pattern remaining the same. From a methodological point of view we learn that our simulation and estimation procedures work straightforwardly and efficiently. We find that the results of the conditional analysis from a classical perspective and the posterior analysis from a Bayesian perspective are very similar apart from the posterior densities of the parameters. So far as computing time is concerned, the calculation of trend and variance of trend for $t = 1, \ldots, n$ took 78 seconds on a Pentium II computer for the classical analysis and 216 seconds for the Bayesian analysis. While the Bayesian time is greater, the time required is still small.

14.3 Heavy-tailed density: outlier in gas consumption in UK

In this example we analyse the logged quarterly demand for gas in the UK from 1960 to 1986 which is a series from the standard data set provided by Koopman *et al.* (2000). We use a structural time series model of the basic form as discussed in §3.2.1:

$$y_t = \mu_t + \gamma_t + \varepsilon_t, \tag{14.4}$$

where μ_t is the local linear trend, γ_t is the seasonal and ε_t is the observation disturbance. The purpose of the investigation underlying the analysis is to study the seasonal pattern in the data with a view to seasonally adjusting the series. It is known that for most of the series the seasonal component changes smoothly over time, but it is also known that there was a disruption in the gas supply in the third and fourth quarters of 1970 which has led to a distortion in the seasonal pattern when a standard analysis based on a Gaussian density for ε_t is employed. The question under investigation is whether the use of a heavy-tailed density for ε_t would improve the estimation of the seasonal in 1970.

To model ε_t we use the t-distribution as in §10.4.1 with logdensity

$$\log p(\varepsilon_t) = \log a\,(\nu) + \frac{1}{2}\log \lambda - \frac{\nu+1}{2}\log\left(1 + \lambda\varepsilon_t^2\right), \qquad (14.5)$$

where

$$a(\nu) = \frac{\Gamma\left(\frac{\nu}{2} + \frac{1}{2}\right)}{\Gamma\left(\frac{\nu}{2}\right)}, \qquad \lambda^{-1} = (\nu - 2)\sigma_\varepsilon^2, \qquad \nu > 2, \qquad t = 1, \dots, n.$$

The mean of ε_t is zero and the variance is σ_ε^2 for any ν degrees of freedom which need not be an integer. The approximating model is easily obtained by the method of §11.5 with

$$h_t^*\left(\varepsilon_t^2\right) = \text{constant} + (\nu + 1)\log\left(1 + \lambda\varepsilon_t^2\right),$$

$$h_t^{*\,-1} = H_t = \frac{1}{\nu + 1}\tilde{\varepsilon}_t^2 + \frac{\nu - 2}{\nu + 1}\sigma_\varepsilon^2,$$

The iterative scheme is started with $H_t = \sigma_\varepsilon^2$, for $t = 1, \dots, n$. The number of iterations required for a reasonable level of convergence using the t-distribution is usually higher than for densities from the exponential family; for this example we required around ten iterations. In the classical analysis, the parameters of the model, including the degrees of freedom ν, were estimated by Monte Carlo maximum likelihood as described in §12.5.3; the estimated value for ν was 12.8.

We now compare the estimated seasonal and irregular components based on the Gaussian model and the model with a t-distribution for ε_t. Figures 14.4 and 14.5 give the graphs of the estimated seasonal and irregular for the Gaussian model and the t-model. The most striking feature of these graphs is the greater effectiveness with which the t-model picks and corrects for the outlier relative to the Gaussian model. We observe that in the graph of the seasonal the difference between the classical and Bayesian analyses is imperceptible. Differences are visible in the graphs of the residuals, but they are not large since the residuals themselves are small. The t-model estimates are based on 250 simulation samples from the simulation smoother with four antithetics for each sample. The number of simulation samples is sufficient because the ratio of the variance due to simulation to the variance never exceeds 2% for all estimated components in the state vector except at the beginning and end of the series where it never exceeds 4%.

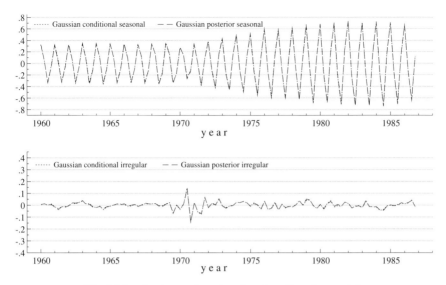

Fig. 14.4. Analyses of gas data based on Gaussian model.

We learn from the analysis that the change over time of the seasonal pattern in the data is in fact smooth. We also learn that if model (14.4) is to be used to estimate the seasonal for this or similar cases with outliers in the observations, then a Gaussian model for ε_t is inappropriate and a heavy-tailed model should be used.

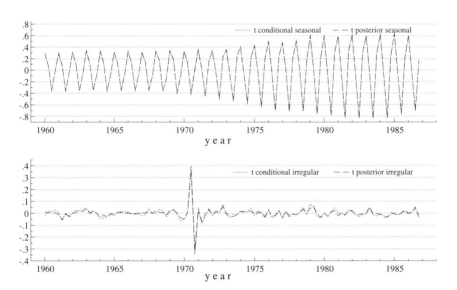

Fig. 14.5. Analyses of gas data based on t-model.

14.4 Volatility: pound/dollar daily exchange rates

The data are the pound/dollar daily exchange rates from 1/10/81 to 28/6/85 which have been used by Harvey *et al.* (1994). Denoting the daily exchange rate by x_t, the observations we consider are given by $y_t = \Delta \log x_t$, for $t = 1, \ldots, n$. A zero-mean stochastic volatility (SV) model of the form

$$y_t = \sigma \exp\left(\frac{1}{2}\theta_t\right) u_t, \qquad u_t \sim N(0, 1), \qquad t = 1, \ldots, n,$$

$$\theta_{t+1} = \phi\theta_t + \eta_t, \qquad \eta_t \sim N\left(0, \sigma_\eta^2\right), \qquad 0 < \phi < 1, \tag{14.6}$$

was used for analysing these data by Harvey *et al.* (1994); see also the illustration of §9.6. The purpose of the investigations for which this type of analysis is carried out is to study the structure of the volatility of price ratios in the market, which is of considerable interest to financial analysts. The level of θ_t determines the amount of volatility and the value of ϕ measures the autocorrelation present in the logged squared data.

To illustrate our approach to SV models we consider the Gaussian logdensity of model (14.6),

$$\log p(y_t | \theta_t) = -\frac{1}{2}\log 2\pi\sigma^2 - \frac{1}{2}\theta_t - \frac{y_t^2}{2\sigma^2}\exp(-\theta_t). \tag{14.7}$$

The linear approximating model can be obtained by the method of §11.4 with

$$\tilde{H}_t = 2\sigma^2\frac{\exp(\tilde{\theta}_t)}{y_t^2}, \qquad \tilde{y}_t = \tilde{\theta}_t - \frac{1}{2}\tilde{H}_t + 1,$$

for which $\tilde{H}_t$ is always positive. The iterative process can be started with $\tilde{H}_t = 2$ and $\tilde{y}_t = \log(y_t^2/\sigma^2)$, for $t = 1, \ldots, n$, since it follows from (14.6) that $y_t^2/\sigma^2 \approx \exp(\theta_t)$. When y_t is zero or very close to zero, it should be replaced by a small constant value to avoid numerical problems; this device is only needed to obtain the approximating model so we do not depart from our exact treatment. The number of iterations required is usually fewer than ten.

The interest here is usually focussed on the estimates of the parameters or their posterior distributions. For the classical analysis we obtain by the maximum likelihood methods of §12.5 the following estimates:

$$\hat{\sigma} = 0.6338, \qquad \hat{\psi}_1 = \log \hat{\sigma} = -0.4561, \qquad SE(\hat{\psi}_1) = 0.1033,$$

$$\hat{\sigma}_\eta = 0.1726, \qquad \hat{\psi}_2 = \log \hat{\sigma}_\eta = -1.7569, \qquad SE(\hat{\psi}_2) = 0.2170,$$

$$\hat{\phi} = 0.9731, \qquad \hat{\psi}_3 = \log \frac{\hat{\phi}}{1 - \hat{\phi}} = 3.5876, \qquad SE(\hat{\psi}_3) = 0.5007,$$

where SE denotes the standard error of the maximum likelihood estimator. We present the results in this form since we estimate the log-transformed parameters, so the standard errors that we calculate apply to them and not to the original parameters of interest. For the Bayesian analysis discussed in Chapter 13 we

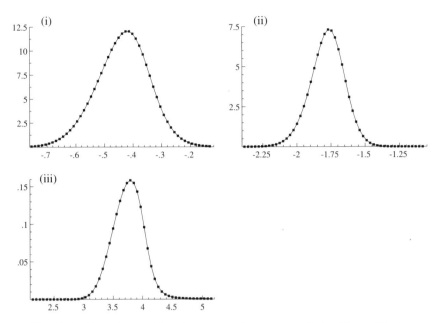

Fig. 14.6. Posterior densities of transformed parameters: (i) ψ_1; (ii) ψ_2; (iii) ψ_3.

present in Figure 14.6 the posterior densities of the parameters. These results confirm that stochastic volatility models can be handled by our methods from both classical and Bayesian perspectives.

14.5 Binary density: Oxford–Cambridge boat race

In the last illustration we consider the outcomes of the annual boat race between teams representing the universities of Oxford and Cambridge. The race takes place from Putney to Mortlake on the river Thames in the month of March or April. The first took place in 1829 and was won by Oxford and, at the time of writing, the last took place in 2000 and was won by Oxford for the first time in eight years. There have been some occasions, especially in the 19th century, when the race took place elsewhere and in other months. In the years of both World Wars the race did not take place and there were also some years when the race finished with a dead heat or some other irregularity took place. Thus the time series of yearly outcomes contains missing observations for the years: 1830–1835, 1837, 1838, 1843, 1844, 1847, 1848, 1850, 1851, 1853, 1855, 1877, 1915–1919 and 1940–1945. We deal with these missing observations as described in §12.4.

The appropriate model is the binary distribution as described in §10.3.2. We take $y_t = 1$ if Cambridge wins and $y_t = 0$ if Oxford wins. Denoting the probability that Cambridge wins in year t by π_t, then as in §10.3.2 we take $\theta_t = \log[\pi_t/(1 - \pi_t)]$.

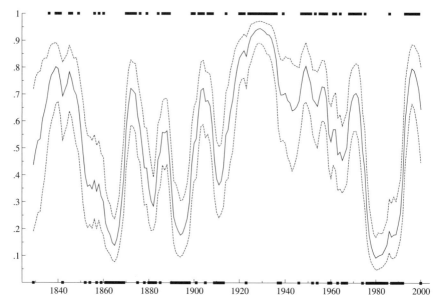

Fig. 14.7. Dot at zero is a win for Oxford and dot at one is a win for Cambridge; the solid line is the probability of a Cambridge win and the dotted lines constitute the 50% (asymmetric) confidence interval.

A winner this year is likely to be a winner next year because of overlapping crew membership, training methods and other factors. Thus we model the transformed probability by the random walk

$$\theta_{t+1} = \theta_t + \eta_t, \qquad \eta_t \sim N(0, \sigma_\eta^2),$$

where η_t is serially uncorrelated for $t = 1, \ldots, n$.

The method described in §11.4 provides the approximating model for this case and maximum likelihood estimation for the unknown variance σ_η^2 is carried out as described in §12.5. We estimated the variance as $\hat{\sigma}_\eta^2 = 0.521$.

The estimated conditional mean of the probability π_t, indicating a win for Cambridge in year t, is computed using the method described in §12.2. The resulting time series of π_t is given in Figure 14.7. The forecasted probability for a Cambridge win in 2001 is 0.67.

14.6   Non-Gaussian and nonlinear analysis using SsfPack

The calculations in this section were carried out using the object oriented matrix programming language *Ox* of Doornik (1998) together with the library of state space functions *SsfPack 2.2* by Koopman *et al.* (1999) and the *Ox* package *SsfNong.ox* for non-Gaussian and nonlinear state space analysis by Koopman *et al.*

(1998). The data and programs are freely available on the Internet at `http://www.ssfpack.com/dkbook`. In each illustration of this section we have referred to relevant *Ox* programs. Documentation of the functions used here and a discussion of computational matters can be found on the Internet at `http://www.ssfpack.com/dkbook`. The *SsfPack* library of computer routines for state space models is discussed in §6.6. The *SsfNong.ox* package can be downloaded from the Internet at `http://www.ssfpack.com`.

References

Akaike, H. and Kitagawa, G. (eds.) (1999). *The Practice of Time Series Analysis*. New York: Springer-Verlag.

Anderson, B. D. O. and Moore, J. B. (1979). *Optimal Filtering*. Englewood Cliffs: Prentice-Hall.

Anderson, T. W. (1984). *An Introduction to Multivariate Statistical Analysis*, 2nd edition. New York: John Wiley & Sons.

Ansley, C. F. and Kohn, R. (1985). Estimation, filtering and smoothing in state space models with incompletely specified initial conditions, *Annals of Statistics*, **13**, 1286–1316.

Ansley, C. F. and Kohn, R. (1986). Prediction mean square error for state space models with estimated parameters, *Biometrika*, **73**, 467–74.

Atkinson, A. C. (1985). *Plots, Transformations and Regression*. Oxford: Clarendon Press.

Balke, N. S. (1993). Detecting level shifts in time series, *J. Business and Economic Statist.*, **11**, 81–92.

Barndorff-Nielsen, O. E. and Shephard, N. (2001). Non-Gaussian OU based models and some of their uses in financial economics (with discussion), *J. Royal Statistical Society B*, **63**. Forthcoming.

Basawa, I. V., Godambe, V. P. and Taylor, R. L. (eds.) (1997). *Selected Proceedings of Athens, Georgia Symposium on Estimating Functions*. Hayward, California: Institute of Mathematical Statistics.

Bernardo, J. M. and Smith, A. F. M. (1994). *Bayesian Theory*. Chichester: John Wiley.

Bollerslev, T., Engle, R. F. and Nelson, D. B. (1994). ARCH Models, In Engle, R. F. and McFadden, D. (eds.), *The Handbook of Econometrics, Volume 4*, pp. 2959–3038. Amsterdam: North-Holland.

Bowman, K. O. and Shenton, L. R. (1975). Omnibus test contours for departures from normality based on $\sqrt{b_1}$ and b_2, *Biometrika*, **62**, 243–50.

Box, G. E. P., Jenkins, G. M. and Reinsel, G. C. (1994). *Time Series Analysis, Forecasting and Control* 3rd edition. San Francisco: Holden-Day.

Box, G. E. P. and Tiao, G. C. (1973). *Bayesian Inference in Statistical Analysis*. Reading, MA: Addison-Wesley.

Box, G. E. P. and Tiao, G. C. (1975). Intervention analysis with applications to economic and environmental problems, *J. American Statistical Association*, **70**, 70–79.

Brockwell, P. J. and Davis, R. A. (1987). *Time Series: Theory and Methods*. New York: Springer-Verlag.

Bryson, A. E. and Ho, Y. C. (1969). *Applied Optimal Control*. Massachusetts: Blaisdell.

Burman, J. P. (1980). Seasonal adjustment by signal extraction, *J. Royal Statistical Society A*, **143**, 321–37.

Cargnoni, C., Muller, P. and West, M. (1997). Bayesian forecasting of multinomial time series through conditionally Gaussian dynamic models, *J. American Statistical Association*, **92**, 640–47.

Carlin, B. P., Polson, N. G. and Stoffer, D. S. (1992). A Monte Carlo approach to nonnormal and nonlinear state-space modelling, *J. American Statistical Association*, **87**, 493–500.

Carter, C. K. and Kohn, R. (1994). On Gibbs sampling for state space models, *Biometrika*, **81**, 541–53.

Carter, C. K. and Kohn, R. (1996). Markov chain Monte Carlo in conditionally Gaussian state space models, *Biometrika*, **83**, 589–601.

Carter, C. K. and Kohn, R. (1997). Semiparameteric Bayesian inference for time series with mixed spectra, *J. Royal Statistical Society B*, **59**, 255–68.

Chu-Chun-Lin, S. and de Jong, P. (1993). A note on fast smoothing, Discussion paper, University of British Columbia.

Cobb, G. W. (1978). The problem of the Nile: conditional solution to a change point problem, *Biometrika*, **65**, 243–51.

Cook, R. D. and Weisberg, S. (1982). *Residuals and Influence in Regression*. New York: Chapman and Hall.

de Jong, P. (1988a). A cross validation filter for time series models, *Biometrika*, **75**, 594–600.

de Jong, P. (1988b). The likelihood for a state space model, *Biometrika*, **75**, 165–69.

de Jong, P. (1989). Smoothing and interpolation with the state space model, *J. American Statistical Association*, **84**, 1085–88.

de Jong, P. (1991). The diffuse Kalman filter, *Annals of Statistics*, **19**, 1073–83.

de Jong, P. (1998). Fixed interval smoothing, Discussion paper, London School of Economics.

de Jong, P. and MacKinnon, M. J. (1988). Covariances for smoothed estimates in state space models, *Biometrika*, **75**, 601–2.

de Jong, P. and Penzer, J. (1998). Diagnosing shocks in time series, *J. American Statistical Association*, **93**, 796–806.

de Jong, P. and Shephard, N. (1995). The simulation smoother for time series models, *Biometrika*, **82**, 339–50.

Doornik, J. A. (1998). *Object-Oriented Matrix Programming Using Ox 2.0*. London: Timberlake Consultants Press.

Doran, H. E. (1992). Constraining Kalman filter and smoothing estimates to satisfy time-varying restrictions, *Rev. Economics and Statistics*, **74**, 568–72.

Doucet, A., deFreitas, J. F. G. and Gordon, N. J. (eds.) (2000). *Sequential Monte Carlo Methods in Practice*. New York: Springer-Verlag.

Duncan, D. B. and Horn, S. D. (1972). Linear dynamic regression from the viewpoint of regression analysis, *J. American Statistical Association*, **67**, 815–21.

Durbin, J. (1960). Estimation of parameters in time series regression models, *J. Royal Statistical Society B*, **22**, 139–53.

Durbin, J. (1987). Statistics and statistical science, (Presidential address), *J. Royal Statistical Society A*, **150**, 177–91.

Durbin, J. (1988). Is a philosophical consensus for statistics attainable? *J. Econometrics*, **37**, 51–61.

Durbin, J. (1997). Optimal estimating equations for state vectors in non-Gaussian and nonlinear state space time series models, in Basawa *et al.* (1997).

Durbin, J. (2000a). Contribution to discussion of Harvey and Chung (2000), *J. Royal Statistical Society A*, **163**, 303–39.

Durbin, J. (2000b). The state space approach to time series analysis and its potential for official statistics, (The Foreman lecture), *Australian and New Zealand J. of Statistics*, **42**, 1–23.

Durbin, J. and Harvey, A. C. (1985). The effects of seat belt legislation on road casualties in Great Britain: report on assessment of statistical evidence, Annexe to Compulsary Seat Belt Wearing Report, Department of Transport, London, HMSO.

Durbin, J. and Koopman, S. J. (1992). Filtering, smoothing and estimation for time series models when the observations come from exponential family distributions, Unpublished paper: Department of Statistics, LSE.

Durbin, J. and Koopman, S. J. (1997). Monte Carlo maximum likelihood estimation of non-Gaussian state space model, *Biometrika*, **84**, 669–84.

Durbin, J. and Koopman, S. J. (2000). Time series analysis of non-Gaussian observations based on state space models from both classical and Bayesian perspectives (with discussion), *J. Royal Statistical Society B*, **62**, 3–56.

Durbin, J. and Quenneville, B. (1997). Benchmarking by state space models, *International Statistical Review*, **65**, 23–48.

Engle, R. F. (1982). Autoregressive conditional heteroskedasticity with estimates of the variance of the United Kingdom inflation, *Econometrica*, **50**, 987–1007.

Engle, R. F. and Russell, J. R. (1998). Forecasting transaction rates: the autoregressive conditional duration model, *Econometrica*, **66**, 1127–62.

Fahrmeir, L. (1992). Posterior mode estimation by extended Kalman filtering for multivariate dynamic generalised linear models, *J. American Statistical Association*, **87**, 501–9.

Fahrmeir, L. and Kaufmann, H. (1991). On Kalman filtering, posterior mode estimation and Fisher scoring in dynamic exponential family regression, *Metrika*, **38**, 37–60.

Fahrmeir, L. and Tutz, G. (1994). *Multivariate Statistical Modelling Based on Generalized Linear Models*. Berlin: Springer.

Fessler, J. A. (1991). Nonparametric fixed-interval smoothing with vector splines, *IEEE Trans. Signal Process*, **39**, 852–59.

Fletcher, R. (1987). *Practical Methods of Optimisation* 2nd edition. New York: John Wiley.

Fruhwirth-Schnatter, S. (1994). Data augmentation and dynamic linear models, *J. Time Series Analysis*, **15**, 183–202.

Gamerman, D. (1998). Markov chain Monte Carlo for dynamic generalised linear models, *Biometrika*, **85**, 215–27.

Gamerman, D. (1997). *Markov Chain Monte Carlo: Stochastic Simulations for Bayesian Inference*. London: Chapman and Hall.

Gelfand, A. E. and Smith, A. F. M. (eds.) (1999). *Bayesian Computation*. Chichester: John Wiley and Sons.

Gelman, A. (1995). Inference and monitoring convergence, in Gilks *et al.* (1996), pp. 131–143.

Gelman, A., Carlin, J. B., Stern, H. S. and Rubin, D. B. (1995). *Bayesian Data Analysis*. London: Chapman & Hall.

Geweke, J. (1989). Bayesian inference in econometric models using Monte Carlo integration, *Econometrica*, **57**, 1317–39.

Ghysels, E., Harvey, A. C. and Renault, E. (1996). Stochastic volatility, In Rao, C. R. and Maddala, G. S. (eds.), *Statistical Methods in Finance*, pp. 119–91. Amsterdam: North-Holland.

Gilks, W. K., Richardson, S. and Spiegelhalter, D. J. (eds.) (1996). *Markov Chain Monte Carlo in Practice*. London: Chapman & Hall.

Godambe, V. P. (1960). An optimum property of regular maximum likelihood estimation, *Annals of Mathematical Statistics*, **31**, 1208–12.

Golub, G. H. and Van Loan, C. F. (1997). *Matrix Computations* 2nd edition. Baltimore: The Johns Hopkins University Press.

Granger, C. W. J. and Newbold, P. (1986). *Forecasting Economic Time Series* 2nd edition. Orlando: Academic Press.

Green, P. and Silverman, B. W. (1994). *Nonparameteric Regression and Generalized Linear Models: A Roughness Penalty Approach*. London: Chapman & Hall.

Hamilton, J. (1994). *Time Series Analysis*. Princeton: Princeton University Press.

Hardle, W. (1990). *Applied Nonparameteric Regression*. Cambridge: Cambridge University Press.

Harrison, J. and Stevens, C. F. (1976). Bayesian forecasting (with discussion), *J. Royal Statistical Society B*, **38**, 205–47.

Harrison, J. and West, M. (1991). Dynamic linear model diagnostics, *Biometrika*, **78**, 797–808.

Harvey, A. C. (1989). *Forecasting, Structural Time Series Models and the Kalman Filter*. Cambridge: Cambridge University Press.

Harvey, A. C. (1993). *Time Series Models* 2nd edition. Hemel Hempstead: Harvester Wheatsheaf.

Harvey, A. C. (1996). Intervention analysis with control groups, *International Statistical Review*, **64**, 313–28.

Harvey, A. C. and Chung, C.-H. (2000). Estimating the underlying change in unemployment in the UK (with discussion), *J. Royal Statistical Society A*, **163**, 303–39.

Harvey, A. C. and Durbin, J. (1986). The effects of seat belt legislation on British road casualties: A case study in structural time series modelling, (with discussion), *J. Royal Statistical Society A*, **149**, 187–227.

Harvey, A. C. and Fernandes, C. (1989). Time series models for count data or qualitative observations, *J. Business and Economic Statist.*, **7**, 407–17.

Harvey, A. C. and Koopman, S. J. (1992). Diagnostic checking of unobserved components time series models, *J. Business and Economic Statist.*, **10**, 377–89.

Harvey, A. C. and Koopman, S. J. (1997). Multivariate structural time series models, In Heij, C., Schumacher, H., Hanzon, B. and Praagman, C. (eds.), *Systematic Dynamics in Economic and Financial Models*, pp. 269–98. Chichester: John Wiley and Sons.

Harvey, A. C. and Koopman, S. J. (2000). Signal extraction and the formulation of unobserved components models, *Econometrics Journal*, **3**, 84–107.

Harvey, A. C. and Peters, S. (1990). Estimation procedures for structural time series models, *J. of Forecasting*, **9**, 89–108.

Harvey, A. C. and Phillips, G. D. A. (1979). The estimation of regression models with autoregressive-moving average disturbances, *Biometrika*, **66**, 49–58.

Harvey, A. C., Ruiz, E. and Shephard, N. (1994). Multivariate stochastic variance models, *Rev. Economic Studies*, **61**, 247–64.

Harvey, A. C. and Shephard, N. (1990). On the probability of estimating a deterministic component in the local level model, *J. Time Series Analysis*, **11**, 339–47.

Harvey, A. C. and Shephard, N. (1993). Structural time series models, In Maddala, G. S., Rao, C. R. and Vinod, H. D. (eds.), *Handbook of Statistics, Volume 11*. Amsterdam: Elsevier Science Publishers.

Hastie, T. and Tibshirani, R. (1990). *Generalized Additive Models*. London: Chapman & Hall.

Holt, C. C. (1957). Forecasting seasonals and trends by exponentially weighted moving averages, Research memorandum, Carnegie Institute of Technology, Pittsburgh, Pennsylvania.

Hull, J. and White, A. (1987). The pricing of options on assets with stochastic volatilities, *J. Finance*, **42**, 281–300.

Jacquier, E., Polson, N. G. and Rossi, P. E. (1994). Bayesian analysis of stochastic volatility models (with discussion), *J. Business and Economic Statist.*, **12**, 371–417.

Jazwinski, A. H. (1970). *Stochastic Processes and Filtering Theory*. New York: Academic Press.

Jones, R. H. (1993). *Longitudinal Data with Serial Correlation: A State-Space Approach*. London: Chapman & Hall.

Kalman, R. E. (1960). A new approach to linear filtering and prediction problems, *J. Basic Engineering, Transactions ASMA, Series D*, **82**, 35–45.

Kim, C. J. and Nelson, C. R. (1999). *State Space Models with Regime Switching*. Cambridge, Massachusetts: MIT Press.

Kim, S., Shephard, N. and Chib, S. (1998). Stochastic volatility: likelihood inference and comparison with ARCH models, *Rev. Economic Studies*, **65**, 361–93.

Kitagawa, G. and Gersch, W. (1996). *Smoothness Priors Analysis of Time Series*. New York: Springer-Verlag.

Kohn, R. and Ansley, C. F. (1989). A fast algorithm for signal extraction, influence and cross-validation, *Biometrika*, **76**, 65–79.

Kohn, R., Ansley, C. F. and Wong, C.-M. (1992). Nonparametric spline regression with autoregressive moving average errors, *Biometrika*, **79**, 335–46.

Koopman, S. J. (1993). Disturbance smoother for state space models, *Biometrika*, **80**, 117–26.

Koopman, S. J. (1997). Exact initial Kalman filtering and smoothing for non-stationary time series models, *J. American Statistical Association*, **92**, 1630–38.

Koopman, S. J. (1998). Kalman filtering and smoothing, In Armitage, P. and Colton, T. (eds.), *Encyclopedia of Biostatistics*. Chichester: Wiley and Sons.

Koopman, S. J. and Durbin, J. (2000). Fast filtering and smoothing for multivariate state space models, *J. Time Series Analysis*, **21**, 281–96.

Koopman, S. J. and Durbin, J. (2001). Filtering and smoothing of state vector for diffuse state space models, mimeo, Free University, Amsterdam, http://www.ssfpack.com/dkbook/.

Koopman, S. J. and Harvey, A. C. (1999). Computing observation weights for signal extraction and filtering, mimeo, Free University, Amsterdam.

Koopman, S. J., Harvey, A. C., Doornik, J. A. and Shephard, N. (2000). *Stamp 6.0: Structural Time Series Analyser, Modeller and Predictor*. London: Timberlake Consultants.

Koopman, S. J. and Hol-Uspensky, E. (1999). The Stochastic Volatility in Mean model: Empirical evidence from international stock markets, Discussion paper, Tinbergen Institute, Amsterdam.

Koopman, S. J. and Shephard, N. (1992). Exact score for time series models in state space form, *Biometrika*, **79**, 823–26.

Koopman, S. J., Shephard, N. and Doornik, J. A. (1998). Fitting non-Gaussian state space models in econometrics: Overview, developments and software, mimeo, Nuffield College, Oxford.

Koopman, S. J., Shephard, N. and Doornik, J. A. (1999). Statistical algorithms for models in state space form using SsfPack 2.2, *Econometrics Journal*, **2**, 113–66. http://www.ssfpack.com/.

Ljung, G. M. and Box, G. E. P. (1978). On a measure of lack of fit in time series models, *Biometrika*, **66**, 67–72.

Magnus, J. R. and Neudecker, H. (1988). *Matrix Differential Calculus with Applications in Statistics and Econometrics*. New York: Wiley.

Makridakis, S., Wheelwright, S. C. and Hyndman, R. J. (1998). *Forecasting: Methods and Applications* 3rd edition. New York: John Wiley and Sons.

McCullagh, P. and Nelder, J. A. (1989). *Generalized Linear Models*. London: Chapman & Hall. 2nd Edition.

Mills, T. C. (1993). *Time Series Techniques for Economists* 2nd edition. Cambridge: Cambridge University Press.

Morf, J. F. and Kailath, T. (1975). Square root algorithms for least squares estimation, *IEEE Transactions on Automatic Control*, **20**, 487–97.

Muth, J. F. (1960). Optimal properties of exponentially weighted forecasts, *J. American Statistical Association*, **55**, 299–305.

Pfefferman, D. and Tiller, R. (2000). Bootstrap approximation to prediction MSE for state-space models with estimated parameters, mimeo, Department of Statistics, Hebrew University, Jerusalem.

Poirier, D. J. (1995). *Intermediate Statistics and Econometrics*. Cambridge: MIT.

Quenneville, B. and Singh, A. C. (1997). Bayesian prediction mean squared error for state space models with estimated parameters, *J. Time Series Analysis*, **21**, 219–36.

Ripley, B. D. (1987). *Stochastic Simulation*. New York: Wiley.

Rosenberg, B. (1973). Random coefficients models: the analysis of a cross-section of time series by stochastically convergent parameter regression, *Annals of Economic and Social Measurement*, **2**, 399–428.

Rydberg, T. H. and Shephard, N. (1999). BIN models for trade-by-trade data. Modelling the number of trades in a fixed interval of time, Working paper, Nuffield College, Oxford.

Sage, A. P. and Melsa, J. L. (1971). *Estimation Theory with Applications to Communication and Control*. New York: McGraw Hill.

Schweppe, F. (1965). Evaluation of likelihood functions for Gaussian signals, *IEEE Transactions on Information Theory*, **11**, 61–70.

Shephard, N. (1993). Fitting non-linear time series models, with applications to stochastic variance models, *J. Applied Econometrics*, **8**, S135–52.

Shephard, N. (1994). Partial non-Gaussian state space, *Biometrika*, **81**, 115–31.

Shephard, N. (1996). Statistical aspects of ARCH and stochastic volatility, In Cox, D. R., Hinkley, D. V. and Barndorff-Nielson, O. E. (eds.), *Time Series Models in Econometrics, Finance and Other Fields*, pp. 1–67. London: Chapman & Hall.

Shephard, N. and Pitt, M. K. (1997). Likelihood analysis of non-Gaussian measurement time series, *Biometrika*, **84**, 653–67.

Shumway, R. H. and Stoffer, D. S. (1982). An approach to time series smoothing and forecasting using the EM algorithm, *J. Time Series Analysis*, **3**, 253–64.

Shumway, R. H. and Stoffer, D. S. (2000). *Time Series Analysis and Its Applications*. New York: Springer-Verlag.

Silverman, B. W. (1985). Some aspects of the spline smoothing approach to non-parametric regression curve fitting, *J. Royal Statistical Society B*, **47**, 1–52.

Smith, J. Q. (1979). A generalization of the Bayesian steady forecasting model, *J. Royal Statistical Society B*, **41**, 375–87.

Smith, J. Q. (1981). The multiparameter steady model, *J. Royal Statistical Society B*, **43**, 256–60.

Snyder, R. D. and Saligari, G. R. (1996). Initialization of the Kalman filter with partially diffuse initial conditions, *J. Time Series Analysis*, **17**, 409–24.

Taylor, S. J. (1986). *Modelling Financial Time Series*. Chichester: John Wiley.

Theil, H. and Wage, S. (1964). Some observations on adaptive forecasting, *Management Science*, **10**, 198–206.

Wahba, G. (1978). Improper priors, spline smoothing, and the problems of guarding against model errors in regression, *J. Royal Statistical Society B*, **40**, 364–72.

Wahba, G. (1990). *Spline Models for Observational Data*. Philadelphia: SIAM.

Watson, M. W. and Engle, R. F. (1983). Alternative algorithms for the estimation of dynamic factor, MIMIC and varying coefficient regression, *J. Econometrics*, **23**, 385–400.

Wecker, W. E. and Ansley, C. F. (1983). The signal extraction approach to nonlinear regression and spline smoothing, *J. American Statistical Association*, **78**, 81–9.

West, M. and Harrison, J. (1997). *Bayesian Forecasting and Dynamic Models* 2nd edition. New York: Springer-Verlag.

West, M., Harrison, J. and Migon, H. S. (1985). Dynamic generalised models and Bayesian forecasting (with discussion), *J. American Statistical Association*, **80**, 73–97.

Winters, P. R. (1960). Forecasting sales by exponentially weighted moving averages, *Management Science*, **6**, 324–42.

Yee, T. W. and Wild, C. J. (1996). Vector generalized additive models, *J. Royal Statistical Society B*, **58**, 481–93.

Young, P. C. (1984). *Recursive Estimation and Time Series Analysis*. New York: Springer-Verlag.

Young, P. C., Lane, K., Ng, C. N. and Palmer, D. (1991). Recursive forecasting, smoothing and seasonal adjustment of nonstationary environmental data, *J. of Forecasting*, **10**, 57–89.

Author index

Akaike, H., 5
Anderson, B. D. O., 5, 67, 71, 128, 184
Anderson, T. W., 10, 37
Ansley, C. F., 62, 63, 71, 75, 81, 101, 139, 172
Atkinson, A. C., 81

Balke, N. S., 13
Barndorff-Nielsen, O. E., 187
Basawa, I. V., 203
Bernardo, J. M., 155, 157, 222, 223
Bollerslev, T., 187
Bowman, K. O., 34
Box, G. E. P., 34, 44, 46, 184
Brockwell, P. J., 5
Bryson, A. E., 71
Burman, J. P., 53

Cargnoni, C., 222
Carlin, B. P., 159, 222
Carlin, J. B., 155, 157, 158, 206, 214, 222,
    223, 225
Carter, C. K., 84, 159, 160, 222
Chib, S., 229
Chu-Chun-Lin, S., 120
Chung, C.-H., 56, 57
Cobb, G. W., 13
Cook, R. D., 81

Davis, R. A., 5
de Freitas, J. F. G., 5
de Jong, P., 22, 71, 75, 81, 84, 101, 105, 115,
    117, 120, 128, 139, 141, 142, 153, 159
Doornik, J. A., 31, 45, 134, 135, 136, 143, 153,
    163, 229, 233
Doran, H. E., 134
Doncet, A., 5
Duncan, D. B., 67
Durbin, J., viii, 3, 44, 51, 53, 56, 57, 101, 103,
    106, 107, 108, 109, 129, 155, 161, 167, 189,
    202, 203, 205, 215, 217, 219, 220, 221, 222,
    227, 230, 231

Engle, R. F., 147, 148, 187, 188

Fahrmeir, L., 5, 128, 193, 202
Fernandes, C., 180
Fessler, J. A., 133

Fletcher, R., 143
Fruhwirth-Schnatter, S., 84, 159, 160

Gamerman, D., 159, 160, 222, 229
Gelfand, A. E., 214
Gelman, A., 155, 157, 158, 159, 206, 214, 222,
    223, 225
Gersch, W., 5
Geweke, J., 156
Ghysels, E., 185, 187
Godambe, V. P., 202, 203
Golub, G. H., 126, 127
Gordon, N. J., 5
Granger, C. W. J., 5
Green, P., 61, 62, 63, 81

Hamilton, J., 5, 150, 151
Hardle, W., 173
Harrison, J., 5, 40, 42, 81, 180
Harvey, A. C., 5, 31, 35, 39, 40, 41, 44, 45,
    53, 56, 57, 60, 81, 101, 123, 124, 133,
    139, 142, 143, 148, 153, 161, 163, 167,
    168, 173, 174, 175, 180, 185, 187, 230,
    233, 236
Hastie, T., 133
Ho, Y. C., 71
Hol-Uspensky, E., 186
Holt, C. C., 50
Horn, S. D., 67
Hull, J., 175
Hyndman, R. J., 169, 171

Jacquier, E., 229
Jazwinski, A. H., 5
Jenkins, G. M., 46
Jones, R. H., 5

Kailath, T., 124
Kalman, R. E., vii, 52
Kaufmann, H., 193
Kim, C. J., 5
Kim, S., 229
Kitagawa, G., 5
Kohn, R., 62, 71, 75, 81, 84, 101, 139, 160,
    222
Koopman, S. J., viii, 3, 31, 35, 45, 71, 72, 75, 81,
    101, 103, 106, 107, 108, 109, 129, 134, 135,

136, 140, 143, 145, 147, 153, 163, 173,
174, 186, 189, 202, 205, 215, 217,
219, 220, 221, 222, 227, 229, 231,
233, 238

Lane, K., 39, 40
Ljung, G. M., 34

MacKinnon, M. J., 81
Magnus, J. R., 112
Makridakis, S., 169, 171
McCullagh, P., 180
Melsa, J. L., 5
Migon, H. S., 180
Mills, T. C., 5
Moore, J. B., 5, 67, 71, 128, 184
Morf, J. F., 124
Muller, P., 222
Muth, J. F., 49, 50

Nelder, J. A., 180
Nelson, C. R., 5
Nelson, D. B., 187
Neudecker, H., 112
Newbold, P., 5
Ng, C. N., 39, 40

Palmer, D., 39, 40
Penzer, J., 153
Peters, S., 148
Pfefferman, D., 152
Phillips, G. D. A., 101, 139
Pitt, M. K., 189, 215, 222, 229
Poirier, D. J., 160
Polson, N. G., 159, 222, 229

Quenneville, B., 56, 152

Reinsel, G. C., 46
Renault, E., 185, 187
Ripley, B. D., 156, 189
Rosenberg, B., 100, 101, 115, 117, 118
Rossi, P. E., 229
Rubin, D. B., 155, 157, 158, 206, 214, 222,
223, 225
Ruiz, E., 175, 235, 236

Russell, J. R., 188
Rydberg, T. H., 188

Sage, A. P., 5
Saligari, G. R., 128
Schweppe, F., 139
Shenton, L. R., 34
Shephard, N., 22, 31, 45, 84, 134, 135, 136,
142, 143, 145, 147, 153, 158, 159, 160,
163, 175, 185, 187, 188, 189, 201, 210,
211, 215, 222, 229, 233, 236, 238
Shumway, R. H., 5, 147
Silverman, B. W., 61, 62, 63, 81, 173
Singh, A. C., 152
Smith, A. F. M., 155, 157, 214, 222, 223
Smith, J. Q., 180
Snyder, R. D., 129
Stern, H. S., 155, 157, 158, 206, 214, 222,
223, 225
Stevens, C. F., 40, 42
Stoffer, D. S., 5, 147, 159, 222

Taylor, R. L., 203
Taylor, S. J., 185
Theil, H., 50
Tiao, G. C., 44, 184
Tibshirani, R., 133
Tiller, R., 152
Tutz, G., 5, 128

Van Loan, C. F., 126, 127

Wage, S., 50
Wahba, G., 53, 61, 63
Watson, M. W., 147, 148
Wecker, W. E., 63, 172
Weisberg, S., 81
West, M., 5, 81, 180, 222
Wheelwright, S. C., 169, 171
White, A., 175
Wild, C. J., 133
Winters, P. R., 50
Wong, C.-M., 62

Yee, T. W., 133
Young, P. C., 5, 39, 40, 71

Subject index

ACD model, 188
airline model, 48
Akaike information criterion, 152
antithetic variables, 205
approximating model, 191
    based on first two derivatives, 193
    based on the first derivative, 195
    exponential family model, 195
    for non-Gaussian state component, 198
    for nonlinear model, 199
    general error distribution, 197
    mixture of normals, 197
    mode estimation, 202
    multiplicative model, 201
    Poisson distribution, 195
    stochastic volatility model, 195
    t-distribution, 197, 199
ARIMA model, 46, 112
    in state space form, 47
ARMA model, 46, 111, 161
    in state space form, 46
    with missing observations, 171
augmented filtering
    loglikelihood evaluation, 140
    regression estimation, 122
augmented Kalman filter and smoother, 115
augmented simulation smoothing, 120
augmented smoothing, 120
autoregressive conditional duration model, 188
autoregressive conditional heteroscedasticity,
    187

Bayesian analysis, 155, 222
    computational aspects, 225
    for linear Gaussian model, 155
    for non-Gaussian and nonlinear model, 222
    MCMC methods, 159, 228
    non-informative prior, 158
    posterior analysis of parameter vector, 226
    posterior analysis of state vector, 155, 222
benchmarking, 54
BFGS method, 143
BIN model, 188
binary distribution, 181, 237
binomial distribution, 181
bivariate models, 161
boat race example, 237

Box-Jenkins analysis, 2, 46, 51, 161
Box-Ljung statistic, 34
Brownian motion, 57

Cholesky decomposition, 14, 22, 97, 131, 133
common factors, 45
concentration of loglikelihood, 31
conditional
    density estimation, 213
    distribution, 11, 16, 65, 84
    distribution function estimation, 213
    mean estimation, 212
    mode estimation, 202
    variance estimation, 212
continuous time, 57, 62
control variables, 219
cycle, 43

data
    exchange rate pound/dollar, 236
    gas consumption, 233
    internet users, 169
    motorcycle acceleration, 172
    Nile, 12, 18, 21, 23, 25, 27, 32, 35, 136
    Oxford/Cambridge boat race, 237
    road accidents, 161, 167, 230
diagnostic checking, 33, 152
diffuse initialisation, 27, 99, 206
diffuse Kalman filter, 27, 101
    augmented filtering, 115
    exact initial Kalman filter, 101
diffuse loglikelihood
    parameter estimation, 149
diffuse loglikelihood evaluation
    using augmented filtering, 140
    using exact initial Kalman filter, 139
diffuse simulation smoothing
    augmented simulation smoothing, 120
    exact initial simulation smoothing, 110
diffuse smoothing
    augmented smoothing, 120
    exact initial smoothing, 106, 109
disturbance smoothing, 19, 73
    augmented smoothing, 120
    covariance matrices, 77
    derivation of, 73
    exact initial smoothing, 109

disturbance smoothing (*cont.*)
    in matrix form, 98
    variance matrix, 75

EM algorithm, 147
estimating equations, 203
EWMA, 49
exact initial filtering
    convenient representation, 105
    derivation of, 101
    regression estimation, 122
    transition to Kalman filter, 104
exact initial Kalman filter, 101
    loglikelihood evaluation, 139
exact initial simulation smoothing, 110
exact initial smoothing, 106
    convenient representation, 107, 109
    derivation of, 106
explanatory variables, 43
exponential distribution, 188
exponential family model, 180
    approximating model, 195
    binary distribution, 181, 237
    binomial distribution, 181
    exponential distribution, 188
    multinomial distribution, 182
    negative binomial distribution, 182
    Poisson distribution, 181, 188, 230
exponential smoothing, 49
exponentially weighted moving average, 49

filtering, 11, 65
financial model, 185
    duration, 188
    trade frequency, 188
    volatility, 185, 187, 236
forecast errors, 13, 33, 49, 68, 152
forecasting, 25, 93, 214

GARCH model, 187
general autoregressive conditional
        heteroscedasticity, 187
general error distribution, 184
    approximating model, 197
Givens rotation, 126
goodness of fit, 152
gradient of loglikelihood, 144

heavy-tailed distribution, 183
    approximating model, 195
    general error distribution, 184
    mixture of normals, 184
    t-distribution, 183, 233
heteroscedasticity diagnostic test, 34

importance resampling, 214
importance sampling, 156, 177, 223
    antithetic variables, 205

computational aspects, 204
    diffuse initialisation, 206
initialisation, 27, 99
innovation model, 69
innovations, 13, 68, 96
intervention variables, 43
irregular, 9

Kalman filter, 11–13, 65, 67
    augmented filtering, 116
    Cholesky decomposition, 14, 97
    derivation of, 65
    exact initial Kalman filter, 101
    in matrix form, 97
    missing observations, 23

least squares residuals, 123
leverage, 186
likelihood evaluation, 138
linearisation
    based on first two derivatives, 193
    based on the first derivative, 195
    for non-Gaussian state component, 198
    for nonlinear model, 199
local level model, 9, 44, 50, 57, 142, 174, 180
local linear trend model, 39, 50, 59, 100, 111,
        118, 128, 133
loglikelihood evaluation, 30, 138
    score vector, 144
    using augmented filtering, 140, 141
    using exact initial Kalman filter, 139
    when initial conditions are diffuse, 139, 140
    when initial conditions are fixed but unknown,
        141
    when initial conditions are known, 138

Markov chain Monte Carlo, 159, 228
maximum likelihood estimation, 30, 138, 142,
        215
mean square error due to simulation, 151, 217
minimum mean square error, 25, 49, 67, 93
missing observations, 23, 59, 92, 136, 169, 214,
        237
mixture of normals, 184
    approximating model, 197
mode estimation, 202
    estimating equations, 203
multinomial distribution, 182
multiplicative models, 185, 201
multivariate models, 44, 167

negative binomial distribution, 182
Newton's method, 142
non-Gaussian models, 179
    binary illustration, 237
    exponential family model, 180
    heavy-tailed distribution, 183
    Poisson illustration, 230

stochastic volatility model, 185
SV model illustration, 236
t-distribution illustration, 233
nonlinear models, 179, 184
normality diagnostic test, 33
numerical maximisation, 142

outliers, 35
Ox, 136, 143, 229, 239

parameter estimation, 30, 142
    effect of errors, 150, 217
    EM algorithm, 147
    for non-Gaussian and nonlinear models, 215
    large sample distribution, 150
    when initial conditions are diffuse, 149
Poisson distribution, 181, 188, 230
    approximating model, 195
posterior analysis of
    state vector, 155

random walk, 9
recursive residuals, 123
regression estimation, 121
    least squares residuals, 123
    recursive residuals, 123
regression lemma, 37
regression model
    with ARMA errors, 54, 114
    with time-varying coefficients, 54
restrictions on state vector, 134

score vector, 144
seasonal, 9, 40, 48, 162, 231
seemingly unrelated time series equations, 44
serial correlation diagnostic test, 34
simulation, 22, 156, 159, 228
    mean square error due to, 217
    variance due to, 213
simulation smoothing, 22, 83, 159
    augmented simulation smoothing, 120
    derivation of, 87
    exact initial simulation smoothing, 110

missing observations, 93
multiple samples, 92
observation errors, 89
recursion for disturbances, 90
state vectors, 91
smoothing, 16, 19, 70
spline smoothing, 61, 115, 133, 161
square root filtering, 124
square root smoothing, 127
SsfPack, 134, 136, 161, 229, 238
STAMP, 45, 143, 163
state estimation error, 15, 68
state smoothing, 16, 70
    augmented smoothing, 120
    covariance matrices, 77
    derivation of, 70
    exact initial smoothing, 106
    fast state smoothing, 75
    in matrix form, 22, 98
    missing observations, 92
    variance matrix, 72
state space model, 1, 38, 65, 99
steady state, 13, 32, 68
stochastic volatility model, 185, 236
    approximate method, 175
    approximating model, 195
structural breaks, 35
structural time series model, 39, 110, 160
    multivariate model, 44, 167
Student's t-distribution, 183
SV model, 185, 236
SVM model, 186

t-distribution, 183, 233
    approximating model, 197, 199
time series, 9
trend, 9, 39

univariate treatment of multivariate series, 128

volatility models, 185, 236

weight functions, 81